China as a Double-Bind Regulatory State

Aifang Ma

China as a Double-Bind Regulatory State

How Internet Regulators' Predicament Produces Regulatees' Autonomy

palgrave
macmillan

Aifang Ma
Peking University
Beijing, China

ISBN 978-981-99-8856-3 ISBN 978-981-99-8857-0 (eBook)
https://doi.org/10.1007/978-981-99-8857-0

Cover illustration: Marina Lohrbach_shutterstock.com

This Palgrave Macmillan imprint is published by the registered company Springer Nature Singapore Pte Ltd.
The registered company address is: 152 Beach Road, #21-01/04 Gateway East, Singapore 189721, Singapore

Paper in this product is recyclable.

To my parents,
for the gift of curiosity

Acknowledgements

In the course of completing this book, I obtained unwavering financial, intellectual, and emotional support. I am grateful for the financial help provided by the Laboratory for Interdisciplinary Evaluation of Public Policies (LIEPP), the Doctoral School of Sciences Po, and the Centre for European Studies and Comparative Politics of Sciences Po. Their research grants and mobility grants enabled me to conduct the two-year field research in Europe and China from 2018 to 2020. I am grateful of the generous help by the Chinese Postdoctoral Science Foundation for my comparative research project on the anti-trust regulation of the digital platforms in China, Europe, and the USA. I take this opportunity to thank the Laura Bassi Foundation for providing the prestigious Laura Bassi Scholarship to help with the editing process of the book.

I would like to warmly thank Professor Nonna Mayer and Professor Colin Hay, two remarkable mentors during my Ph.D. studies at the Centre for European Studies and Comparative Studies of Sciences Po. Nonna crucially contributed to the initiation and the development of this research project in 2019. From 2019 to 2022, she took on her own time to read what I wrote and discussed with me every two weeks. Even the Covid-19 did not break down her rhythm, although it did oblige us to meet virtually for most part of 2020. My warmest thanks also go to Colin Hay. He was an exceptional supervisor for both my master and doctoral studies at Sciences Po Paris. All the Ph.D. candidates supervised by Colin remain amazed by the speed with which he replied to our emails: his responsiveness is absolutely unequalled. I sincerely thank Colin for having made very detailed comments on each chapter of my work. The challenging questions he asked often become the starting point of new intellectual discoveries.

I am indebted to Dominique Cardon of the Médialab of Sciences Po for his generous help. As the Director of Médialab since 2016, Dominique organised a series of seminars called "Transnum". With the help of Dominique, I had the opportunity to present my research several times at Médialab. The suggestions I collected from the participants had helped me to sharpen my arguments over time. I am also indebted to Cyril Benoît and Caterina Froio, two brilliant young scholars at the Centre for European Studies and Comparative Politics at Sciences Po. Without their kind advice, I could not have been able to find out my way when I felt confused about my research at the end of the first year of my PhD studies. Since this book is based on my Ph.D. thesis, the comments and suggestions of my thesis defence committee are so helpful for improving the argumentation of this monograph. Therefore, I would like to present my sincerest acknowledgements to Guobin Yang, Jean-Louis Rocca, Colin Hay, Chloé Froissart, Rongbin Han, and Séverine Arsène. Their stimulating remarks have pushed me to rethink the definition of key concepts and the robustness test of the regulatory approach conceptualised in this book. I thank my colleagues at the School of Journalism and Communication at Peking University. The discussions with Yong Hu, Gang Chen, Jinnan Tang, Kaihe Chen, Weijia Wang, Zengzhi Shi, Huiyu Zhang, Jing Xu, Jincan Xu, Shijie Guan, Di Lu, Shu He, Jinhe Liu, and Mengyao Wang have provided me with new elements of reflections. I am also grateful of the continuous help and encouragement of Jacob Dreyer, the most efficient editor I have seen. I sincerely thank him for having accepted to publish this monograph and for having invested so many efforts during the editing process.

My gratitude also goes to my family. My grand-parents, who worked on the land for their whole life, taught me the values of diligence and perseverance since my childhood. My parents have shown unreserved love to me. The optimism, patience, resilience, and kindness of my mother remind me that there are no conflicts between working as a respectable researcher, enjoying one's life to the fullest, and being considerate to others. My father keeps telling me to bear some ambition in my life and to try my best when realising each objective. I thank my two lovely nephews, Ma Kai and Liang Yu, for having enlightened my boring life with their chuckles and jokes of joyful adolescents. Last but not least, I express my deepest gratitude for my partner Christian. His unlimited love, tenderness, and tolerance have been indispensable to reassure me. They allowed me to keep focused on completing this project without having to prepare or wash dishes. Love makes miracles happen.

Contents

Abbreviations

CAC	Cyberspace Administration of China
CAITC	China Academy of Information and Communications Technology
CATR	China Academy of Telecommunication Research
CCCCP	Central Committee of the Chinese Communist Party
CC-CCYL	Central Committee of Chinese Communist Youth League
CCIC	Central Cybersecurity and Informatisation Commission
CCI-SLG	Central Cybersecurity and Informatisation Small Leading Group
CCTV	China Central Television
CDO	Central Department of Organisation
CDP	Central Department of Propaganda
CERNET	China Education and Research Network
CFIS	China Federation of Internet Societies
CNCERT/CC	National Computer Network Emergency Response Technical Team/Coordination Centre of China
CNNIC	China Internet Network Information Centre
CPPCC	Chinese People's Political Consultative Conference
CSTNET	China Science and Technology Network
EDR	Economics-Driven Regulation
EISI	Emergent Industries with Strategic Importance
FDI	Foreign Direct Investment
FYP	Five-Year Plan
GAQSIQ	General Administration of Quality Supervision, Inspection and Quarantine
GFW	Great Firewall of China

ICP	Internet Content Provider
ICTs	Internet and Communication Technologies
ISC	Internet Society of China
ISM	Internet and social media
ISP	Internet Service Provider
ISPC	Internet Sector Party Committee
MAU	Monthly Active Users
MEI	Ministry of Electronics Industry
MII	Ministry of Information Industry
MIIT	Ministry of Industry and Information Technology
MPS	Ministry of Public Security
MPT	Ministry of Posts and Telecommunications
MSS	Ministry of State Security
NASSP	National Administration of State Secrets Protection
NCFC	The Networking and Computing Facility of China
NDRC	National Development and Reform Commission
NIISC	National Information Infrastructure Steering Committee
NPC	National People's Congress
NPCPC	National Public Complaints and Proposals Commission
NRTA	National Radio and Television Administration
PAIC	Provincial Administration for Industry and Commerce
PBOC	People's Bank of China
PDR	Politics-Driven Regulation
RFO	Regime-Friendly Objectives
RNR	Real Name Registration
SAIC	State Administration for Industry and Commerce
SAMR	State Administration for Market Regulation
SAPPRFT	State Administration of Press, Publications, Radio, Film and Television
SASAC	State-owned Assets Supervision and Administration Commission
SCIO	State Council Information Office
SCIWSC	State Council Informatisation Work Leading Small Group
SFO	Society-Friendly Objectives
SIIO	State Internet Information Office
SNS	Social Network Service
SOEs	State-Owned Enterprises
TOS	Terms of Services
UGCs	User-Generated Content
VAS	Value-Added Services

List of Figures

List of Tables

CHAPTER 1

Introduction

I took interest in the internet and social media (ISM) in China in September 2016, when I returned from Paris to Beijing and started working at the Embassy of France in China. At that time, after more than 20 years of development, the ISM was already an indispensable component in most Chinese people's daily life. In the morning, when I took the Beijing metro line 6, well known for its crowdedness, to go to the Embassy, almost everybody inside the metro was looking at their smartphone. People who just succeeded in jumping into the train only had very tiny space to put their feet. However, such inconvenience seldom discouraged them from checking smartphones. With one hand grabbing the handrail and supporting the bag on the forearm, they used the other hand to support the smartphone on the screen of which a heated discussion was already under way. Some people apparently discussed with their co-workers, some with their kids to make sure that they arrived at school on time, some with delivery men, some chuckled at the posts they were consulting, and some busy at typing content on the screen without raising their eyes for three or four stations in a row.

Normally, one well-dressed young lady would struggle to move forward in the train, always one glittering smartphone in the hand. She walked slowly, went to everybody one by one, gently showing the QR code of a WeChat public account on her smartphone and trying to convince people to scan it. Most of the time, it was intended to sell

A. Ma, *China as a Double-Bind Regulatory State*,
https://doi.org/10.1007/978-981-99-8857-0_1

products or services. It also happened that the person claimed to collect money for disabled people. Nonchalant passengers, busy with "phubbing" (shortened from "phone snubbing"), did not raise their eyes when talked to. Whatever ISM apps people used, and whatever they did on it, everybody looked so concentrated. I felt like that they simply forgot everything else around them. They paid attention to other people only when they had to rush in or out of the train during the peak hours. Unlucky sorts, utterly grabbed by what they were doing on the smartphone, even forgot to get off at the right station.

I have kept asking myself why Chinese people are so enthusiastic to accept new things, even though they are aware of the dark side of using these fanciful applications, including large-scale data collection. One straightforward explanation is that in China, the ISM makes life easier. WeChat combines the functions of several apps used in Europe and the USA: Facebook, PayPal, WhatsApp, Snapchat, and even Zoom. Most needs of an individual living in China can be satisfied by using WeChat. Another explanation is that the use of the ISM makes individuals feel less frustrated and less helpless in an increasingly consumerist and authoritarian China. The feeling of being connected, which stems from the use of the ISM, works as balm—even though temporary—against the ubiquitous discipline by the state, the capital, and, increasingly, the algorithms.

The destiny of one group of Chinese netizens attracted my attention: entrepreneurs in the ISM sector. Generally graduated from the departments of computer science at prestigious American universities, they returned back to China to create their own business in the late 1990s and the early 2000s. Ebullient and feisty, they jumpstarted the Chinese IT industry and created the first internet boom of the country. Within one decade, their business acquired a global stardom, on a par with their American counterparts GAFAM. The success stories of these private ISM firms and their founders are particularly intriguing in authoritarian China where the party-state's vigilance towards private entrepreneurship is well known (Dickson, 2007).

More than one observer has contributed to build and perfect the image of a coercive, ruthless, and control-hungry Chinese party-state. The message is pretty clear: Chinese society, private businesses included, lives in the shadow of the party-state and does not have any space of autonomy, because there is no limit to the party-state's power and that it can intervene wherever and whenever it sees fit. In her book *Rouge Vif*, Ekman highlights the CCP's seamless social control. For her, the internet and social media serve above all as modern tools of social and ideological control in China. She mentions several facts to back up her argument, including recruiting government-paid trolls to spread pro-regime voices and promoting the learning of Xi Jinping's thoughts among party members (Ekman, 2020: 74, 135). Xi's China is compared to a "surveillance state" and a "dictatorship 2.0", in which the CCP mobilises the ISM to establish a comprehensive social control and to emasculate potential opponents (Strittmatter, 2020). In a nutshell, these writings concur to show a Chinese society enjoying scant (or nonexistent) autonomy from the party-state.

However, empirical evidence does not lend significant support to this argument of an all-powerful party-state. Since the existing literature widely views Xi's presidency as the most authoritarian period in China in the post-Mao era, I will precisely focus on this period to empirically refute the argument of "omnipotent party-state". If it can be shown that private ISM firms and netizens under Xi enjoy some manoeuvring room, it can be logically inferred that it existed under Xi's less authoritarian predecessors, too.

The Private ISM Firms' Room to Manoeuvre

...continues to exist under Xi Jinping, as shown, among others, by their growing influence. From 2017 to 2021, the percentage of the digital economy over the national GDP has risen from 32.9% to 39.8%,[1] dwarfing the 9% in the USA (Nicholson, 2020: 2) and the 7% in the EU (European Central Bank, 2020). Apart from the BATHX (Baidu, Alibaba, Tencent, Huawei, Xiaomi), the ISM firms created after these five behemoths are no less competitive. In 2020, the Research Centre of China's Private Enterprises (*Zhongguo minying qiye yanjiu zhongxi*) of Renmin University released the ranking of the world's top 500 unicorn firms.[2] 6 of the top 10 unicorns were from China, and all operated in the ISM

[1] From 2017 to 2021, the digital economy represents respectively 32.9%, 34.8%, 36.2%, 38.2%, and 39.8% of the Chinese GDP. See the *Digital China Development Report* released by the Cyberspace Administration of China in 2018, 2019, 2020, 2021, and 2022.

[2] In the 47th *China Statistical Report on Internet Development*, the CNNIC defines a unicorn firm in the ISM sector as a "firm the value of which is estimated at 1 billion dollars as of the latest financing activity and which is not yet listed on the stock exchange market" (p. 4).

sector. ByteDance ranked first, *ex-aequo* with the American manufacturer of driverless cars Wagmo.[3] From 2013 to 2020, Chinese E-commerce firms have been holding the largest online retail market in the world.

Chinese citizens' exceptional enthusiasm in creating ventures in the ISM sector also challenges the argument of an all-powerful party-state. In December 2020, the capitalisation of Chinese ISM firms, listed at home and at abroad combined, achieved roughly $2588.6 billion, an increase of 51.2% compared with December 2019.[4] At the end of 2019, China counted 187 unicorn firms among its ISM industry, and this number rose to 207 by December 2020.[5] The great majority of Chinese ISM firms are privately owned.

Given the party-state's vigilance towards private entrepreneurs (McGregor, 2010; Clark, 2016; Dickson, 2003, 2007). The success stories of private ISM firms are truly intriguing. What matters is not that they now face greater restrictions (embodied by the anti-trust regulatory move since November 2020) than before. What deserves investigation is why, in the past 20-odd years, they have been able to prosper in a communist regime. The CCP's power is seldom checked in China. In order to ensure the regime's survival, the party-state instinctively controls the society—business and citizens included. Control is the nature of the Chinese party-state (Ringen, 2016). Therefore, it would be more logical to see state-owned firms, far more trusted by the party-state than private firms, prosper in the profitable and increasingly strategic ISM industry. Why is it that an authoritarian regime like China, where strict state control is supposed to stifle creativity, has managed to become a big power in the creativity-demanding ISM industry? Is the success of private ISM firms voluntarily nurtured by the party-state or do they fortuitously escape from its grip in their rise to national, and definitely international, stardom?

In China, the coexistence between political authoritarianism and economic liberalism, the latter being epitomised by the creation of a genuine tech boom sustained by private tech giants, deserves special attention. It is not often present in other places than in China: either an autocracy is unable to nurture a booming tech industry, as in Russia;

[3] "Ranking of the Top 500 Unicorn Firms in the World in 2020," China Industry News, 4 January 2021, https://www.cinn.cn/dfgy/202101/t20210104_237346.html.

[4] CNNIC, *The 47th China Statistical Report on Internet Development*, February 2021.

[5] Ibid., p. 4.

or a genuine tech boom turns up in democracies, as in the USA. China is not only the biggest autocracy, but also the most digitally connected one in the world. What is even more striking is that the more the ISM-driven economy gallops, the more authoritarianism deepens in China (Bandurski, 2018). According to some preeminent political scientists, although economic growth alone does not necessarily lead to democratisation, it is generally a favourable factor to democratisation (Przeworski and Limongi, 1993). This statement does not hold true in China: the economic growth sustained in part by the ISM industry has not generated democratisation, and the possibility of this perspective is slimmer. If the Chinese case is an exception to confirm the rule, then why is it in China, rather than elsewhere, that such an exception exists and continues to grow?

The Puzzle

As shown above, although the party-state disposes strong capacities enabling it to tame private ISM firms in myriad ways if it so chooses (McNight et al., 2021: 13), private ISM firms in China have managed to prosper in their own way. The coexistence between political authoritarianism and economic prowess of private ISM firms turns China into a *sui generis* autocracy. It is *sui generis* because in the bilateral interactions between the party-state and ISM firms, the party-state often gains ground, but does not win definitively over firms. It does not matter how much ground the party-state gains each time or at which frequency. What is astonishing is that it never triumphs definitively. Is it because the party-state voluntarily chooses not to win, so as to keep the game going in China? Or is it simply not able to win, being outwitted by agile and highly adaptive ISM firms?

These questions lump together to form the core research question of the book: *in an authoritarian China where the apparently all-powerful and unchecked party-state has been trying to consolidate its grip on power, how can one explain the existence of a relatively autonomous space for private ISM firms, where the latter have grown into the mainstay of national economy, to the extent that they are now even perceived to be a threat to the party-state's authority*? The relationship of the party-state with private ISM firms and netizens constitutes the two faces of the same coin. Investigating these relations will clarify why the Chinese society continues to enjoy a significant manoeuvring room against the Chinese authoritarian backdrop.

The "Double-Bind Regulatory State" Hypothesis

This book scrutinises the "manufacturing" of the manoeuvring room for private ISM firms in authoritarian China. Its central argument is that in the regulation of the ISM industry, the Chinese party-state acts as a "***double-bind regulatory state***". In state-firm relations, the party-state as the meta-regulator pursues two contradictory sets of objectives that I call respectively as RFOs and SFOs. I define RFOs as the objectives that ensure the regime's survival and the realisation of which may imply that the party-state does not act in a rule-abiding manner. For instance, arbitrary power and non-justified violence are mobilised to serve RFOs. RFOs include, above all, national security, social stability, and ideological leadership of the party-state. SFOs stand opposite to RFOs. I define SFOs as the objectives that benefit more the society than the party-state and the realisation of which implies that the party-state acts in a rule-abiding manner. In this article, SFOs include principally, but not exhaustively, technological innovation, economic upscaling, the possibility for private firms to "voice" criticisms (Hirschman, 1970), and citizens' enjoyment of high-quality internet services (Austin, 2014).

Although both sets of objectives are indispensable for the regime's survival, the party-state has to simultaneously pursue contradictory logics and courses of actions to realise them. When it moves to RFOs, the party-state tends to enact centralised regulation and exercises more control over ISM firms. In contrast, when it moves to SFOs, the regulatory process features greater decentralisation, with local governments being delegated greater competences. The control of the party-state over ISM firms lessens accordingly. Since the regulators pursuing RFOs and SFOs are not the same and that they compete for resources and competences, the party-state must arbitrate between them, so that the realisation of one set of objectives does not damage the realisation of the other set of objectives.

This book argues that the contradiction between RFOs and SFOs is indispensable to set DBR in motion, for it creates a governance predicament for the party-state. RFOs and SFOs do not always clash (e.g., the development of digital surveillance technologies benefits both the party-state and firms). However, more often than not, they are incompatible, for they imply opposite rationales and methods in the course of being achieved. I argue that the party-state takes on a *split personality* when pursuing RFOs and SFOs: while it behaves like an unpredictable and self-serving "dictator" when seeking after RFOs, it becomes a reliable and public-regarding "democrat" when pursuing SFOs. In addition, while pursuing RFOs often implies tighter and more arbitrary control from the party-state, pursuing SFOs pushes it to leave greater autonomy to firms. Given the importance of both SFOs and RFOs for the regime's survival, the party-state cannot pursue RFOs all the time, for this would throw private firms into total uncertainty and discourage them from delivering high-quality services to citizens. The party-state must pursue SFOs, too. Its constant alternation between RFOs and SFOs opens autonomy-enabling opportunities for private firms.

Why national security and social stability are put into RFOs, not SFOs? When construed broadly, both can also be categorised into SFOs, for they create a propitious environment for the blossoming of private businesses. The reason why I include them in RFOs is that for the party-state, the fact that national security and social stability can simultaneously secure its power and serve public interest matters differently. Ensuring the regime's survival is the top concern of autocrats (see Olson, 1993). Therefore, putting "national security" and "social stability" into RFOs is coherent with Chinese political authoritarianism, although this classification does not mean that they are opposite to SFOs.

Depending on the party-state's priorities in specific periods (e.g., emphasis on informatising the national economy under Jiang Zemin; taming tech firms to realise "common prosperity" under Xi Jinping), it tends to periodically favour SFOs or RFOs. However, when examined as a whole, the party-state has kept alternating between the two sets of objectives, causing the level of autonomy of private firms to fall and rise accordingly. I argue that the double-bind logic, being inherent in the party-state's governance logic, opens a window of opportunities through which private ISM firms can seek, and have sought, margins of manoeuvring. The very fact that the party-state cannot afford to abandon either set of objectives makes it safe to say that ISM firms in China continue to enjoy manoeuvring room, however authoritarian the party-state may appear.

The DBR approach does away with actor-centric and incident-centric approaches, which take single actors or single incidents as units of analysis. The analytical approaches centred around individual actors or incidents have produced biased understanding of the power relations in the regulation of private ISM firms in China. Since they easily and/or selectively exaggerate the influence of certain actors or incidents, misrepresentation of the power relations between stakeholders into regulatory processes emerges. DBR differs from the existing approaches, for it sees the party-state and firms as forming a dialectic and interactive ensemble, and takes the interplays between actors as units of analysis. In DBR, the margins of manoeuvring of regulatees stem from their continuous wrestling with regulators. Invalidating the conventional wisdom portraying the Chinese party-state as omnipotent, omniscient, and omnipresent, DBR reveals that structural-level dynamics do exist in China and put limits on the growth of political authoritarianism in China.

Key Concepts

Three terms are essential to understand the unfolding of DBR in China: party-state, autonomy, and ISM (internet and social media) firms. This section defines each of them to pinpoint the scope of the discussion in the book. I interchangeably use "party-state", "China", and the "Chinese government", although "party-state" appears more frequently. The latter is conceptually more appropriate to capture the essential features of the Chinese political system: one-party rule with weak counter-powers. Therefore, "party-state" is frequently used by preeminent scholars in Chinese studies (Shambaugh, 2008; Ringen, 2016; Cabestan, 2018). This reflects a well-known fact: the party rules the state (Ringen 2016: 1). This peculiar institutional design is closely related to the centrality of the CCP in China: "the CCP is the cornerstone of the Chinese political system, and, in many ways, is the political system" (Cabestan, 2017). Therefore, describing China not simply as a "state" but as a "party-state" is more accurate.

The centrality of the CCP in China justifies the use of "party-state". The CCP is different from political parties in democratic countries in two aspects: party status and inter-party relations. In terms of party status, the CCP is a permanent ruling party in China. Its leadership is enshrined in the Chinese constitution. In contrast, political parties in democracies must compete with each other before being enabled to govern. On inter-party relations, the Chinese party system is a hierarchical one: political parties don't have the same status. Although China counts 9 political parties, only the CCP is constitutionally allowed to rule. It would be anti-constitutional for any of the eight democratic parties to become the ruling party. In democracies, every political party can become the ruling party, as long as it wins the elections.

The second term is "autonomy". Firms' autonomy from the party-state's control is at the heart of the book. Dworkin is a leading scholar on theories and practices of autonomy. He defines autonomy as being represented, "in desperate brevity, by the formula autonomy = authenticity + independence" (1976: 24). Authenticity and independence relate to each other in the following way: "by identifying with one's motives a person makes them hers; but only when this identification is made in a procedurally independent way are these motives her own. They are 'hers' in that they have become aspects of her identity" (Haworth, 1991: 131). These philosophical reflections help pinpoint the meaning of "autonomy" in China. This book does not simply understand "autonomy" as describing people's acting freely. Instead, it considers autonomy as a relational concept which denotes firms' manoeuvring room in their wrestling with the party-state. Firms are considered as acting "autonomously" when they defend their "autonomy" from being gnawed by external actors (e.g., autocrats).

Third, ISM (the internet and social media) firms refer to the profit-seeking entities providing two types of ISM services: content services (e.g., online gaming, instant messaging, live-streaming, video-sharing, etc.) and non-content-services (e.g., E-commerce, ride-hailing, food delivery). It is worth pointing out that the scope of the book is limited to private firms, rather than state-owned enterprises (SOEs). This choice is motivated by two considerations. On the one hand, SOEs operate in the economic sectors with commanding heights or the sectors which do not attract private capital due to long payback periods for investments. Such sectors include energy, aviation, telecommunications, railways, non-ferrous metals, etc. In contrast, although the internet infrastructure is controlled by SOEs (China Mobile, China Telecom, and China Unicom), most firms in the downstream sectors of the internet industry are privately owned. This book considers the business activities in these downstream sectors as forming the ISM industry.

On the other hand, the close linkage between SOEs and the party-state makes it conceptually erroneous to qualify their relationship as "regulation". The SASAC, affiliated to the State Council and enjoying full control over the core company in each SOE group, has direct and formal oversight on SOEs in terms of their business strategies (e.g., overseas investment, pricing, initial public offering, mergers and acquisitions). SOEs are not governed by Chinese corporate law which applies to private companies only (Lin and Milhaupt, 2013). Top-level leaders of SOEs, including CEOs, vice-CEOs, and party secretaries, are not designed by the board of directors, as private firms do. They are appointed by the Central Department of Organisation (CDO) and the SASAC. Therefore, there is no organisational separation between SOEs and the party-state, whereas organisational separation is one component in the definition of regulation (Noll, 1980; Mitnick, 1980: 7). Given the greater independence of private businesses from the party-state's interference, this book focuses on the regulation of private firms.

It is equally essential to define what types of firms are considered as ISM firms. The definition of both the internet and social media entails three dimensions: technological, social, and economic. I define the internet as an infrastructural platform where information circulates and social interactions (including business activities) take place. Social media are defined as "techno-social platforms" (Surowiec and Stetka, 2018: 5) that "increase our ability to share, to co-operate, with one another, and to take collective action, all outside of the framework of traditional institutions and organisations" (Shirky, 2008: 20–21). The definition of Shirky is mainly drawn from the perspective of citizen empowerment. Obar and Wildman enrich the definition of social media by introducing the commercial dimension (2015: 746–747). They consider "social media" services as composed of four commonalities: (1) social media services are (currently) Web 2.0 internet-based applications; (2) user-generated content is the lifeblood of social media; (3) individuals and groups create user-specific profiles for a site or app designed and maintained by a social media service; and (4) social media services facilitate the development of social networks online by connecting a profile with those of other individuals and/or groups.

Given that ISM firms' economic activities fully fall in the regulatory scope in China, this book considers the commercial dimension as an integral part of the definition of social media. ISM firms studied here are profit-seeking entities and involve four types of firms which (1) provide internet and social media services (e.g., Baidu, Bytedance, Tencent, Sina), (2) rely upon the internet and social media to provide services (e.g., Didi, Meituan, Alibaba, JD.com, Ctrip), (3) manufacture essential parts of the internet and social media (e.g., Xiaomi, Huawei), or (4) in cases where a firm provides both online and offline services, its revenues principally come from ISM-based services (e.g., Ping An Group, etc.).

Contributions to Existing Literature

The conceptualisation of DBR draws from four bodies of literature: regulation studies, communication studies, Chinese political economy, and the political sociology of social movements. In return, DBR contributes to each of these four parental bodies of literature in its own way.

Regulation Studies

The overwhelming majority of regulation studies focus on the regulation practices in democracies. Hsueh (2011) and Yeo (2020) partially address this problem by extending regulation studies to authoritarian regimes like China. DBR further tailors existing regulation studies to increase their adaptability to China. The tailoring is conducted by incorporating three interrelated factors that have been insufficiently discussed in the existing literature: the authoritarian context, the ISM sector, and netizens. The importance of using regulation theories to study the ISM sector in authoritarian contexts is that this sector has both crucial political and economic stakes for autocrats. Since the existing literature principally studies the sectors representing above all economic stakes, research into the ISM sector helps extend the scope of regulation studies.

While existing regulation studies mostly focus on the regulation of firms, DBR incorporates netizens as a regulable entity, too. It conceptualises the regulation as covering state-society relations as a whole, rather than state-market relations only. This is justified by the particularity of authoritarian contexts: interest divergence between autocrats and citizens. The regulation of citizens is no less important than the regulation of private firms under authoritarian settings. In autocracies, the regulation serves to influence and correct regulatees' behaviour (as in democracies), on the one hand, and to consolidate autocrats' hold on to power, on the other hand. Therefore, no one is allowed to escape from the net of control, firms and citizens alike. This book takes netizens as a stakeholder in their own rights in the regulation of private ISM firms. It studies netizens as a game-changer in influencing state-firm power relations.

The three leading regulatory approaches (the neoliberal approach, the statist approach, and the "market-institutional" approach, see Yeo 2020) construe the state regulation as how far the state position itself from the market. What distinguishes DBR is that it does not concern itself solely or principally with the state-market distance. Likewise, it is not mainly interested in regulators' selectiveness in deciding which sectors to regulate more intrusively than other sectors (Hsueh, 2011). Instead, it is akin to a theory of "wrestling" in which the distance between involved parties is irrelevant. Instead of measuring how far regulators stand next to regulatees, DBR captures the constraints pushing regulators to step in and out of regulatees' zone of autonomy. The originality of DBR is that it associates state regulation with state-society power relations in authoritarian regimes, a dimension under-analysed in existing studies. DBR will portray a dialectic movement: while authoritarian regulators mobilise regulation to sustain authoritarian rule, regulatees exploit regulators' predicament to resist encroachments on their autonomy.

Communication Studies

Existing Chinese communication studies are characterised by a dichotomous reading on who is empowered by the ISM. One set of scholars argues that the ISM empowers citizens, enabling them not only to develop a nascent civil society (Tai, 2006; Tai and Sun 2007; Hu, 2008; Wang and Shi 2018; Esarey & Xiao 2011), but also to push progressive democratisation in China (Xiao, 2011). The second set of scholars argues that the ISM empowers the party-state, enabling it to have sharpened tools to control the society (MacKinnon, 2011; Noesselt, 2014; Bandurski, 2018; Economy, 2018). This dichotomous reading on the role of the internet in Chinese politics interestingly echoes that on the functions of the internet in world politics in general. While scholars like Benjamin Barber (1984, 1998) and Henri Oberdorff (2010) argue how information and communication technologies (ICTs) are a boon for democracy and democratisation (e.g., lowering the costs of citizens' participation in democratic affairs), other scholars point out that ICTs do not necessarily play out this role, and that internet diffusion does not automatically generate political changes (Norris, 2001). Captured into what Rongbin Han calls a "liberalisation-control" paradigm (2018), this dichotomous reading has provided precious insights into the respective advantages of the party-state and the Chinese society in their wrestling. However, this paradigm has become increasingly problematic, for predicting the victory of either the party-state or firms or netizens is overly hasty. The changeability of state-society power relations over time and stakeholders' agency in reversing established power relations have been under-studied.

Different from these two directions, a more nuanced approach has attracted attention. Han highlights the "pluralisation" of online expression in Chinese cyberspace. As he argues, "the impact of the internet on Chinese politics is much more mixed and complicated than a dyadic model in which either the society or the state dominates" (2018: 5). DBR goes to a similar direction. Its contributions to communication studies are that it analyses state-firm relations as an ongoing cyclical movement, rather than a one-off game with clear-cut outcomes. Instead of focusing on either the state or firms, which sustains some single-sided accounts of either powerful state or agile internet firms, DBR puts the two actors in an arena where a mix of different combinations awaits them. The pursuit of competing objectives pushes the party-state to alternate between mobilising more control and less control over ISM firms, causing the latter's level of autonomy to fall and rise. Firms can increase their autonomy by pushing the party-state to accelerate its move to SFOs and to stay there longer.

Portraying state-firm relations as a constant game of wrestling, DBR recognises a multitude of possible outcomes. In so doing, it avoids being tempted to predict the victory or the failure of any of the involved parties. More concretely, it is not because the party-state reinforces its control over private internet firms that it can shrink their manoeuvring room. The efficiency of internet control is influenced by how firms react to it. Likewise, it is not because ISM firms have significant economic power that their manoeuvring room grows. Given the selectivity of the structure (party-state) vis-à-vis the strategies of agents (private ISM firms) (Hay and Wincott, 1998: 954; Hay, 2002, 2009: 267), whether the economic power of firms translates into their tangible autonomy is influenced by the party-state's priority concerns at specific moments.

This book brings another contribution to communication studies: since the double-bind logic stems from regulators' built-in constraints, the manoeuvring room opened for ISM firms will persist as long as the regime remains unchanged. It is built-in in China. The double-bind logic functions like a curse casted to the regime and will only break down when the necessity of pursuing competing objectives disappears. For instance, if the ISM can enable China to transition from an autocracy to a democracy, the double-bind logic is expected to become dysfunctional. The reason is that in a democracy, since power transition is inscribed into constitutional arrangements, the incumbent political party can/does not use the regulation as a policy instrument to maintain itself in power. In this sense, DBR not only transcends the dichotomous reading of the control-liberalisation paradigm, but also reconciliates the two sides of arguments constituting it.

Chinese Political Economy Studies

Chinese political economy studies are characterised by the emphasis on the party-state's capacity to bring business, private and public alike, under control. The control over SOEs is realised through SASAC. For private firms, their private ownership does not protect them from the encroachments by public actors. The party-state can ensure its oversight over private firms through capital control (Pairault, 2013), cooptation (Dickson, 2003, 2007; Milhaupt and Zheng, 2015), subsidies (Milhaupt and Zheng, 2015: 685), creation of party cells (Xue, 2018), or "extralegal control" (Milhaupt and Zheng, 2015: 685–688). Therefore, scholars concur that ownership is a poor indicator for evaluating state-business relations in China. However, although it is true that the party-state has diverse means to control private business, it is regretful that scholars have not deployed equal endeavours to reflect on the means by which private firms can protect themselves. In addition, it is important to examine whether firms' overt obedience is intended to help their covert disobedience.

The contributions of DBR to Chinese political economy studies are twofold. First, instead of studying the party-state's powerfulness and inventiveness in controlling private firms, DBR examines how private ISM firms are not powerless in resisting encroachments. More importantly, I argue that Chinese ISM firms' manoeuvring room is embedded in the Chinese political system, because the source of their manoeuvring room (namely the party-state's predicament) is built in the regime's governance logic. Concurring that ISM firms do commit disobedience for pursuing profits, DBR goes one step further by wrapping up firms' non-compliance for economic concerns into a larger explanatory framework. It identifies the root causes of corporate disobedience in the governance logic of the regime itself.

Second, DBR highlights that the ownership structure of private firms does matter in influencing state-firm relations. The existing literature largely agrees that in China, the ownership structure of firms poorly explains how they are regulated (Pairault, 2013; Milhaupt and Zheng, 2015). DBR provides an alternative interpretation. It argues that the influence of firms' private ownership differs depending on whether it is the EDR (economics-driven regulation) or the PDR (politics-driven regulation) at stake. For the EDR, this books joins the existing literature: the way in which the party-state controls SOEs and private ISM firms is similar. However, concerning the PDR, firms' private ownership structure matters in explaining state-firm relations. The analysis of the content regulation rules in the book shows that private firms are much more limited in their rights to collect and report hard news, in contrast to state-owned media. While the latter can report self-collected information, private firms must, at least in principle, parrot hard news of state media.

Social Movement Theories

If the contestation movements have happened in democracies more frequently than in autocracies, they have been "normalised", acquiring a "planetary dimension" (Mayer, 2009: 433). In authoritarian contexts, investigating the resistance of firms is equally important than investigating the resistance of citizens. DBR makes three contributions to social movement studies. First, by examining the margins of manoeuvring of private ISM firms in China, DBR helps clarify the resistance of firms against unfavourable regulations and their chances of outsmarting authoritarian regulators. Second, it challenges the two-end polarised reading of state-society relations in China. The polarisation, captured into "repression everywhere or resistance everywhere", is illustrated by the obsession of absolutely identifying a winner and a loser in state-society interactions, neglecting the temporality and the multiplicity in the acts of winning or losing over someone. The literature highlighting the party-state's all-powerfulness is particularly attracted by this obsession. DBR mitigates the polarisation by incorporating the party-state's governance logic into the factors that influence protesters' political opportunity structure. It does not provide one-off diagnostics about who wins over whom on the sole basis of the factors with high visibility. It dives into the invisible and analyses state-society relations as an endless game in which the outcomes of one round participate in shaping the game configuration for the next round. In DBR, there are no winners and losers. There are only opportunities and constraints that the participants of the game try to turn in their favour.

Third, it provides a more structuring analytical framework to "anarchic" and "individualistic" "forms of resistance" (Hibou, 2011: 18) or "everyday forms of resistance" (Scott, 1985: 36). When studied separately, these diverse forms of piecemeal resistance illustrate protesters' inventiveness in authoritarian settings. However, when studied within DBR, piecemeal resistance is no longer as simple as a contingent choice, but an inevitable one by citizens and firms. When the party-state moves from SFOs to RFOs, the cost of not mobilising tight control overtakes the benefits of using less control, thus shrinking protesters' political opportunity structure. The repression of overt deviance intensifies to secure the regime. Piecemeal resistance under such contexts (e.g., mockery, dual discourses, coded languages) is an inevitable strategy to survive coercion. When such seemingly inoffensive resistance erodes the legitimacy of the regime, it pressures autocrats to move away from RFOs to SFOs, thus increasing the manoeuvring room of firms and citizens.

Methodology and Data

This book represents the results of five years of extensive field research in China and France from 2019 to 2022. It rests upon six sources of empirical data (Table 1.1).

The collection methods of three sources of data, including interviews, ISM-related regulations, and biannual reports of the CNNIC, need to be explained at length because they are mobilised in a general way in the book. For the other types of data used for more specific purposes, I explain their collection methods in corresponding chapters.

Interviews

Using the snowballing strategy, I conducted 52 in-depth semi-directive interviews from 2018 to 2022 in Paris, Beijing, Qingdao, three rural villages of Qingdao, and on the internet. I interviewed six types of people: government officials, firm managers and employees, scholars, users, lawyers, and journalists. Interviews lasted between 40 minutes and 2 hours. Since the Covid-19 paralysed international travels and that China

Table 1.1 Empirical data and collection period

Data types	*Quantity*	*Period covered by the data*	*Other remarks*
Interview	52	–	Beijing, Qingdao, three rural villages of Qingdao, Paris, Zoom, WeChat, phone, email; data collected from 2018 to 2022
Biography analysis	143	2001–2020	128 leaders of the ISC (Internet Society of China); 15 leaders of the CFIS (China Federation of Internet Societies)
Laws and statutes	473	1987–2022	National-level laws, statutes, legal interpretation, and ministerial decisions related to the regulation of the ISM sector
CNNIC reports	51	1997–2022	Biannual statistical reports on Chinese internet development
Party building	100	2020	Creation of party cells in the top 100 Chinese ISM firms in 2020
Online contentious cases	170	2002–2022	16 analytical reports of the public opinion on the Chinese internet, compiled by the Public Opinion Supervision Bureau of people.com.cn

applied the zero-covid policy, I interviewed most China-based informants through WeChat, Zoom (zoom is not blocked in China), or phones.

Since my research related to the control of the party-state over private internet firms, it was not always easy to make people talk. Directly mentioning the word "censorship" or "control" often panicked interviewees. Therefore, I used several techniques to reassure them. First, when I felt that the interviewee was unwilling to talk, I suggested that she talk about the experiences of her friends or colleagues. This method worked well, because it allowed the interviewee not to take personal responsibilities for her words. Second, I used the interview method "focus group", allowing several interviewees who knew each other to be interviewed at the same time, often at the behest of interviewees. Since the interaction between participants "stimulates the identification and sharing of various perspectives on the same topic" (Carter et al., 2014: 545; Morgan, 1996), my interviewees became more talkative when interviewed as focus groups.

Third, each interviewee was given full guarantee of anonymity, including no mentioning of her personal information in the book, no showing of her face during recorded Zoom or WeChat calls, no recording of the interviews without consent, and the use of the interview data for no other purposes than the present research. In the end, 38 out of 52 interviews were recorded. For the other 14 interviews, I took notes during or after the interview, depending on whether the fact of taking notes in front of interviewees discouraged them from talking.

I went through a learning curve when conducting interviews on the field. Like Pierre Bourdieu who tended to talk more than his interviewees, I also "break the rhythm of the interview and encourage the interviewees to passivity, waiting for the next question" (Mayer, 1995: 363), at least for the initial interviews I effectuated. I remembered having interviewed one eminent French scholar in internet law in his office in Paris. Since my questions were too precise, I talked much more than him. He was bored and told me that he had an appointment. The interview finished much earlier than scheduled. After that, I adjusted my interview methods, leaving interviewees with more time to speak. The interviews in 2020 and 2021 were more informative than those in 2018. I learnt to "listen, not be afraid of silence, put myself in the other person's shoes and help him explore his thoughts without influencing him, without imposing my thoughts on him" (Mayer, 2018: 12).

Since the research question touches upon internet censorship, which is a highly sensitive topic in China, guaranteeing interviewees' full anonymity is necessary to ensure that they cannot be identified and that they don't get into trouble for their participation in this research project. The anonymisation method goes as follows: each interviewee is presented according to the city and the date of interview. The individuals interviewed more than once count more than once as well. Since the on-site interviews were conducted in Beijing, Qingdao (including the 3 rural villages of Qingdao), and Paris, I represent these three cities respectively by the capital letters B, Q, and P. Therefore, one example of marking an on-site interview can be "B-24/12/2019". Regarding the interviews conducted on WeChat, Zoom, or Email, they are all coded as "O" followed by the date of interview. One example of a coded online interview can be "O-23/10/2020". If more than one interview is effectuated within one day, according to the order, I put a number behind the initial letter of the city. For instance, during my stay in Beijing in December 2019, since many interviews had been planned before my arrival to the field, I often conducted several interviews in one day. Therefore, terms like "B1-21/12/2019", "B2-21/12/2019", and "B3-21/12/2019" mean three interviews conducted in one day, but with different persons.

Laws and Statutes on ISM Regulation

I collect all the laws, statutes, decisions, legal interpretation, and notices on the ISM regulation that Chinese national-level regulators have adopted from 1987 to 2022. Given regional discrepancies in regulation priorities (e.g., provinces in the west of China regulate much less than provinces in the east), policies adopted by local governments significantly vary from region to region. To prevent regional differences from hindering the generalisability and representativeness of research findings, all the collected documents in the database are released by the central government. Since the regulations approved by national regulators reflect the party-state's priority concerns and that they are taken into account in the policymaking at the local level, they constitute reliable and representative sources for studying the national-level ISM regulation. Collected regulations are adopted by two types of regulators: economic regulators and political regulators.

Economic regulators studied in this book are in charge of planning and deciding economic policies in the ISM sector, ranging from investment policies, anti-trust policies, technological innovation policies, and labour policies. Economic regulators include:

- State Council;
- Ministry of Industry and Information Technology (MIIT);
- National Development and Reform Commission (NDRC);
- Ministry of Science and Technologies (MST);
- State Administration for Market Regulation (SAMR, after 2018).

Political regulators studied in this book take charge of content regulation and ideological guidance. They also ensure the security of the key internet infrastructure of China (e.g., international gateways, backbone networks). They include

- Cyberspace Administration of China (CAC);
- Ministry of Culture;
- National Administration for Radio and Television (NART);
- Ministry of Public Security (MPS);
- Central Department of Propaganda (CDP).

Laws and decisions adopted by the National People's Congress (NPC) also count as national-level regulations. I do not classify NPC into economic regulators or political regulators for two reasons. First, the regulation of the ISM industry in China is mainly effectuated by low-hierarchical rules, such as ministerial and departmental rules. The mobilisation of high-hierarchy rules, like laws, is much less frequent (Miao et al., 2021). Second, while some of the laws and decisions of NPC relate to the economic regulation of the ISM sector, others relate to its political regulation. Therefore, NPC can be considered as both an economic regulator and a political regulator. The compiled database includes 473 regulations.

This database helps reveal the regulation patterns under each leadership team in China, including (1) regulatory agencies and their principal responsibilities, (2) regulatory priorities, and (3) strictness/looseness of the regulatory environment. It also informs how the central government has been designing the regulation context from the late 1980s onwards. Frequent amendments and adjustments to the same policy (e.g., three adjustments to the licensing system in 2000, 2005, and 2017) often reveal the importance of the issue to Beijing. In addition, the number of laws and statutes issued each year reveals the evolution of the regulation intensity in China. That the Xi Jinping administration has issued the highest number of regulations since 1987 (284 out of 473) shows its high attention to the ISM sector and signals the creation of a stricter regulatory environment.

CNNIC Reports 1997–2022

To ensure the systematicity of the findings, I collect all the 51 statistical reports that the CNNIC has released from November 1997 to September 2022. The 51st report released in March 2023 reflects the internet development in China until December 2022, namely the end point of the book. Created on 3 June 1997, the CNNIC is affiliated to the MIIT. Since November 1997, it has been releasing statistical reports in the middle and the end of each year. Although the first report is issued in November 1997, its data also cover 1994, 1995, and 1996. The CNNIC reports are the only official sources of information on Chinese internet development that have been released without discontinuity from the late 1990s until today. They report, among others, on the internet penetration rate, urban-rural discrepancy, digital economy, netizens' demographic structure, and their habits in using ISM platforms. CNNIC reports provide a precious source enabling researchers to obtain a systematic view of the Chinese internet development.

The CNNIC reports are generally reliable data. As the builder, operator, and manager of the fundamental internet resources in China, the CNNIC has the privileged position for keeping track of the Chinese internet development. Since it is affiliated to an economic regulator (the MIIT) and not an ideology-leaning political regulator, it faces less political pressure for data manipulation or concealment. That CNNIC reports have been widely quoted in Chinese communication studies is also a sign of data reliability. Lastly, these reports are highly relevant to public authorities and private businesses for their decision-making. For instance, the information on the rural-urban digital discrepancy is important for the central government to decide whether or not to move public services online, in what regions to do so, and what to do for the regions with a lower internet coverage. Given the importance of its data for the policymaking authorities, it goes to the CNNIC's interest to increase, not weaken, the reliability of its reports.

Organisation of the Book

The book contains eight chapters. After this introductory chapter, Chapter 3 conceptualises the DBR approach. The conceptualisation daws insights from regulation studies, communication studies, political economy, and social movement theories. Chapters 4 and 5 adopt a historical perspective. They respectively investigate the economic and political particularities of the ISM regulation in China.

Backed up by extensive empirical evidence, Chapters 5, 6, and 7 focus on the "manufacturing" of the party-state's double-bind logic and its implication on the ISM regulatory landscape in China. Each chapter develops an in-depth analysis of one specific pair of competing objectives of the party-state.

Chapter 5 discusses the contradiction between political and economic goals. Given that the pursuit of these two sets of objectives mobilises different regulators, rationales, and approaches, the party-state has to navigate between them. That it is caught in-between empowers ISM firms. Aware of their economic importance, firms engage themselves into "inverse cooptation", a process where firms exploit their status of "coopted entrepreneurs" to seek favourable policies.

Chapter 6 discusses "fragmented authoritarianism" (Lieberthal and Oksenberg, 1988) unfolding at the horizontal and vertical levels of the Chinese political system. Both levels of fragmentation generate manoeuvring room for firms, enabling them to ally with one group of regulators to resist other groups of regulators.

Chapter 7 shows that the party-state's double-bind logic also stems from the contradiction between its aspirations to global leadership and the continuation of domestic information control. The more the party-state aspires for global leadership, the more it becomes important to woo the international audience. The relevance of the international public opinion to the party-state opens a window of opportunity for private ISM firms, because they can ally themselves with international audiences to reduce the costs of committing disobedience.

Chapter 8 answers the question raised in the introduction. That the party-state faces competing objectives in the ISM regulation explains how the manoeuvring room of private ISM firms emerges: since the party-state can rule neither as an arbitrary autocrat nor as a rule-abiding democrat, it must muddle through SFOs and RFOs all the time and scrupulously control the intensity of authoritarianism so as not to be swallowed by it. Its constant navigation between SFOs and RFOs enables private ISM firms to seek opportunities to manoeuvre. Chapter 8 introduces external robustness test and sectoral robustness test of DBR. On the one hand, it tests the validity of DBR in Singapore, the ISM regulation of which significantly resembles that of China. On the other hand, it tests the validity of DBR in another sector than the ISM in China, namely the cosmetic industry. The dual robustness test sheds light on the generalisability of the DBR approach.

References

Austin, Greg. 2014. *Cyber Policy in China*. Cambridge, UK: Polity.

Bandurski, David. 2018. "Can the Internet and Social Media Change the Party?" In *Routledge Handbook of the Chinese Communist Party*, edited by W.W.-L. Lam, 372–390. Oxon, UK: Routledge.

Barber, Benjamin. 1984. *Strong Democracy : Participatory Politics for a New Age*. Berkeley, CA: University of California Press.

Barber, Benjamin. 1998. "Three Scenarios for the Future of Technology and Strong Democracy." *Political Science Quarterly* 113(4): 573–589.

Cabestan, Jean-Pierre. 2018. *Demain la Chine: démocratie ou dictature ?* Paris: Gallimard.

Cabestan, Jean-Pierre. 2017. "The Party Runs the Show: How the CCP Controls the State and Towers over the Government, Legislature and Judiciary." In *Routledge Handbook of the Chinese Communist Party*, edited by W.W.-L. Lam, 75–91. Oxon, UK: Routledge.

Carter, Nancy, Denise Bryant-Lukosius, Alba Dicenso, Jennifer Blythe, and Alan J. Neville. 2014. "The Use of Triangulation in Qualitative Research." *Oncology Nursing Forum* 41(5): 545–547.

Clark, Duncan. 2016. *Alibaba: The House that Jack Ma Built*. New York: HarperCollins Publishers.

Dickson, Bruce. 2003. *Red Capitalists in China: The Party, Private Entrepreneurs, and Prospects for Political Change*. Cambridge, UK: Cambridge University Press.

Dickson, Bruce. 2007. "Integrating Wealth and Power in China: The Communist Party's Embrace of the Private Sector." *The China Quarterly* 192: 827–854.

Dworkin, Gerald. 1976. "Autonomy and Behavioural Control." Hastings Center Report 6: 23–28.

Economy, Elisabeth. 2018. *The Third Revolution: Xi Jinping and the New Chinese State*. Oxford: Oxford University Press.

Ekman, Alice. 2020. *Rouge Vif: l'Idéal Communiste Chinois*. Paris: Editions de l'Observatoire.

Esarey, Ashley, and Qiang Xiao. 2011. "Digital Communication and Political Change in China." *International Journal of Communication* 5: 298–319.

European Central Bank. 2020. "The Digital Economy and the Euro Area." *Economic Bulletin* 8.

Han, Rongbin. 2018. *Contesting Cyberspace in China: Online Expression and Authoritarian Resilience*. New York: Columbia University Press.

Haworth, Lawrence.1991. "Dworkin on Autonomy". *Ethics* 102(1): 129–139.

Hay, Colin, and Daniel Wincott. 1998. "Structure, Agency and Historical Institutionalism." *Political Studies* 46: 951–957.

Hay, Colin. 2002. *Political Analysis: A Critical Introduction*. Basingstoke: Palgrave.

Hay, Colin. 2009. "King Canute and the 'Problem' of Structure and Agency: On Times, Tides and Heresthetics." *Political Studies* 57: 260–279.

Hibou, Béatrice. 2011. *Anatomie Politique de la Domination*. Paris: La Découverte.

Hirschman, Albert. 1970. *Exit, Voice and Loyalty: Responses to Declines in Firms, Organisations, and States*. Cambridge, MA: Harvard University Press.

Hsueh, Roselyn. 2011. *China's Regulatory State: A New Strategy for Globalisation*. Ithaca, NY: Cornell University.

Hu, Yong. 2008. *Zhong Sheng Xuan Hua: Wang Luo Shi Dai De Ge Ren Biao Da Yu Gong Gong Tao Lun* [The Rising Cacophony: Personal Expression and Public Discussion in the Internet Age]. Guilin: Guang Xi Shi Fan Da Xue Chu Ban She [Guangxi Normal University Press].

Lieberthal, Kenneth, and Michel Oksenberg. 1988. *Policy Making in China: Leaders, Structures and Processes*. Princeton, NJ: Princeton University Press.

Lin, Li-wen, and Curtis Milhaupt. 2013. "We Are the (National) Champions: Understanding the Mechanisms of State Capitalism in China". *Stanford Law Review* 65(4): 697–759.

MacKinnon, Rebecca. 2011. "Liberation Technology: China's 'Networked Authoritarianism'." *Journal of Democracy* 22(2): 32–46.

Mayer, Nonna. 1995. "L'entretien selon Pierre Bourdieu, analyse critique de La misère du monde". *Revue Française de Sociologie* 36(2): 355–370.

Mayer, Nonna. 2009. "Opinion Publique". In *Dictionnaire des mouvements sociaux*, edited by Olivier Fillieule, Lilan Mathieu and Cécile Péchu, 429–435. Paris : Presses de Sciences Po.

Mayer, Nonna. 2018. "Qualitatif ou quantitatif ? Plaidoyer pour l'éclectique méthodologique." *Bulletin de Méthodologie Sociologique* 139: 7–33.

McGregor, Richard. 2010. *The Party: The Secret World of China's Communist Rulers*. New York: HarperCollins Publishers.

McNight, Scott, Martin Kenney, and Dan Breznitz. 2021. "Platformizing the Economy? Building and Regulating Chinese Digital Platforms." Available at SSRN https://papers.ssrn.com/sol3/papers.cfm?abstract_id=3885190.

Miao, Weishan, Min Jiang, and Yunxia Pang. 2021. "Historicizing Internet Regulation in China: A Meta-Analysis of Chinese Internet Policies (1994–2017)". *International Journal of Communication* 15: 2003–2026.

Milhaupt, Curtis, and Wentong Zheng. 2015. "Beyond Ownership: State Capitalism and the Chinese Firm." *The Georgetown Law Journal* 103: 665–722.

Mitnick, Barry M. 1980. *The Political Economy of Regulation: Creating, Designing and Removing Regulatory Forms*. New York: Columbia University Press.

Morgan, David L. 1996. "Focus Groups." *Annual Review of Sociology* 22: 129–152.

Nicholson, Jessica R. 2020. "New Digital Economy Estimates." Bureau of Economic Analysis, U.S. Department of Commerce, 1–32. https://www.bea.gov/system/files/2020-08/New-Digital-Economy-Estimates-August-2020.pdf.

Noesselt, Nele. 2014. "Microblogs and the Adaptation of the Chinese Party-Stat's Governance Strategy." *Governance* 27(3): 449–468.

Noll, Roger G. 1980. "What Is Regulation?" Social Science Working Paper 324, California Institute of Technology.

Norris, Pippa. 2001. *Digital Divide : Civic Engagement , Information Poverty, and the Internet Worldwide*. New York: Cambridge University Press.

Obar, Jonathan A. and Steve Wildman. 2015. "Social Media Definition and the Governance Challenge: An Introduction to the Special Issue." *Telecommunications Policy* 39(9): 745–750.

Oberdorff, Henri. 2010. *La démocratie à l'ère numérique*. Grenoble: Presses Universitaires de Grenoble.

Olson, Mancur. 1993. "Dictatorship, Democracy and Development." *American Political Science Review* 87(3): 567–576.

Pairault, Thierry. 2013. "Les entreprises chinoises sous la tutelle directe du gouvernement illustré par leur investissement en Afrique." *Revue de la Régulation* 13, https://journals.openedition.org/regulation/10195

Przeworski, Adam, and Fernando Limongi. 1993. "Political Regimes and Economic Growth." *The Journal of Economic Perspectives* 7(3): 51–69.

Ringen, Stein. 2016. *The Perfect Dictatorship: China in the 21st Century*. Hong Kong: HKU Press.

Scott, C. James. 1985. *Weapons of the Weak: Everyday Forms of Peasant Resistance*. New Haven: Yale University Press.

Shambaugh, David. 2008. *China's Communist Party: Atrophy and Adaptation*. Washington, DC: Woodrow Wilson Center Press.

Shirky, Clay. 2008. *Here Comes Everybody: The Power of Organizing Without Organizations*. New York: The Penguin Press.

Strittmatter, Kai. 2020. *Dictature 2.0 : Quand la Chine surveille son people (et demain le monde)*. Translation by Olivier Mannoni. Paris: Tallandier.

Surowiec, Pawel, and Vaclav Stetka. 2018. "Introduction: Social Media, Politics and Democracy in Post-Transition Central and Eastern Europe." In *Social Media and Politics in Central and Eastern Europe*, edited by V. Stetka & P. Surowiec, 1–20. London: Routledge.

Tai, Zixue. 2006. *The Internet in China: Cyberspace and Civil Society*. New York: Routledge.

Tai, Zixue, and Tao Sun. 2007. "Media Dependencies in a Changing Media Environment: The Case of the 2003 SARS Epidemics in China." *New Media & Society* 9(6): 987–1009.

Wang, Hongyu, and Fayong Shi. 2018. "Weibo Use and Political Participation: The Mechanism Explaining the Positive Effect of Weibo Use on Online Political Participation Among College Students in Contemporary China." *Information, Communication & Society* 21(4): 516-530.

Xiao, Qiang. 2011. "The Rise of Online Public Opinion and Its Political Impact." In *Changing Media, Changing China*, edited by S. Shirk, 202–224. Oxford: Oxford University Press.

Xue, Xiaorong. 2018. *Hu Lian Wang Qi Ye Dang Jian: Ji Shu Zi Ben Yu Zheng Zhi Luo Ji Zhang Li Xia De Zheng Dang Zhi Li* [Party-Building in Internet Firms: Party Governance under the Logics of Technology, Capital and Politics]. Beijing: Current Affairs Press.

Yeo, Yukyung. 2020. *Varieties of State Regulation: How China Regulates Its Socialist Market Economy*. Cambridge, MA: Harvard University Press.

CHAPTER 2

A Conceptual Scheme for the Double-Bind Regulatory State

Introduction

Coined initially by Anglophone researchers, regulation "has become a multi-disciplinary field, with substantial contributions to regulatory debates being made by political scientists, lawyers, sociologists, anthropologists, and others" (Baldwin et al., 2010: 4). Although regulation has reached maturity in both intellectual and practical senses of the term (Baldwin et al., 2010: 5), it continues to cause debates about its definition, involved actors, intentionality, objectives, and organisational separation between regulators and regulatees. Persistent academic disagreement on these issues explains the difficulties in building a widely accepted definition. Therefore, the definition of regulation adopted in this book is voluntarily broad and includes only the strictly necessary features of regulation. The rationale is that detailed prescribing of the patterns required for qualifying a public action as "regulation" will inevitably disqualify quite many autocracies, because the more the definition is granular, the more it becomes circumscribed to democracies where regulation studies originate.

Those strictly necessary features of regulation are identified and entailed in the essence-based definition of regulation and a pattern-based one (Koop and Lodge, 2017). The former defines regulation as "the intentional intervention in the activities of a target population", while the latter defines it as "intentional intervention in the activities of a

A. Ma, *China as a Double-Bind Regulatory State*,
https://doi.org/10.1007/978-981-99-8857-0_2

target population, where the intervention is typically direct – involving binding standard-setting, monitoring and sanctioning – and exercised by public-sector actors on the economic activities of private-sector actors" (Koop and Lodge, 2017: 104–105). The pattern-based definition is abandoned in this book, because Chinese ISM regulators regulate both economic and non-economic activities. Moreover, in regulator-netizen relations, non-economic activities are the priority concern of regulators. The pattern-based definition is too strict. It excludes from the cluster of regulatory states the countries whose public actions fail to live up to its expectations. Therefore, this book uses the essence-based definition which allows greater conceptual adaptability and creativeness.

Two Typologies of Regulation Studies

Regulation studies originate in advanced economies the economic activities of which have been the principal objects of study for a long period of time. Based on regulation objectives, Benoît (2018) establishes the typology of three leading regulation theories that have hugely influenced regulation studies from the 1930s onwards: the public interest theory, the capture theory, and various scholarly endeavours using a more eclectic view of the relations between the actors involved in the regulation (Laffont and Tirole, 2012; Carpenter and Moss, 2014). The public interest theory (also called market failure approach) considers that regulators' primordial objectives are to urge private firms to deliver public goods because firms' competition and pursuit of profits are prone to damaging public wellbeing. Therefore, influencing economic actors to maximise social wellbeing is believed to be the goal of all regulation (Coglianese and Lazer, 2003). The capture theory turns up in the 1970s. Instead of considering regulators as altruistic, it contends that powerful firms corrupt regulatory agencies to maximise their profits (Stigler, 1971). Therefore, the results of one regulatory intervention represent the outcomes of bargaining between regulators and powerful firms (Benoît, 2018).

Since the 1990s onwards, scholars have been trying to overcome the opposition between the public interest approach and the capture theory. For instance, among several criticisms, Carpenter and Moss (2014) highlight that both approaches remain blind to the permanent reconfigurations between the public and private in any regulatory process. They propose that the variety of institutional configurations from one economic sector to another be taken into account in regulation studies. The emphasis on whether a regulator is independent or is captured by powerful firms has reduced, because a regulator is seldom entirely independent or entirely captured in the regulation. In addition, the role of ideas started being granted greater importance in regulation studies.

Yeo (2020) provides a second typology which differs from the typology by Benoît in two respects. First, Yeo's typology is based on regulation patterns, not objectives. Second, the empirical basis in Yeo's typology covers both democratic and authoritarian settings. She conceptualises the market-institutional approach to overcome the inadaptability of the neoliberal and the statist approaches to China. She then applies it to Vietnam to test its robustness. She argues that although both China and Vietnam have socialist institutions, Vietnam is still different, due to "the fragile regulatory authority of Vietnam's planning agency and the Communist Party over the business of leading sectors" (Yeo, 2020: 119). Therefore, highlighting the spearheading role of socialist politico-economic institutions in designing the regulation, Yeo incorporates autocracies into the big family of regulatory states.

The neoliberal reforms in the UK and the USA since the 1970s onwards provide empirical materials for developing the neoliberal approach (Yeo, 2020: 134). Besides privatisation and deregulation, national governments create independent regulatory agencies for promoting fair competition, and there is no differentiated treatment of firms. In so doing, regulators act as independent referees and do not confound with players, to use the metaphor of Pearson (2005). The statist approach recognises the selectiveness in national governments' boosting of economic sectors. Japan and South Korea, for instance, voluntarily supported big business corporations, represented respectively by "keiretsu" and "chaebol". National governments "orchestrated" market competition to increase the international competitiveness of selected actors and to nurture national champions. Although the statist approach captures Chinese state-market relational patterns, since it "is grounded largely on advanced liberal-market economies (such as Japan or France), it is devoid of socialist politico-economic institutions, such as a planning bureau or the Communist Party" (Yeo, 2020: 139). Consequently, the statist approach "fails to properly explain how those resilient socialist institutions have complicated and even reproduced statist efforts in establishing regulatory institutions for a market-oriented economy" (Yeo, 2020: 139).

As one variation of state regulation, the market-institutional approach differs from the neoliberal approach because it considers market competition as orchestrated, not genuine. It differs from the statist approach because socialist politico-economic institutions, which are absent in the statist approach, spearheads the market competition in the market-institutional approach. The latter analyses the state intervention as a two-layered process. First, at the cross-sectoral level, the strategic value that one economic sector represents for the state shapes the regulatory intensity. The more a sector is perceived as strategically important, the more the regulation is intense. Second, within the same sector, the regulation intensity depends on the ownership level of firms. Firms owned at the central level are subject to more intensive regulation than firms owned at the local level.

Extensive empirical explorations have been done in regulation studies. Sectors as diverse as health (Benoît, 2020; Carpenter, 2010), ticket-pricing (Finez, 2014), aviation (Woll, 2008), telecommunications (Woll, 2008; Hsueh, 2011; Yeo, 2020), automobile (Yeo, 2020), and textile (Hsueh, 2011) have been thoroughly studied. Sociological research on regulation is related to long-distance transport, manufacturing, or university safety management (Gray and Silbey, 2014). These writings, some of which are cross-country (Lodge, 2002; Benoît, 2020) or cross-sector studies (Hsueh, 2011; Yeo, 2020; Woll, 2008), have generated important insights on how regulators, regulatees, and market conditions interact with and shape each other.

Four Weaknesses of Current Regulation Studies

Despite theoretical and methodological maturity, regulation studies have four weaknesses. First, most regulation studies focus on democracies. Originating from and largely tested in democracies, regulation remains under-studied in autocracies like China, although this research agenda is attracting attention. For instance, brilliant research has been done on the market regulation in China (Yeo, 2020). Nevertheless, while regulation studies in democracies have achieved a high level of conceptual systematicity, allowing the theories to travel beyond single countries, regulation studies on autocracies lag behind in terms of conceptualisation and generalisation. On conceptualisation, most existing research touches upon the regulation of economic activities, leaving the regulation of the activities with political importance underexploited. On generalisation, the existing research tends to consider China as an exceptional case, the particularity of which prevents it from discussing with the non-Chinese world. Lack of comparison and insufficient "banalisation" of the Chinese case (Rocca, 2006: 19) explain the continuous accumulation of particularity-centred narratives. It hinders the pan-authoritarian systematisation of regulation studies.

Second, the ISM representing both high economic and political stakes for national governments has attracted less attention than the sectors whose economic importance clearly prevails over the political one. The bias of economic centrality is particularly harmful for regulation studies in autocracies. Since the grip onto power is one of the major concerns of autocrats, regulation serves to realise autocrats' political objectives as much as (if not more than) economic ones. The academic lacunae in this field are currently particularly acute, because existing regulation studies on authoritarian regimes largely focus on other sectors than the ISM. In addition, although the existing research does scrutinise the relationship between national governments and ISM firms (Jiang, 2012; Negro, 2017), scholars tend to use "governance" and "regulation" interchangeably, without fully considering the differences between them. As a result, the association between the ISM and regulation studies remains tenuous.

"Governance" and "regulation" mean different things. While "governance" emphasises "different modes of action coordination", regulation is related to various "forms of deliberative collective actions in matters of public interest" (Mayntz, 2009: 121–122). Thus, regulation is narrower than governance in its scope. Lastly, national governments, regardless of being democratic and authoritarian, are currently undergoing similar processes: unleashing the economic potential of the ISM sector on the one hand and increasing state intervention into ISM firms on the other hand (Ma, 2023c). The cross-regime convergence in the regulation of the digital market makes it necessary to mobilise a more systematic approach than an exploratory one to make sense of what we are living today. Regulation studies provide an ideal explanatory framework for doing so, for they specifically address how national governments influence private actors.

The third weakness relates to the inability of existing regulation studies to capture who are regulated in autocracies. They consider firms as the principal actors on which regulation acts upon, and tend to neglect individuals as stakeholders in the regulation in their own rights. Unequal attention to firms and netizens stems from the fact that regulation studies take roots in the West where there exists interest alignment between regulators and citizens. Since citizens in democracies enjoy plentiful institutionalised channels (e.g., regular elections, media supervision, inter-party supervision, demonstrations, strikes) to hold regulators accountable, the costs of being caught violating citizens' interests are particularly high for regulators. Interest alignment between regulators and citizens makes the latter theoretically non-regulable under democratic rules.

In autocracies, regulators being accountable to autocrats more than to citizens, interest alignment between regulators and citizens is tenuous. Regulation becomes one of the policy instruments enabling autocrats to self-serve. Moreover, since citizens have few effective channels to hold autocrats accountable, the costs for regulators to deviate from citizens' interests are relatively low. Interest divergence between regulators and citizens, on the one hand, and a low cost for regulators to maintain such divergence, on the other hand, make it both necessary and feasible to regulate citizens, such that autocrats' interests take precedence over citizens' interests. Abundant empirical data on autocrat-citizen struggles in autocracies (Yang, 2009a; Han, 2018) show that treating citizens as regulatees in their own rights is fully legitimate.

Fourth, alongside the negligence of citizens as regulatees, existing research has granted stronger emphasis on the role of the party-state or specific political leaders in spearheading a policy process. This focus is partly due to the scholarly interest in the *sui generis* nature of the one-party rule in China and the high visibility of the party-state's actions in media. It is also explained by the fact that data on firms and netizens in autocracies are not always available and reliable. Therefore, a complete picture on the interactions among all stakeholders in the ISM sectors is lacking. Based on extensive first-hand empirical data (see Chapter 1), this study delves into the Chinese regulation of the ISM industry which is comparable to a miniaturised Chinese society grouping all the major stakeholders. This systematic study is expected to bring valuable insights, for it represents the views of three actors (the party-state, firms, and citizens), rather than two actors or one actor only.

State-society relations vary considerably within and between democracies and autocracies. Therefore, it is necessary to build a more eclectic regulatory approach, rather than an approach solely or principally centred on democracies. If the objective to build such an encompassing regulation theory workable in both democracies and autocracies can hardly be achieved in the foreseeable future, scholarly endeavours should at least by deployed to work out an approach well suited to autocracies in general or to one specific autocracy. That is a more constructive and feasible undertaking than trying to figure out a too generalised approach with universal explanatory power for all types of political regimes. This is where this book hopes to bring its theoretical contributions. By focusing on one specific industry in one specific autocracy, it narrows the gap between existing regulation theories and the reality of regulation in an authoritarian regime, and tailors a China-specific regulation theory which can be further tested in other authoritarian settings. This book allows to apprehend the regulation of the ISM sector not through an entrenched Western prism, which still prevails, but with a more eclectic analytical framework.

Bearing the aforementioned lacunae in mind, this book investigates the Chinese regulation of the ISM sector from 1987 to 2022. As the start point of this study, 1987 is the year in which China used the internet to send the first email out of the country. As the end point, 2022 marks the end of Xi Jinping's second presidential term. The time span of 35 years covers four full leadership teams: Deng Xiaoping (1987–1989), Jiang Zemin (1989–2002), Hu Jintao (2002–2012), and Xi Jinping (2012–2022). The length of the chosen period allows to obtain systematic views on the ISM regulatory patterns under each leadership. Theoretically, this book takes up the dual challenge of identifying regulatory patterns specific to China, on the one hand, and banalising the Chinese case through integrating it into a pan-authoritarian explanatory framework, on the other hand. By applying regulation studies to the ISM sector in autocracies and considering individual citizens as regulates in their full rights, this book "tailors" existing literature and adapts it to authoritarian regulation practices. In what follows, I will build a new regulatory approach, called "Double-Bind Regulation" (DBR) approach.

Nurturing the Double-Bind Logic in an Autonomised State

Scholars inscribed into Marxist traditions tend to deny the state its autonomy, considering the state as an interest-maximising machine for elites. Contradicting Marxist scholars, Mann (1984) and Carpenter (2001) argue for the autonomous power of the state, autonomy reflected in the independence of the state from being captured by ruling elites (Mann, 1984) and the powerfulness of entrenched bureaucrats vis-à-vis elected politicians (Carpenter, 2001). Inspired by Mann and Carpenter, this part conceptualises the Chinese party-state as a political entity the increasing institutionalisation of which leads to its autonomisation. Instead of dictating their interests and objectives to the state, ruling elites see the realisation of their objectives conditioned by the priorities of the state. The autonomisation of the state power is an essential building block of the double-bind regulatory approach, because, with autonomisation, the state acquires its own objectives.

Autonomisation of the State Power

For scholars in Marxist traditions, the state is a contingent superstructure determined by the economic basis, and its vocation is to disappear when the private property of the means of production and the antagonism of the classes cease existing (Mendras and Etienne, 1996: 58). Marxists portray the state as a machine that the ruling class uses to maximise its interests. This is "perhaps the most prevalent conception of the state within Marxist theory" (Hay, 1999: 154). In the *Manifesto of the Communist Party*, Marx and Engels wrote that "the history of all hitherto existing society is the history of class struggles" (1848). Class confrontation, central to Marxism, implies that the exploitation is realised through oppressors' instrumentalisation of the state as an interest-maximisation machine. The elite's control of the state is both material and ideological. The Gramscian notion of "ideological hegemony" (Gramsci, 1979) denotes how the ruling class appropriates the state and legitimises it by changing their ideology into the ideology of the whole society. In so doing, the exploited class "actually accepts its situation as a normal, even justifiable part of that social order" (Scott, 1985: 39).

That the Marxist and, to a lesser extent, the neo-Marxist views do not recognise the state as autonomous explains the "parasitic" nature of the state (Jessop, 1977: 354) in which the state depends on the ruling class for being set in motion. This view has attracted wide academic controversies and has been rejected by the scholars arguing for the opposite direction (Mann, 1984; Skocpol, 1979; Hintze, 1975; Tilly, 1981; Giddens, 1981). The state extracts resources from the society and bases its autonomous power upon such resources. The autonomy of the state enables it to run "against either the dominant class, or against domestic war or peace factions and foreign states" (Mann, 1984: 187). In the famous Miliband-Poulantzas debate between the "instrumentalist" and the "structuralist" theories of the state (Barrow, 2002: 4), for instance, Poulantzas argue that "if the function of the state in a determinate social formation and the interests of the dominant class coincide, it is by reason of the system itself: the direct participation of members of the ruling class in the state apparatus is not the cause but the effect" (Poulantzas and Miliband, 1972: 247).

Autonomisation of the State Power in Democracies and Autocracies

In contrast to Marxist scholars who tend to be shy on the issue of state autonomy, some scholars recognise that the state does acquire autonomy from the ruling class. Mendras and Etienne argue that the state can gain its autonomy if the ruling class is in decline, or if the conflicts between classes or between factions of the classes end up with neutralising the competing classes, which will allow the state to temporally become an arbitrator (1996: 59). To illustrate the autonomy of the state, the two scholars quote the example of France under Napoléon III, a period in which the French bureaucracy squeezed, controlled, regulated, and supervised the civil society (Mendras and Etienne, 1996: 59). Michael Mann goes one step further, arguing that the state has always been autonomous from elites' control. He divides the power of the state into despotic power and infrastructural power. The despotic power of the state concerns the "range of actions which the elite is empowered to undertake without routine, institutionalised negotiations with civil society groups" (1984: 188). The infrastructural power of the state means the "capacity of the state to actually penetrate civil society, and to implement logistically political decisions throughout the realm" (1984: 188). While the despotic power is more related to the privileges of the elite class, the infrastructural power denotes the autonomy of the state. To illustrate the autonomy of the state from the elites, Mann wrote:

> President Nixon or M. Chaban-Delmas may have paid no taxes; political leaders may surreptitiously amass wealth, infringe the civil liberties of their opponents, and hold onto power by slyly undemocratic means. But they do not brazenly expropriate or kill their enemies or dare to overturn legal traditions enshrining constitutional rule, private property or individual freedoms. On the rare occasions this happens, we refer to it as a coup or a revolution, an overturning of the norms. If we turn from elected politicians

> to permanent bureaucrats, we still do not find them exercising significant autonomous power over the civil society. (1984: 190)

This quote illustrates that the state is autonomous because the "despotic power" of the elites is locked into the institutionalisation of the "infrastructural power". The locking leads above all to the autonomisation of the state, the functioning of which is dictated more by institutionalised rules than by elites' (dis)likings. Although the autonomisation unfolds differently depending on democracies and autocracies, what remains unchanged is the relative independence of the state from the instrumentalisation by elites. The autonomisation in democracies is reflected into the bureaucratic independence from politicians' control. Taking the example of the rural free delivery developed in the USA at the end of the nineteenth century, Carpenter argues that the initiation and the coming into fruition of this project is in large part thanks to the entrepreneurship of bureaucratic agencies, rather than congressional majorities. For him, "moving with the veneer of efficiency and neutrality, and backed by coalitions that support this reputation and the agency's program innovations, autonomous bureaucrats take command over policy. Politicians who oppose their designs are able to resist them only at great cost: rationally, they usually defer to the agency's wishes" (Carpenter, 2001: 113). That a bureaucratic official in the USA is often affiliated to a political party is "an asset to his bureaucratic entrepreneurship, not a liability" (Carpenter, 2001: 117). It implies that bureaucrats defend the turf of their agency before defending their party.

The autonomisation of the state in China unfolds differently, because, except at the grassroots level, China has no democratically elected officials in the strict sense of the term. Therefore, the autonomisation is reflected above all by the institutionalisation of the party-state. In China, although the regime's functioning is still based on the rule by law, the institutionalisation allows the party-state certain autonomy from elites, for institutionalisation makes it difficult for elites, grouped into the *nomenclatura* system, to arbitrarily exercise their power. The "authoritarian dilemma of modernisation" (Lei, 2018: 6), term coined by Ya-wen Lei, captures the predicament of Chinese elites in their course of institutionalising the functioning of the party-state. After building the legal, economic, and political institutions for the sake of pursuing socio-economic development of the country, it becomes difficult for elites to exercise arbitrary control over the party-state. The dilemma facing them is that they cannot reap the benefits of being the dominating class without at least appearing to subject themselves to the order of the law (Lei, 2018: 8). The freedom of action of the elites is constrained by the rules they themselves have fixed. It is certainly true that the elites spearhead the creation of the party-state from 1921 to 1949. However, upon the termination of the creation process, instead of being an appendage of the elites, the party-state acquires its own *raison d'être*.

This institutionalisation process and its demonstration have been widely studied (Nathan, 2003; Shirk, 2018; Cabestan, 2018). Calling the peaceful power succession as the "centerpiece of the effort to prevent dictatorship and regularise political competition" (2018: 22), Shirk writes that the power transition from Jiang to Hu in 2002 "was the first time that any ruler of a communist nation had left office without dying or being deposed by a coup" (2018: 23). Likewise, power transition from Hu to Xi in November 2012 also features peacefulness in Chinese succession politics. What is even better in the 2012 transition is that Hu achieves a complete retirement, enabling Xi to take all the three core functions of the party-state: General Secretary of the CCP, President of China, and President of the Central Military Commission.

Nathan (2003) highlights four dimensions in the institutionalisation of the party-state in the post-1989 China: (1) the increasingly norm-bound nature of its succession politics, (2) significant increase in meritocratic as contrasted to factional considerations in the promotion of cadres, (3) differentiation and functional specialisation of institutions of the regime, and (4) the establishment of channels for political participation to strengthen the party's legitimacy. Cabestan analyses the institutionalisation of the party and the state separately. For him, while the party remains opaque and weakly institutionalised, significant modernisation and professionalisation do characterise the functioning of the state (2018: 30–35). He recognises that "generally speaking, the administrative machine works, and works well in general, and even works better and better, if we compare it with the end of the Maoist era" (2018: 31). The institutionalisation creates the party-state's stability, which contrasts with continuous changes among elites. As Perry emphasises, "while the composition of the political elite has changed dramatically since Mao's day (reflecting, among other things, an exponential growth in its educational credentials), its organisational structure remained remarkably stable" (Perry, 2007: 9).

Three Implications of the Autonomisation of the State Power

The autonomisation has three implications on the interpretation of state-society relations. First, it enables the state to run against elites' interests and makes it difficult for elites to instrumentalise the state. An autonomised state is a non-capturable prey and, to some extent, is not a prey. Autonomisation implies that despite their dominant socio-economic position in the society and the privileges coming along, elites cannot realise their objectives by changing their personal objectives/interests into the objectives/interests of the state. The automatic association between the state and ruling elites disappears. Instead, elites must realise their interests by operating within the state structure and continuously negotiating and battling with the ruled (Mann, 1984). Autonomisation "normalises" or banalises state-elite relations, tearing apart the conventional idea according to which the state "automatically" serves ruling classes. Applied to China, the autonomisation of the party-state explains the party-state's development of its own objectives, which are not necessarily in line with the objectives of the ruling *nomenklatura*. The matching between the party-state's objectives and those of elite becomes a contingency, not a fatality.

Portraying the party-state as automatically serving the ruling elites' interests constitutes the pitfall that the existing literature frequently falls prey to, as if the interests of the two were indistinguishable from each other. Considering that China has been exercising the one-party rule since 1949, the attempts to establish an automatic association between the two are particularly strong in the existing Chinese studies. The literature on the preeminent role of the CCP in driving Chinese political and economic life is particularly characterised by the tropism of interest alignment between the party-state and its elites. In her analysis of the CCP's genuine adherence to communism, Ekman (2020) treats the elites not only as constituting a homogeneous group, but also as indistinguishable from the party-state. Xi Jinping, probably the most frequently used term in her book, seems to be able to dictate the party-state on the sole basis his personal will. Ekman's analysis leaves readers with the impression that Xi Jinping is the Chinese party-state itself, while this is not true. Despite the richness of her book, it exaggerates the capacity of the ruling elites to dictate the functioning of the party-state. It neglects the constraints that the institutionalised party-state imposes on the elites' exercise of power.

It also neglects the impact of internecine struggles among the elites on their capacity to dictate the party-state, if that capacity does exist.

Second, the autonomisation of the state produces the autonomisation of the society, making sense of citizens' victories in outwitting the elites for defending their interests, regardless of regime types where elite-citizen battles unfold. The reason is that the techniques mobilised by the elites for consolidating their power are aptly appropriated by citizens. In China, the power of citizens co-evolves with that of the ruling class, and citizens excel in identifying counter-policies in response to higher-level authorities (Rocca, 2006: 46). Likewise, in the ideological domain, the capacity of the ruling class to transform its own values into the values of the whole society is challenged. While Gramsci argues that the ruling class "engineer consent for their rule" by controlling both material and symbolic means of production (e.g., culture, religion, education and media) (Scott, 1985: 39), Scott rejects the capacity of the ruling class to do so. Taking the example of the peasant resistance against landlords in Malaysia, he illustrates how landlords have failed to generate consent:

> If bandits and poachers are made into folk heroes, we can infer that transgressions of elite codes evoke a vicarious admiration. If the forms of outward deference are privately mocked, it may suggest that peasants are hardly in the thrall of a naturally ordained social order. If those who try to curry the personal favour of elites are shunned and ostracised by others of their class, we have evidence that there is a lower-class subculture with sanctioning power. (Scott, 1985: 41)

The autonomisation of the state transforms the state into a neutral "arena" (Mann, 1984: 210). In it, despite their control of overwhelming material and symbolic means of production, elites do not automatically win over citizens. Instead of staying in the grandstand, elites must descend into the "arena" and wrestle there to defend their privileges. The autonomisation of the state power opens plenty of possibilities of power configuration between the ruling and the ruled, one of which is that the ruled win over the ruling. The dictatorship of the proletariat that Lenin and Trotsky instate in Russia after the 1917 Revolution, for instance, illustrates the state as a neutral wrestling arena between opposing social groups. As an arena, the state does not automatically favour one group over the other, and the power relations between the ruling and the ruled become less preordained and less Manichean than as portrayed by Marxist and neo-Marxist scholars.

Third, the autonomisation of the state implies that the ruling class is conditioned by the configuration of opportunities and constraints inherent in the state structure when they try to champion their interests. The objectives compatible with the state structure are more likely to be realised than those incompatible with it. Bureaucratisation is an important sign of the rise of the modern state (Mann, 2012; Weber, 1946). Based on the ten constituent elements of bureaucracy (Weber, 1978: I:220–221), Mann encapsulates them into five in his definition of bureaucracy:

> Bureaucrats are officials (1) separated from ownership of office by an employed, salaried status and (2) appointed, promoted and dismissed according to impersonal criteria of competence. Bureaucratic offices are (3) organised within departments each of which is centralised and embodies a functional division of labour; (4) departments are integrated into a single overall administration, also embodying functional division of labour and centralised hierarchy. Finally, bureaucracy presupposes (5) insulation from the wider society's struggle over values.

Bureaucratisation replaces kinship with impersonal links, thus squeezing the arbitrary power of the ruling elites. Bureaucratisation has a long history in China. Since the creation of the Qin dynasty in BC 221, China has always been an administrative state (Cabestan, 2018). Bureaucratisation and its improvements in later historical periods create a stable structure in China which is relatively impervious to external changes. Therefore, while the way in which the ruling elites change is often brutal and sudden, the changes to the state structure have been much more incremental. The ruling elites go, and the state remains. The coming onto stage of the CCP is no exception, because the realisation of the interests of the party elites depends on the opportunities and constraints that the bureaucratised state presents.

Recognising the party-state as autonomised is the foundation for any objective analysis of state-society relations in China, because conceptually liberating the party-state from automatic instrumentalisation by the CCP opens diverse configurations of power relations. "Powerful state and powerless society" are only one of these configurations. Most crucially, it pushes scholars to watch, with equal attention, the actors other than the

CCP itself, to relativise the latter's all-powerfulness, and to portray citizens and firms as dynamic actors capable of negotiating their manoeuvring room on an equal footing with the ruling elites. In contrast, not recognising the autonomisation of the state precludes other possibilities than the most visible one: the activism of citizens and firms is doomed to fail. It puts scholars in two lethal tropisms: the tropism of interest alignment between the state and the elites, on the one hand, and the tropism of powerful state subduing the powerless society, on the other hand.

From an Autonomous Party-State to a "Double-Bind State"

After explaining the theoretical foundations of the autonomisation, this section further pinpoints the nature of the Chinese party-state. If the latter is autonomised, what does it look like and how does it work? It is certainly true that China is an authoritarian regime. Nevertheless, simply associating China with authoritarianism is too broad to convey any precise information. This section examines several adjectives scholars have worked out to qualify Chinese authoritarianism and advances the approach of "double-bind authoritarianism".

Scholars have found several adjectives to pinpoint authoritarianism in the Chinese way: "consultative" (Teets, 2013: 35), "bargaining" (Lee and Zhang, 2013), "resilient" (Nathan, 2003), "adaptive" (Shambaugh, 2008), "pragmatic" (Lai, 2016), and "populist" (Tang, 2016). Important attention has been paid to qualify political authoritarianism under Xi Jinping. Scholars concur that China under Xi is more authoritarian than before. Xi is portrayed as applying a "personalistic rule" (Shirk, 2018), a "more closed, disciplinary type of authoritarianism" (Howell and Pringle, 2018), and taking China down on a "neo-totalitarian" path (Béjà, 2019). Power concentration under Xi motivates the British scholar Elisabeth Economy to term Xi' era as the "third revolution" in China (2018). Compared with his predecessors, "Xi and his team have set out to run the race differently... They have moved away from a collective leadership to elevate Xi as the preeminent leader, deepened the role of the Communist Party and state in society and in the economy..." (2018: 10).

Despite these diverse adjectives, scholars largely recognise that China has its particularity that established typologies can hardly capture. Nathan famously declares that China remains "an authoritarianism of a still poorly understood new type" (2009: 38). Although taking inspiration from the

Marxist-Confucianist ideological register, China has crafted hybrid categories both vis-à-vis the Chinese culture and the political models known in the West. The literature on "consultative", "bargaining", or "deliberative" authoritarianism, for instance, puts China in-between democracy and autocracy. Scholars agree that the party-state does take the initiative of enlarging public participation. However, it holds the agenda-setting power and controls the venues of "consultation", "bargaining", or "deliberation", so as to ward off the risks of regime change (He and Warren, 2011). Similarly, the literature on the party-state's adaptiveness finds the Chinese particularity is that the regime is constrained neither by ideological dogma nor by its roots of being a revolutionary mass party. Instead, it incorporates capitalism into socialism to create not a "socialism with Chinese characteristics", but a "capitalism with Chinese characteristics" (Huang, 2008). The diversity of the adjectives for pinpointing which authoritarianism China practices shows that the only certainty concerning the political nature of China is the uncertainty of finding a workable definition for it.

This book focuses on one of the Chinese particularities. It is the fact that numerous internal contradictions (e.g., statism and entrepreneurship, political monopoly and individual liberty, etc.), instead of shortening the regime, have contributed, at least currently, to its longevity. China has survived these contradictions thanks to its resilience and adaptability. I partially agree with these two explanatory frameworks, because since their publication, they have shown much greater explanatory power than competing narratives that predict the regime decay in China or even its collapse. Nevertheless, the weakness with the "resilience" and "adaptability" theses is that neither of them fully considers the party-state as autonomised from the CCP. Both portray the party-state as an appendage of the CCP. The title of Shambaugh's book, *China's Communist Party: Atrophy and Adaptation*, is particularly telling in this regard. Its message is clear: that the CCP has succeeded in overcoming atrophy is automatically taken as an explanation of the success of China in the past few decades. Despite the insightfulness and originality of Shambaugh's book, no distinction between the party-state and the CCP has been provided. I don't fully agree with this practice, because while the term "party-state" can be used to designate China, that of "CCP" cannot. Therefore, this book carefully distinguishes the party-state from the CCP.

The two elements in each contradiction reflect the two faces of the party-state: a society-friendly face and a regime-friendly face. The party-state with the society-friendly face leashes an autonomous zone to the society and acts in a rule-abiding way. In contrast, the party-state with the regime-friendly face takes the regime's survival as its top priority. It allows itself to gnaw the autonomous zone of firms and netizens for the sake of ensuring the regime's security. Take the contradiction between statism and entrepreneurship as an example: while the party-state may mobilise statism to ensure its control over private business, it has to simultaneously champion entrepreneurship to secure the economic basis of the regime. The same can be said when it comes to the coexistence of political monopoly and individual freedom, political arbitrary power and legal certainty, state domination of the market and respect of market rules.

This book calls the dynamic pushing the same actor to seek after contradictory objectives as a "double-bind logic". It qualifies China as practising "double-bind authoritarianism". Compared with the existing literature, "double-bind authoritarianism" captures wide state-society interactive fields. It covers, and is not limited to, the consultation/deliberation/bargaining between the government and the society. It denotes state-society power struggles at a larger and more structural scale, arguing that the Chinese party-state is continuously split by competing forces: one being the self-serving force and the other being the public-serving force. As will be shown in this book, that the party-state is driven by two competing forces significantly impacts state-society interactive patterns, in particular in the field of state regulation of private ISM firms.

"Double-Bind Authoritarianism"

That China practices "double-bind authoritarianism" importantly influences regulatory patterns of the country, captured into the term of "double-bind regulatory state". The typology of regulations by Yeo (2020), including the neoliberal approach, the statist approach, and the market-institutional approach, is based on how far the party-state positions itself vis-à-vis the market. While the state remains farthest to the market within the neoliberal approach, the distance between the two is the smallest within the statist approach. Since the latter recognises that the state selectively intervenes into the market based on the strategic value of regulated actors and sectors (Hsueh, 2011) as well as firms' ownership level (Yeo, 2020), it has the merit of recognising the variations in state-market distances. The distance can be close when firms are perceived as strategically important or when they are controlled by central regulators. In contrast, the distance is far when concerned firms are perceived as strategically non-important or when they are locally owned. The market-institutional approach complements the statist approach. However, what distinguishes the two is that the market-institutional approach highlights in particular the spearheading role of socialist politico-economic institutions in deciding how far the state positions itself vis-à-vis the market and when it should take such positions.

DBR conceptualised here does not inform of the state-market distance. It does not inform either of the regulation selectivity. Instead, it is akin to a "wrestling" theory. Instead of informing how far the state should stand next to the market, it captures the constraints obliging the former to step in or out of regulatees' autonomous sphere. This approach differs from existing regulatory approaches, for it considers the constraints not as deliberately created or desired by the party-state, but as located beyond its control. Portraying the constraints of the party-state as exogenous helps breathe regulatees into powerful actors capable of pressuring the party-state into compromises or choosing specific courses of actions. In this sense, DBR challenges the traditional wisdom which frequently features the opposition between an all-mighty party-state versus a powerless society in China (Ekman, 2020; Ringen, 2016).

DBR works bidirectionally, not unidirectionally. It recognises that the regulation from regulators and the reactions of regulatees are co-evolving and co-shaping. The quintessence of DBR is that regulatees cause dilemmas to the party-state. For firms, they can both boost economic upscaling and compete with propaganda organs; for citizens, they can both legitimise and destabilise the regime. Given these dilemmas, the party-state has to alternate between contradictory courses of actions in dealing with regulatees: centralisation versus decentralisation, more control versus less control. Without alternation, the party-state could hardly simultaneously garner the benefits and control the risks that regulatees create. The alternation puts the party-state at two contrasting poles, termed as "regime-friendly pole" (RFP) and "society-friendly pole" (SFP) in this book. The party-state's objectives differ depending on which side it finds itself in: while it has to stifle destabilising factors at the RFP, it must unleash a greater autonomous zone at the SFP. Despite their contradiction, since both sides of interests are crucial to the regime's survival, the party-state must shift from one pole to the other, because the exclusive pursuit of one side damages the other side.

Given the coercive and manipulative capacities of the regime, some may object that it is not necessary for the party-state to move to the SFP: Is it not evident that the regime could simply neglect the popular requests and pursue its own agenda? Is it not counterintuitive to say that regulatees in China can "oblige" the party-state to do this or that? Such objections are legitimate when examined from a state-centred perspective, for the latter denies that the Chinese society could hold the government accountable. However, this reasoning is simplistic, because it neglects that the party-state also faces considerable constraints in exercising its power. Actors face two types of constraints in their behaviour: formal and informal (North, 1990). In China, both formal (e.g., institutionalisation of the functioning of the regime) and informal constraints (e.g., traditional political philosophy of China) exist for the party-state and put limits to its freedom of action.

North particularly highlights informal constraints, which "come from socially transmitted information and are a part of the heritage that we call culture" (1990: 37). Informal constraints tangibly influence people's behaviour, the evidence being that "the same formal rules and/or constitutions imposed on different societies produce different outcomes" (North, 1990: 36). Different cultural codes make people reason and behave differently. The mental processing of information depends "upon the brain's ability to learn by being programmed with one or more elaborately structured natural languages that can code for perceptual, attitudinal and moral (behavioural) as well as factual information" (Johansson, 1988; quoted in North, 1990: 37). Since the Chinese culture emphasises benevolence, virtue, and responsibility of rulers (Tong, 2011: 147–149), it is expected that the party leadership has the moral pressure of being accountable even in the absence of formal constraints.

DBR differs from the existing approaches, in that its scope of applicability is both larger and smaller. Three points deserve special attention. First, DBR deals with state-market relations as well as state-society relations, for it includes citizens as regulable actors in their own rights. In so doing, DBR is well adapted to the political and social realities in the regulation of the ISM, a sector in which citizens are no less important than private firms for influencing the party-state's chances of survival. While the existing approaches mostly tackle state-market relations, DBR is more encompassing. It provides an explanatory framework for state-society relations as a whole, and not simply state-market relations.

Second, although the conceptualisation of the DBR approach is based on the study of the ISM sector, DBR holds heuristic value for studying the regulation of other economic sectors than the ISM in China. The cross-sectorial applicability of DBR is related to the fact that it explains how the pursuit of contradictory objectives goads the party-state to deploy contradictory courses of action, a pattern which is not limited to the ISM industry. Since the regime's survival stands at the core of the party-state, it is expected that the political concerns are present in the conception of regulatory frameworks for a wide range of industrial sectors. Although the ISM sector is the sector where the alternation between more control and less control is fairly evident, the same pattern exists in some futuristic economic sectors which represent high political stakes for the party-state. The elaboration of regulatory rules for artificial intelligence, for instance, is expected to be similarly influenced by the double-bind logic, because AI is an engine for driving economic upscaling as well as a tool of surveillance of citizens' behaviour (Ma, 2018).

Third, DBR has a limited scope of applicability when it comes to the types of political regimes where it unfolds, because it is at play in priority in authoritarian settings, not democratic settings. The double-bind logic seldom appears in the regulation of democratic countries, because power preservation and power legitimisation don't enter into conflict there. Since constitutional arrangements in democracies make it illegal for politicians to preserve power without legitimising it (mostly through elections), regulation in democracies is seldom intended to enable politicians to stay in power indefinitely. The independence of regulatory agencies in democracies further hinders politicians' attempts to instrumentalise the regulation. In contrast, autocrats frequently mobilise the regulation as a way to preserve their power and the privileges coming along.

State-Firm Relations in the Double-Bind Regulatory State

Lengthy academic discussions have been held on how the Chinese party-state exercises effective control over private firms. Above all, the party-state neutralises the threats of private business by coopting successful private entrepreneurs (Dickson, 2003, 2007). The integration of "wealth and power" disincentivises private entrepreneurs from challenging the *status quo*, because cooptation closely ties the interests of private business to those of the regime (Dickson, 2003, 2007). Imbalanced power relations between regulators and firms are deepened by the party-state's selective cooptation: the more a private firm is successful, the more its owners are likely to be coopted (McGregor, 2010). That private entrepreneurs depend on the party-state leads to their failure to form meaningful counter-power groups in China. As a result, they are not perceived as a reliable force for democratising China (Dickson, 2003, 2007; Tsai, 2005; Strittmatter, 2020), against the predictions of some enthusiastic scholars (Glassman, 1991; White, 1994; Sautman, 1992). Tsai suggests that Chinese private entrepreneurs do not form a political class in the Marxist and Weberian senses of the term (Weber, 1998; Marx, 1926). The root cause is that they have no awareness of their distinctive identity (Tsai, 2005: 1135). In addition, they will not challenge the party-state unless an alternative political system proves better in protecting their interests (Tsai, 2005).

In the specific ISM sector, scholars reach the similar conclusion of imbalanced state-firm relations. The party-state is perceived as having all necessary techniques and resources to bring firms under control (Tsui, 2003). Influential firms like Sina Weibo cooperate with content regulators in removing "harmful" content (Roberts, 2018; Economy, 2018). Content regulators issue directives to ISM firms each day and firms conduct censorship in accordance with these directives (Roberts, 2018: 123). The threats of punishments incur greater corporate obedience (Roberts, 2018: 153). Mechanisms of rewards are also mobilised. The enthusiasm of private firms in creating grassroots cells of the CCP, for instance, is explained by potential rewards, including the possibility of receiving the visit of national-level leaders (McGregor, 2010). Cooptation, another form of reward, serves to protect private businesses from being preyed upon local governments (Dickson, 2007). To sum up, "China's large internet companies and the state enjoy a symbiotic relationship where the former enjoys much-needed state investment and market protectionism, while granting the latter compliance with censorship directives, and even proactively working to exercise 'self-discipline' in ensuring the spread of pro-Party messages online" (Cairns, 2016: 14).

Operating under the party-state's carrots and sticks, ISM firms have been found to contribute to realise the party-state's objectives. For instance, tech giants act as the party-state's policy arms on social management. Huawei has spearheaded the installation of surveillance equipment in the Xinjiang autonomous region (Strittmatter, 2020). Sesame Credit, financial arm of Alibaba, replies to the call of the People's Bank of China by rolling out the pilot project of the social credit system in 2015, together with 7 other chosen fintech firms (Strittmatter, 2020). The readiness of private ISM firms to serve the party-state creates the impression that they are "parasitic actors" in Chinese political economy that are deprived not only of the possibility of pursuing their own interests, but also of the willingness to do so. Given the parasitic nature of private businesses, their level of autonomy seems to largely depend on the goodwill of the party-state.

However, despite political authoritarianism in China, the image of "powerful state, powerless business" is too reductionist to be an objectivised account of state-business relations. It obliterates by one stroke of pen the multiple loci of interest struggles couched in state-business interactions. In what follows, I will show that there exist "multiple shades of powerfulness" for the party-state and "multiple shades of powerlessness" for firms. In parallel to the party-state's jumping from one to another shade of powerfulness, firms change their positions accordingly across the spectrum of powerlessness. This, however, does not mean that firms' changes of shades are invariably in lockstep with the party-state's changes. Instead, I argue that in a double-bind regulatory state, firms can accelerate or slow down the party-state's move by participating in changing the circumstances facing regulators. The movements of the two types of actors can happen in a desynchronised way. The state-firm struggles mean that it is not because the state becomes more powerful that firms will be more powerless. Instead, it is very possible that the more the state is powerful, the less firms are powerless. The same logic applies to state-netizens interactions.

Media Commercialisation in China

Polanyi argues that there exist a variety of political economy. In his words, "a range of alternatives in political economy are possible, because markets can be embedded in a range of different ways. Many varieties of political economy have been developed as a result of different social and political choices implemented through state intervention" (quoted in Yeo, 2020:

10). Therefore, social and political arrangements chosen by the state influence the "making" of the market, and the state is always present in the process of market-making. The state not only "guarantees the respect of the property rights. It is also the guarantor of means of exchange...and the regulator of the market. Therefore, the state makes the market possible" (Hay and Smith, 2018: 27). To extrapolate the decisiveness of the state in the market creation to the Chinese case, the objectives of the party-state in initiating the ISM sector in the late 1980s deserve special attention for making sense of DBR. The Chinese internet development serves academic purposes in the first stage (1980s), economic purposes in the second stage (1990s–2000s), and political purposes in the third stage (since the end of the 2000s). Nevertheless, if the pursuit of academic purposes does not lead to the confrontation of interests, the pursuit of economic and political objectives pits economic elites against political elites, because the pursuit of the two sets of objectives implies opposite courses of actions.

The party-state is the engineer-in-chief of the ISM industry in China. After the Cultural Revolution from 1966 to 1976, economic development became the priority concern of the Chinese leadership in the 1980s and the 1990s. Reformist leaders around Deng decentralised competences to local governments, devolved significant economic property rights to enterprises, and promoted technological innovation to boost economic recovery (Lynch, 1999: 74). Media commercialisation, deliberately designed by the party-state (Zhao, 2000; Jiang, 2012), was initiated against this backdrop. Zhao found two reasons behind media commercialisation. First, the expansion of the market economy nourished the press boom and enabled the newspapers controlled by the state to be financially independent. Second, given the increase of the state expenditure and the weak national GDP growth (in particular in the 1980s), the party-state actively pushed the press to market and reduced the provision of direct subsidies and other financial incentives (Zhao, 2000: 5). To develop an information economy, the State Science and Technology Commission listed newspapers and broadcasting as "information commodification industries" in 1987. In late 1992, the State Press and Publication Administration announced a "landmark policy", requiring "all major newspapers, apart from a few central party organs such as the People's Daily, to achieve financial independence by 1994" (Zhao, 2000: 6). The party-state regarded the press as part of the tertiary sector to be promoted as a new point of economic growth (Zhao, 2000: 6).

The zeitgeist of commercialisation extended to the internet industry. Upon the kick-off of the Chinese ISM industry in the mid-1990s, the party-state introduced the two-pronged regulatory approach based on firms' "strategic value" (Hsueh, 2011). Since economic catch-up was one fundamental motivation behind the Chinese government's investments into the IT sector (Tai, 2006; Negro, 2017), the party-state consciously let private firms rather than awkward SOEs spearhead the downstream sectors of the IT industry, namely email services, music-sharing, E-commerce, etc. (Jiang, 2012). Senior party leaders thought that private entrepreneurs were able to "fill up the gaps" left by an ailing public sector and bloated state firms (Lagerkvist, 2012: 2630). However, this layout did not mean that SOEs were unimportant. In the telecommunications sector, located at the upstream of the IT industry and considered as constituting one of the national economic lifelines (Yeo, 2020; Hsueh, 2011), SOEs enjoyed a quasi-monopolistic control. This industrial layout provided security guarantee to the party-state: since SOEs controlled the internet infrastructure, and that private firms could not operate without infrastructure, the party-state could hold a decisive leverage over private ISM firms by controlling SOEs.

Corporate Disobedience and Commercialised Censorship

Commercialisation of the print media and the nurturing of private ISM firms did not lead to the same results. Since the print media remained controlled by the party-state both before and after commercialisation, the inclusion of critical elements within these newspapers was mild and orchestrated (Zhao, 2000). On the contrary, dominated by privately-owned firms, the development of the ISM industry has "grown out of plan" (Naughton, 1995). The "wild growth" of the ISM industry in the 2000s and the 2010s, nurtured partially by the tolerant attitude of the Chinese government, created antagonistic relations between the party-state and private ISM firms (Chen and Gong, 1997: 162). Since Chinese netizens developed greater reliance on private ISM platforms for information and entertainment (Tai and Sun, 2007), private firms acquired importance in shaping people's mind and became a rival of the party-state in the ideological field (Lynch, 1999; Xue, 2018). As the initiator of the ISM industry, the party-state saw its creatures develop their agendas. In the words of Susan Shirk, "by choosing to give up some degree of control over the media, the rulers of authoritarian countries like China make a trade-off. Most obviously, they gain the benefit of economic development; the market operates more efficiently when people have better information...On the other hand, surrendering control over information creates severe political risks. It puts new demands on the government that it may not be able to satisfy, and it could reveal to the public the divisions behind the façade of party unity" (Shirk, 2011: 5).

One straightforward sign of the "growing out of plan" of private ISM firms is the boldness with which private ISM firms disobeys to state regulations. The commercialisation wave in the 1980s and the 1990s (Zhao, 1998; Yang, 2003, 2009a; Lynch, 1999) deeply transforms the nature of the media sector and increases the difficulties of the propaganda apparatus in conducting the thought work (Lynch, 1999). From serving exclusively propagandistic purposes in the Maoist era, the media sector has progressively been treated as a sector for driving economic growth in a similar way to other economic sectors. For the ISM industry, the progressive decoupling between its economic functions and its propaganda functions makes the satisfaction of citizens' needs as important as the observance of propaganda authorities' instructions (Zhao, 1998).

Since users, not the party-state, are the principal source of revenues for private firms, firms tend to satisfy users' needs in priority (Tai, 2006; Negro, 2017). The imperative of making profits explains ISM firms' half-hearted implementation of regulations (Jiang, 2014). In addition, that private firms make higher profits by carrying explosive news further incentivises them to side with consumers, not with regulators (Yang, 2009a, 2009b). Although the licensing system prohibits private ISM firms from reporting self-collected hard news, firms frequently bypass these rules and create their own teams of reporters (Bandurski, 2018). To increase their revenues, private ISM firms avoid being considered by netizens as the party-state's "throat and tongue", a reputation assumed by state-backed media (Negro, 2017). When implementing content regulations, firms ensure that their implementation is more lenient than their competitors, such that they look more independent from the party-state's influence (Miller, 2018).

The principal-agent framework is mobilised to explain disobedience of private ISM firms in China (Miller, 2018; Lagerkvist, 2012). Miller finds that the delegation of content regulation to private ISM firms creates firms' "deliberate and strategic shirking", which causes agency loss for the government (Miller, 2018: 10). For him, "because there is often a misalignment of preferences between the government principal and the ICP agent, private Internet companies may ignore the directive or partially implement it" (Miller, 2018: 6). Firms develop agency discretion which "occurs when the agency succeeds in choosing a policy in line with agency goals, when those goals differ from what the executive and legislature expected at the appointment stage" (Calvert et al., 1989: 605).

Corporate disobedience illustrates how the development of the ISM industry has deviated from the party-state's initial plans (Naughton, 1995). Governmental investment in the ISM sector since the 1980s is initially intended to drive up the Chinese economic catch-up with advanced economies in the West. However, since the 2000s onwards, the ISM has shown its potential to cause political and social destabilisation in China. The party-state is left with two choices. First, it pursues either political or economic goals by abandoning the other side. More concretely, it can dissipate political risks by taking command of private ISM firms, namely through nationalisation; or it pursues economic objectives by neglecting political risks. Since both economic growth and political stability are indispensable to the regime's survival, this option is unfeasible. Therefore, the party-state is left with the second option: pursuit of both sides of objectives, however difficult the undertaking seems to be.

Balancing Political and Economic Goals

The party-state pursues both economic (e.g., technological innovation, economic upscaling, job creation) and political objectives (e.g., ideological security, regime legitimacy, social stability) when regulating the ISM sector. However, not all these objectives are compatible with each other. For instance, to ensure ideological security implies that the party-state must micro-manage firms, while micro-management very probably stifles firms' technological innovation. The bifurcated regulation of the ISM—"hard regulation" over political issues and "soft regulation" over economic issues (Yeo, 2020: 78, 51)—finds its roots in the party-state's retrospective reflections on the crackdown of the pro-democracy movements in 1989. After the blood-shedding repression in 1989, "the party's slogan for the new era was simple: 'on economic matters, relaxed controls; for political matters, tight controls'" (McGregor, 2010: 42). The party-state has faithfully applied this maxim to the regulation of the ISM sector, although well-entrenched political conservatism in China makes some observers question why the ISM would have been developed at all in an authoritarian regime like China (Clark, 2016).

Quite many observers have argued that the party-state holds an ambivalent attitude towards the ISM industry (Qian and Bandurski, 2011; Xiao, 2011; Cullen and Choy, 2005; McGeary, 2001): on the one hand, it uses the ISM to realise economic development; on the other hand, it tries to limit citizens' access to information. However, scholars disagree on whether political and economic objectives constitute a dilemma for the party-state. Some scholars contend that simultaneously seeking after the two sets of objectives do not cause any predicament for the Chinese government. Liao argues that China has largely achieved both sides of objectives (Liao, 2016: 245). The Xi Jinping administration has acquired the capacity to simultaneously pursue economic and political objectives with greater ease than under its predecessors (Bandurski, 2018). That the regime concurrently pursues economic liberalisation and preserves political conservatism has nothing contradictory, and the combination between the two is believed to be the basis of Chinese economic success (Shirk, 1993).

Other scholars, however, emphasise the negative impact of mismatched political conservatism and economic liberalism. Goldstone highlights four negative consequences of the mismatch: the split within the ranks of the party leadership, the rift between party leaders and China's other elites, the decline in the Party's direct control of the Chinese society, and discontent among peasants and workers (Goldstone, 1995: 41–46). The conflicts between political elites and economic elites, and those between the party-state and Chinese netizens on the freedom of speech corroborate Goldstone's predictions. Lagerkvist highlights the "uneasy coexistence between the animated discussions on Chinese microblogging platforms, such as Sina Weibo and Tencent Weibo, and the party-state's fixation on social and political stability" (2012: 2629). The party-state's contradictory attitude towards the ISM sector has provoked "uneven and sometimes unintended policy outcomes" (Kalathil and Boas, 2003, p.14), one of which is that citizens' freedom of speech continues to be controlled despite substantially increased social connectivity. Nevertheless, netizens' creation of online public spheres (Rauchfleisch and Schäfer, 2015; Hu, 2008) and their continuous acquisition of circumvention skills (Yang 2003, 2009a) illustrate another "unintended policy outcome": when censorship cannot produce its expected results, it can "over time facilitate the formation of a semi-autonomous, critical public opinion" (Lynch, 1999: 74).

Autonomy of Private ISM Firms with Fragmented Authoritarianism

Fragmented authoritarianism, conceptualised by Lieberthal and Oksenberg (1988), deepens the party-state's predicament in the regulation of private ISM firms. Based on their study of the energy sector in the 1980s, they point out vertical and horizontal fragmentation in the Chinese authoritarian system. They argue that the bargaining and negotiations among bureaucracies at the horizontal and vertical levels make Chinese decision-making processes anything but linear and streamlined.

Although it is commonly believed that China is a unitary state, it is in fact one of the most decentralised countries in the world (Birney, 2013; Landry, 2008; Fu, 2018: 51). Fragmentation of the Chinese political system has led to wide discussions in academia (O'Brien, 1994; Montinola et al., 1995; Lieberthal, 1995; Jin et al., 2005; Zheng, 2006; Stern and O'Brien, 2012; Mertha, 2009). The governance system features a "*de facto* federalism" (Zheng, 2006), with localities having veto power when implementing the policies adopted by the central government (Miller, 2018: 7). With decentralisation, "Beijing now could only pretend to rule the provinces, while the provinces only pretended to be ruled by Beijing" (Zheng 2006: 3; Segal, 1994). To reduce Beijing's budgetary burden, since 1979 onwards, China has encouraged local governments to finance the introduction of foreign capital and technologies on their own and to innovate the provision of public goods (Dollar, 2018). Since well-to-do foreign investors have the capital that local governments vie for, local governments have engaged themselves into a wave of "competitive liberalisation", in which economic liberalisation in one locality pressures other localities to follow suit (Yang, 1997: 46).

Inter-jurisdictional competition damages the authority of Beijing due to local governments' strategic shirking and selective policy implementation. Decentralisation shortens Beijing's power reach also because local governments jealously keep information in their own hands (Mertha, 2009; McGregor, 2010; Wong, 1985). Under the pressure of outperforming each other, local governments are as entrepreneurial as private entrepreneurs. They compete with each other to innovate the methods for attracting investors and promising firms (McGregor, 2010). This is what Oi termed "local state corporatism" in China, defined as the merger of state and economy into a new institutional development (1992: 100). Oi argues that the Chinese fiscal reform in the 1990s grants local governments property rights over increased income and incentivises them to pursue local economic development. In this process, local governments increasingly resemble a business corporation, with local officials working as the equivalent of a board of directors (Oi, 1992: 101). In a similar way as private firms, local governments compete with each other for foreign investment, policy privileges, and exemptions of obligations, the objective being to outperform their neighbours.

It is safe to say that local governments have largely dwarfed Beijing on policy innovation. Instead of initiating new policies, Beijing is more often than not a codifier and disseminator of good practices worked out by local authorities beforehand. Using the example of laid-off workers in the mid-1990s, Rocca illustrated how the policymaking in Beijing was driven by local policymaking, not the other way around. He wrote that the wave of laid-off workers started in the years 1994–1996 in provinces like Liaoning before the central government institutionalised this policy in September 1997 (Rocca, 2006: 111).

The Strategic Value of ISM Firms

Fragmented authoritarianism has deep implications on the level of autonomy of private ISM firms in China. Since the 1980s onwards, the regulation of private firms has followed the principle of "jurisdiction-based regulation" (*shu di guan li*). This principle means that a private firm is regulated by the local government where the firm is administrative registered. The consequence of the "jurisdiction-based regulation" is that private firms change into one of the instruments by which local governments realise their socio-economic objectives. Private firms acquire "strategic value" for local governments (Hsueh, 2011). Firms' "strategic value" influences how they are regulated. Hsueh highlights that "[T]he strategic value logic accounts for the government's subjective assessment and the structural features of the sector in question to explain why the state exercises deliberate control in certain sectors and subsectors and incidental control in others" (2011: 23). Hsueh argues that industries with high strategic value will be closely regulated, implying less autonomy for this sector. However, when applied to locally registered private firms, the picture is different. Private firms with strategic value at the local level tend to be protected and enjoy high influence over local governments' decision-making. Therefore, I argue that the strategic value of a private firm certainly attracts more regulation, but it also grants the firm higher leverage over local governments. For firms, their strategic value transforms from a constraint at the sectorial level and the national level into an autonomy-generating opportunity at the firm level and the local level.

Local governments' dependence on private firms explain their ambivalent attitude characterised by the mixing of paternalism and utilitarianism. For paternalism, development-oriented local governments protect private firms in their jurisdiction (Wong, 1985; McGregor, 2010). Their protectionism goes as far as to lobby the central government to facilitate the business of private firms (Chai, 2018). For utilitarianism, local governments are largely selective in choosing which firms to support. Financial support is given in priority to the firms with the highest potential to benefit local economy (Boullenois, 2020; McGregor, 2010). That local officials support local private firms in return reinforces their control over firms, because it "entrenches local officials as the most prominent holders of local power, and encourages strong interventionist policies on their part" (Boullenois, 2020: 54).

Paternalism and utilitarianism of local governments are interrelated, because private firms with scant utility are less likely to receive paternalistic protection. Nevertheless, local governments' utilitarianism creates their vulnerability, for they must make compromises to important firms in exchange for their utility. The strategic value of private ISM firms for local governments increases after the Chinese leadership fixes the ambition of building China into an information society in 2000 (Austin, 2014: 156). Against this backdrop, differentiated treatment of big firms and small firms, strategic firms and non-strategic ones, intensifies. Big firms are found to have more frequent interactions with local governments (Dickson, 2007), and their needs are satisfied in priority. They are also powerful enough to directly lobby local governments, often bypassing semi-official industrial associations (Kennedy, 2011).

Fragmented authoritarianism enables private ISM firms to enlarge their manoeuvring room by creating alliances with local governments, the precondition being that they have "utility" to local authorities. Nevertheless, one cautionary note is important. The risk for private firms is that when they acquire national importance—BATHX (Baidu, Alibaba, Tencent, Huawei, and Xiaomi)—they will attract the attention of the central government. The entering of central regulators into regulator-firm interactions at the local level changes the configuration of opportunities and constraints for firms. At such moments, the pressure from Beijing is likely to overtake that of local governments. The anti-trust regulatory wave, started by the abrupt interruption of the IPO (Initial Public Offering) of Ant Financial in November 2020, illustrates the risks facing Chinese tech magnates.

Fragmented Authoritarianism at the Horizontal Level: "Divide to Manoeuvre"

Fragmented authoritarianism in China unfolds at both the vertical and horizontal levels of governance. While "*de-facto* federalism" (Zheng, 2006) captures vertical fragmentation, horizontal fragmentation denotes the competition of regulatory agencies with overlapping competences and their lack of coordination. In their bureaucratic politics model, Allison and Halperin argue that "the 'maker' of governmental policy is not one calculating decision-maker, but rather a conglomerate of large organisations and political actors who differ substantially about what their government should do on any particular issue and who compete in attempting to affect both governmental decisions and the actions of their government" (1972: 42). Therefore, each organisation or bureaucracy involved in policymaking processes tries to maintain and extend its turf: "maintaining autonomy and organisational morale, protecting the organisation's essence, maintaining or expanding roles and missions, and maintaining or increasing budgets" (Allison and Halperin, 1972: 48). The final decision achieved represents the result of bureaucratic bargaining of different agencies (Allison, 1971; Lieberthal and Oksenberg, 1988).

A similar process of bureaucratic bargaining exists in China. The image of a unified government capable of streamlining the interests of competing agencies is anything but true. Political governance in China is characterised by "a fragmented bureaucratic structure of authority" and a "protracted, disjointed, and incremental" policy process (Lieberthal and Oksenberg, 1988: 22). Although top political leaders could wield significant pressure over agencies, without consensus building, "progress can easily bog down in the bowls of Chinese various bureaucracies" (Lieberthal and Oksenberg, 1988: 23). Since the 1980s, from the debates on how to boost economic growth (Harding, 1987; Riskin, 1987), to ministerial struggles for imposing their respective approach to tackling inflation (Solinger, 1987; Shirk, 1985), decision-making in China has been fraught with bureaucratic turf battles.

Bureaucratic fragmentation in China enables ISM firms to enlarge their manoeuvring room by exploiting inter-agency conflicts. Firms can create alliances with some agencies to resist other agencies. For instance, Sina Weibo allies itself with content regulators of Beijing municipality. While it implements the instructions of the Beijing government with diligence, it poorly implements the instructions of other local governments (Miller, 2018). The shirking of ISM firms is also facilitated by the coexistence between a large number of regulators and the geographic concentration of ISM firms in China. The most important Chinese internet companies are located in Beijing, a few in Shanghai and Guangdong, and almost none elsewhere (Cairns, 2016: 14). That regulators outnumber regulatees creates the problem of "common agency", defined as a situation where "many government agencies have the power to issue directives to single Internet companies" (Miller, 2018: 7). It helps private firms obtain an important advantage over regulators, allowing them the discretion to decide which agency's directives to follow, in particular when regulators hold divergent preferences (Miller, 2018: 7).

Conclusion

This chapter lays down the conceptual scheme of the book and advances the building blocks of the double-bind regulation (DBR) approach. The conceptualisation of DBR comes in two steps: (1) autonomisation of the state vis-à-vis political elites and (2) creation of the double-bind logic in state-firm relations. First, inspired by Mann (1984) and Carpenter (2001), I argue that the Chinese party-state has a certain level of autonomy vis-à-vis the party leadership. Given that the party-state constitutes the structure where political elites operate, an autonomised party-state with its own interests and objectives conditions the realisation the interests and objectives of political elites.

As the ruling party, the Chinese Communist Party (CCP) has important agenda-setting power and policymaking power. One of the particularities of Chinese politics is that while the party runs the state (Cabestan, 2018), the party and the state are also inseparable and embedded into each other. Given the embeddedness between them, when extrapolated to China, the thesis "autonomisation of the state" precisely signifies the autonomisation of the party-state. The gist for explaining the Chinese case is that the party-state acquires autonomy from its ruling elites: while ruling elites go, the party-state remains. More importantly, the autonomisation means that the party-state, as an autonomous political entity, can override the interests of the elites, irrespective of whether they work inside state bureaucracies (e.g., Ministry of Industry and Information Technology) or party institutions (e.g., Central Department of Propaganda).

The party-state's double-bind logic in the regulation of private ISM firms comes from the contradiction of its sought-after objectives. The particularity of the ISM sector in China is that it represents both vital economic and political stakes for the regime's survival. However, the party-state has to mobilise opposite approaches to handle them: centralised regulation (or more control) for political issues and decentralised regulation (or less control) for economic issues. That the political and economic regulations of the ISM sector are taken up by two different groups of regulators deepens the party-state's predicament. As the meta-regulator, the party-state cannot let either group of regulators go as far as it sees fit and has to pull it back when the pursuit of one set of interests damages the other. The party-state's inevitable alternation between political and economic objectives opens autonomy-enabling opportunities for ISM firms. To obtain greater manoeuvring room, firms can capitalise on their economic leverage to resist unfavourable regulations.

Fragmented authoritarianism at the vertical and horizontal levels of governance also causes predicament for the party-state. At the vertical level, China has been practising "*de facto* federalism" (Zheng, 2006) since the beginning of the 1980s. The decentralisation of significant economic competences to localities creates two interrelated problems: "corporatisation" of local governments (Oi, 1992; McGregor, 2010) involved in fierce interjurisdictional competition and incentivisation of local governments to deviate from the instructions of the central government. Private ISM firms being all registered at the local level (municipalities and counties), their reliance on local governments far overtakes their reliance on Beijing. In the localities where ISM firms represent "strategic value" (Hsueh, 2011) to local development, local governments "spoil" them and even lobby Beijing on their behalf (Chai, 2018; McGregor, 2010). The caveat here is that private firms with national influence are likely to attract the attention of central regulators, thus shortening the protective arm of local governments.

At the horizontal level, the party-state's double-bind logic stems from lack of coordination among regulators with overlapping competences. In China, more than 50 content regulators, local and central regulators combined, can instruct content providers to moderate contents (Miller, 2018). Fragmented bureaucratic authority often plays in favour of regulatees (Rahman and Thelen, 2019). Weak coordination among regulators enables private internet firms to selectively implement content regulation. More precisely, they implement the regulation representing the least costs to them (Miller, 2018).

DBR represents a major difference when compared with the existing neoliberal, statist, and market-institutional approaches. Instead of measuring how far regulators are positioned vis-à-vis the market, DBR measures the regulator-regulatee wrestling and its implications on the autonomy of private ISM firms. More concretely, it measures the manoeuvring opportunities enabling the allegedly powerless regulatees to outwit the allegedly powerful regulators in authoritarian China. These opportunities are endogenous, for the party-state's double-bind logic is inherent in its governance logic. The implication of the endogeneity of the predicament is that the party-state is by nature a chameleon autocrat jumping back and forth among different shades of authoritarianism. Although it is possible that it can reach the extreme shade characterised by harshest authoritarianism, the double-bind logic makes the party-state's stay at this extreme point temporary rather than permanent. DBR gives Chinese citizens the reason for staying optimistic, because the tight social control under the Xi era may very well be temporary. As a sign, wasn't the stubborn zero-covid policy, practised in China for almost 3 years (2020–2022) and expected to permanently exist in the country by more than one observer, well abandoned at the beginning of December 2022?

References

Allison, Graham. 1971. *Essence of Decision: Explaining the Cuba Missile Crisis*. Boston: Little Brown.

Allison, Graham, and Morton H. Halperin. 1972. "Bureaucratic Politics: A Paradigm and Some Policy Implications." *World Politics* 24: 40–79.

Austin, Greg. 2014. *Cyber Policy in China*. Cambridge, UK: Polity.

Baldwin, Robert, Martin Cave, and Martin Lodge. 2010. "Introduction: Regulation—The Field and the Developing Agenda." In *The Oxford Handbook of*

Regulation, edited by R. Baldwin, M. Cave, and M. Lodge, 3–16. Oxford: Oxford University Press.

Bandurski, David. 2018. "Can the Internet and Social Media Change the Party?" In *Routledge Handbook of the Chinese Communist Party*, edited by W. W.-L. Lam, 372–390. Oxon, UK: Routledge.

Barrow, Clyde W. 2002. "The Miliband-Poulantzas Debate: An Intellectual History." In *Paradigm Lost: State Theory Reconsidered*, edited by S. Aranowitz and P. Bratsis, 3–52. Minneapolis, MN: University of Minnesota Press.

Béjà, Jean-Philippe. 2019. "Xi Jinping's China: On the Road to Neo-Totalitarianism." *Social Research* 86(1): 203–230.

Benoît, Cyril. 2018. "Régulation." In *Dictionnaire d'économie politique: capitalisme, institutions, pouvoir*, edited by C. Hay and A. Smith, 415–420. Paris: Presses de Sciences Po.

Benoît, Cyril. 2020. *Réguler l'accès aux médicaments*. Grenoble: Presses Universitaires de Grenoble.

Birney, Mayling. 2013. "Decentralisation and Veiled Corruption Under China's 'Rule of Mandates'." *World Development* 53: 55–67.

Boullenois, Camille. 2020. "Poverty Alleviation in China: The Rise of State-Sponsored Corporate Paternalism." *China Perspectives* 3: 47–56.

Cabestan, Jean-Pierre. 2018. *Demain la Chine: démocratie ou dictature ?* Paris: Gallimard.

Cairns, Christopher. 2016. "Fragmented Authoritarianism? Reforms to China's Internet Censorship System Under Xi Jinping." Presented at the American Political Science Association Annual Meeting in Philadelphia, PA on 3 September 2016.

Calvert, Randall L., Mathew D. McCubbins, and Barry R. Weingast. 1989. "A Theory of Political Control and Agency Discretion." *American Journal of Political Science* 33(3): 588–611.

Carpenter, Daniel P. 2001. "The Political Foundations of Bureaucratic Autonomy: A Response to Kernell." *Studies in American Political Development* 15: 113–122

Carpenter, Daniel. 2010. *Reputation and Power: Organisational Image and Pharmaceutical Regulation at the FDA*. Princeton, NJ: Princeton University Press

Carpenter, Daniel, and David Moss. 2014. *Preventing Regulatory Capture: Special Interest Influence and How to Limit It*. New York: Cambridge University Press.

Chai, Yanfei. 2018. *Zhejiang Gaige Kaifang 40 Nian Koushu Lishi (1978–2018)* (Oral History of 40 Years of Reform and Opening-Up in Zhejiang). Hangzhou: Zhejiang Science and Technology Press.

Chen, Feng, and Ting Gong. 1997. "Party Versus Market in Post-Mao China: The Erosion of the Leninist Organisation from Below." *Journal of Communist Studies and Transition Politics* 13: 148–166.

Clark, Duncan. 2016. *Alibaba: The House that Jack Ma Built*. New York: HarperCollins Publishers.

Coglianese, Cary and David Lazer. 2003. "Management-Based Regulation: Prescribing Private Management to Achieve Public Goals." *Law & Society Review* 37: 691–730.

Cullen, Richard, and D. W. Choy. 2005. "China's Media: The Impact of the Internet." *San Diego International Law Journal* 6(323): 323–339.

Dickson, Bruce. 2003. *Red Capitalists in China: The Party, Private Entrepreneurs, and Prospects for Political Change*. Cambridge, UK: Cambridge University Press.

Dickson, Bruce. 2007. "Integrating Wealth and Power in China: The Communist Party's Embrace of the Private Sector." *The China Quarterly* 192: 827–854.

Dollar, David. 2018. "The Highs and Lows of Chinese Decentralisation." East Asia Forum, Brookings Institution, 1 November.

Economy, Elisabeth. 2018. *The Third Revolution: Xi Jinping and the New Chinese State*. Oxford: Oxford University Press.

Ekman, Alice. 2020. *Rouge Vif: l'Idéal Communiste Chinois*. Paris: Editions de l'Observatoire.

Finez, Jean. 2014. "La construction des prix à la SNCF une socio-histoire de la tarification: de la péréquation au yield management (1938–2012)." *Revue Française de Sociologie* 55(1): 5–39.

Fu, Diana. 2018. *Mobilizing Without the Masses: Control and Contention in China*. Cambridge: Cambridge University Press.

Giddens, Anthony. 1981. *A Contemporary Critique of Historical Materialism*. London: Macmillan.

Glassman, Ronald M. 1991. *China in Transition: Communism, Capitalism, and Democracy*. New York: Praeger.

Goldstone, Jack. 1995. "The Coming Chinese Collapse." *Foreign Policy* 99: 35–52.

Gramsci, Antonio. 1979. *Letters from Prison*. London: Quartet Books

Gray, Garry C., and Susan S. Silbey. 2014. "Governing Inside the Organisation: Interpreting Regulation and Compliance." *American Journal of Sociology* 120(1): 96–145.

Harding, Harry. 1987. *China's Second Revolution*. Washington DC: Brookings Institution.

Han, Rongbin. 2018. *Contesting Cyberspace in China: Online Expression and Authoritarian Resilience*. New York: Columbia University Press.

Hay, Colin. 1999. "Marxism and the State." In *Marxism and Social Science*, edited by A. Gamble, D. Marsh, and T. Tant, 152–174. Champaign, IL: University of Illinois Press.
Hay, Colin, and Andy Smith. 2018. "Introduction: Le Rapport Capitalisme-Politique". In *Dictionnaire d'économie politique*, edited by C. Hay, and A. Smith, 19–49. Paris: Presses de Sciences Po.
He, Baogang, and Mark Warren. 2011. "Authoritarian Deliberation: The Deliberative Turn in Chinese Political Development." *Perspectives on Politics* 9(2): 269–289.
Hintze, Otto. 1975. *The Historical Essays of Otto Hintze*. New York: Oxford University Press.
Howell, Jude, and Tim Pringle. 2018. "Shades of Authoritarianism and State-Labour Relations in China." *British Journal of Industrial Relations* 57(2): 223–246.
Hsueh, Roselyn. 2011. *China's Regulatory State: A New Strategy for Globalisation*. Ithaca, NY: Cornell University.
Hu, Yong. 2008. *Zhong Sheng Xuan Hua: Wang Luo Shi Dai De Ge Ren Biao Da Yu Gong Gong Tao Lun* [The Rising Cacophony: Personal Expression and Public Discussion in the Internet Age]. Guilin: Guang Xi Shi Fan Da Xue Chu Ban She [Guangxi Normal University Press].
Huang, Yasheng. 2008. *Capitalism with Chinese Characteristics: Entrepreneurship and the State*. Cambridge: Cambridge University Press.
Jessop, Bob. 1977. "Recent Theories of the Capitalist State." *Cambridge Journal of Economics* 1: 353–373.
Jiang, Min. 2012. "Internet Companies in China: Dancing Between the Party Line and the Bottom Line." *Asie Vision* 47: 1–50.
Jiang, Min. 2014. "The Business and Politics of Search Engines: A Comparative Study of Baidu and Google's Search Results of Internet Events in China." *New Media & Society* 16(2): 212–233.
Jin, Hehui, Yingyi Qian, and Barry R. Weingast. 2005. "Regional Decentralization and Fiscal Incentives: Federalism, Chinese Style." *Journal of Public Economics* 89(9): 1719–1742.
Johansson, S. Ryan. 1988. "The Computer Paradigm and the Role of Cultural Information in Social Systems." *Historical Methods* 21: 172–188.
Kalathil, Shanthi, and Taylor C. Boas. 2003. *Open Networks, Closed Regimes: The Impact of the Internet on Authoritarian Rule*. Washington, DC: Carnegie Endowment for International Peace.
Kennedy, Scott. 2011. "Fragmented Influence: Business Lobbying in China in Comparative Perspectives." In *Beyond the Middle Kingdom: Comparative Perspectives on China's Capitalist Transformation*, edited by S. Kennedy, 113–135. Stanford, CA: Stanford University Press.

Koop, Christel, and Martin Lodge. 2017. "What Is Regulation? An Interdisciplinary Concept Analysis." *Regulation & Governance* 11: 95–108.

Laffont, Jean-Jacques, and Jean Tirole. 2012. *Théories des incitations et réglementations*. Paris: Economica.

Lagerkvist, Johan. 2012. "Principal-Agent Dilemma in China's Social Media Sector? The Party-State and Industry Real-Name Registration Waltz." *International Journal of Communication* 6: 2628–2646.

Lai, Hongyi. 2016. *China's Governance Model: Flexibility and Durability of Pragmatic Authoritarianism*. Abingdon: Routledge.

Landry, Pierre F. 2008. *Decentralised Authoritarianism in China: The Communist Party's Control of Local Elites in the Post-Mao Era*. Cambridge: Cambridge University Press.

Lee, Ching Kwan, and Zhang, Yonghong. 2013. "The power of instability: unravelling the microfoundations of bargained authoritarianism in China." *American Journal of Sociology* 118(6): 1475–1508.

Lei, Ya-wen. 2018. *The Contentious Public Sphere: Law, Media and Authoritarian Rule in China*. Princeton, NJ: Princeton University Press.

Liao, Zhixiong. 2016. "An Economic Analysis on Internet Regulation in China and Proposals to Policy and Law Makers." *International Journal of Technology Policy and Law* 2(2, 3, 4): 242–256.

Lieberthal, Kenneth, and Michel Oksenberg. 1988. *Policy Making in China: Leaders, Structures and Processes*. Princeton, NJ: Princeton University Press.

Lodge, Martin. 2002. *On Different Tracks: Designing Railways Regulation in Britain and Germany*. Westport: Praeger.

Lynch, Daniel. 1999. "Dilemmas of 'Thought Work' in Fin-De-Siècle China." *The China Quarterly* 157: 173–201.

Ma, Aifang. 2018. *L'intelligence artificielle en Chine: Un état des lieux*. Paris: Fondation pour l'Innovation Politique.

Ma, Aifang. 2023c. *La Régulation du numérique: Chine, Etats-Unis, France*. Paris: Fondation pour l'Innovation Politique.

Mann, Michael. 1984. "The Autonomous Power of the State: Its Origins, Mechanisms and Results." *European Journal of Sociology* 25(2): 185–213.

Mann, Michael. 2012. *The Sources of Social Power 2: The Rise of Classes and Nation-States, 1760–1914*. New York: Cambridge University Press.

Marx, Karl. 1926. *The Eighteenth Brumaire of Louis Bonaparte*. Translated by P. Eden. London: Allen and Unwin.

Marx, Karl, and Frederick Engels. 1848. *Manifesto of the Communist Party*.

Mayntz, Renate. 2009. "The Changing Governance of Large Technical Infrastructure Systems." In *Über Governance: Institutionen und Prozesse politischer Regelung*, edited by Renate Mayntz, 121–150. Frankfurt: Campus Verlag.

McGeary, Aaron D. 2001. "China's Great Balancing Act: Maximizing the Internet's Benefits While Limiting Its Detriments." *The International Lawyer* 35(1): 219–230.

McGregor, Richard. 2010. *The Party: The Secret World of China's Communist Rulers*. New York: HarperCollins Publishers.

Mendras, Henri, and Jean Etienne. 1996. *Les Grands Auteurs de la Sociologie: Tocqueville, Marx, Durkheim, Weber*. Paris: Hatier.

Mertha, Andrew. 2009. "'Fragmented Authoritarianism 2.0': Political Pluralization in the Chinese Policy Process." *The China Quarterly* 200: 995–1012.

Miller, Blake. 2018. "The Limits of Commercialized Censorship in China." Working Paper, http://eprints.lse.ac.uk/101417/1/miller_limits_of_commercialized_censorship_in_china.pdf.

Montinola, Gabriella, Yingyi Qian, and Barry R. Weingast. 1995. "Federalism, Chinese Style: The Political Basis for Economic Success in China." *World Politics* 48(1): 50–81.

Nathan, Andrew. 2009. "China Since Tiananmen: Authoritarian Impermanence." *Journal of Democracy* 20(3): 37–40.

Nathan, Andrew. 2003. "China's Changing of the Guard: Authoritarian Resilience." *Journal of Democracy* 14(1): 6–17.

Naughton, Barry. 1995. *Growing Out of the Plan: Chinese Economic Reform 1978–1993*. Cambridge: Cambridge University Press.

Negro, Gianluigi. 2017. *The Internet in China: From Infrastructure to a Nascent Civil Society*. Cham: Palgrave Macmillan.

North, Douglass. 1990. *Institutions, Institutional Change and Economic Performance*. New York: Cambridge University Press.

Oi, Jean C. 1992. "Fiscal Reform and the Economic Foundations of Local State Corporatism in China." *World Politics* 45: 99–126.

O'Brien, Kevin. 1994. "Implementing Political Reform in China's Villages." *The Australian Journal of Chinese Affairs* 32: 33–59.

Pearson, Margaret M. 2005. "The Business of Governing Business in China: Institutions and Norms for the Emerging Regulatory State." *World Politics* 57: 296–322.

Perry, Elisabeth J. 2007. "Study Chinese Politics: Farewell to Revolution?" *China Journal* 57(1): 1–22.

Poulantzas, Nicolas, and Ralph Miliband. 1972. "The Problem of the Capitalist State". In *Ideology in Social Science: Readings in Critical Social Theory*, edited by Robin Blackburn, 238–262. New York: Pantheon Books.

Qian, Gang, and David Bandurski. 2011. "China's Emerging Public Sphere: The Impact of Media Commercialization, Professionalism, and the Internet in an Era of Transition." In *Changing Media, Changing China*, edited by S. Shirk, 38–76. Oxford: Oxford University Press.

Rahman, K. Sabeel, Thelen, Kathleen. 2019. "The Rise of the Platform Business Model and the Transformation Twenty-First-Century Capitalism." *Politics & Society* 47(2): 177–204.

Rauchfleischer, Adrian and Schäfer, Mike. 2015. "Multiple Public Spheres of Weibo: A Typology of Forms and Potentials of Online Public Spheres in China." *Information, Communication & Society* 18(2): 139–155.

Ringen, Stein. 2016. *The Perfect Dictatorship: China in the 21st Century*. Hong Kong: HKU Press.

Riskin, Carl. 1987. *China's Political Economy: The Quest for Development since 1949*. New York: Oxford University Press.

Roberts, Margaret. 2018. *Censored: Distraction and Diversion Inside China's Great Firewall*. Princeton: Princeton University Press.

Rocca, Jean-Louis. 2006. *La Condition chinoise: la mise au travail capitaliste à l'âge des réformes (1978- 2004)*. Paris: Karthala.

Sautman, Barry. 1992. "Sirens of the Strongman: Neo-Authoritarianism in Chinese Political Theory." *The China Quarterly* 129: 72–102.

Scott, C. James. 1985. *Weapons of the Weak: Everyday Forms of Peasant Resistance*. New Haven: Yale University Press.

Segal, Gerald. 1994. "China Changes Shape: Regionalism and Foreign Policy: Introduction." The Adephi Papers 34(287): 3–5.

Shambaugh, David. 2008. *China's Communist Party: Atrophy and Adaptation*. Washington, DC: Woodrow Wilson Center Press.

Shirk, Susan. 1993. *The Political Logic of Economic Reform in China*. Berkeley, CA: University of California Press.

Shirk, Susan. 1985. "The Politics of Industrial Reform." In *The Political Economy of Reform in Post-Mao China*, edited by E. J. Perry, and C. Wong, 195–221. Cambridge, MA: Harvard Council on East Asian Studies.

Shirk, Susan. 2011. "Chapter 1: Changing Media, Changing China." In *Changing Media, Changing China*, edited by S. Shirk, 1–37. Oxford: Oxford University Press.

Shirk, Susan. 2018. "China in Xi's 'New Era': The Return to Personalistic Rule." *Journal of Democracy* 29(2): 22–36.

Skocpol, Theda. 1979. *States and Social Revolutions*. Cambridge: Cambridge University Press.

Solinger, Dorothy J. 1987. "The 1980 Inflation and the Politics of Price Control in the PRC." In David M. Lampton (ed.), *Policy Implementation in Post-Mao China*. Berkeley: University of California Press, pp. 106–110.

Stern, Rachel E., and Kevin O'Brien. 2012. "Politics at the Boundary: Mixed Signals and the Chinese State." *Modern China* 38(2): 174–198.

Stigler, George J. 1971. "The Theory of Economic Regulation." *Bell Journal of Economics and Management Science* 2: 3–21.

Strittmatter, Kai. 2020. *Dictature 2.0: Quand la Chine surveille son people (et demain le monde)*. Translation by Olivier Mannoni. Paris: Tallandier.

Tai, Zixue. 2006. *The Internet in China: Cyberspace and Civil Society*. New York: Routledge.

Tai, Zixue, and Tao Sun. 2007. "Media Dependencies in a Changing Media Environment: The Case of the 2003 SARS Epidemics in China." *New Media & Society* 9(6): 987–1009.

Tang, Wenfang. 2016. *Populist Authoritarianism: Chinese Political Culture and Regime Sustainability*. Oxford: Oxford University Press.

Teets, Jessica C. 2013. "Let many civil societies bloom: the rise of consultative authoritarianism in China." *The China Quarterly* 213: 19–38.

Tilly, Charles. 1981. *As Sociology Meets History*. New York: Academic Press.

Tong, Yanqi. 2011. "Morality, Benevolence, and Responsibility: Regime Legitimacy in China from Past to the Present." *Journal of Chinese Political Science* 16: 141–159.

Tsai, Kelley S. 2005. "Capitalists Without a Class: Political Diversity Among Private Entrepreneurs in China." *Comparative Political Studies* 38(9): 1130–1158.

Tsui, Lokman. 2003. "The Panopticon as the Antithesis of a Space of Freedom: Control and Regulation of the Internet in China." *China Information* 17(2): 65–82.

Weber, Max. 1946. *Politics as a Vocation*. New York: Oxford University Press.

Weber, Max. 1978. *Economy and Society*. New York: Bedminster Press.

Weber, Max. 1998. "Class, Status and Party." In *From Max Weber: Essays in Sociology*, edited and translated by H. H. Gerth, and C. W. Mills, 180–195. New York: Oxford University Press.

White, Gordon. 1994. "Democratization and Economic Reform in China." *Australian Journal of Chinese Affairs* 31: 73–92.

Woll, Cornelia. 2008. *Firm Interests: How Governments Shape Business Lobbying on Global Trade*. Ithaca, NY: Cornell University Press.

Wong, Christine. 1985. "Material Allocation and Decentralization: Impact of the Local Sector on Industrial Reform." In *The Political Economy of Reform in Post- Mao China*, edited by E. Perry, and C. Wong, 253–278. Cambridge, MA: Harvard University Press.

Xiao, Qiang. 2011. "The Rise of Online Public Opinion and Its Political Impact." In *Changing Media, Changing China*, edited by S. Shirk, 202–224. Oxford: Oxford University Press.

Xue, Xiaorong. 2018. *Hu Lian Wang Qi Ye Dang Jian: Ji Shu Zi Ben Yu Zheng Zhi Luo Ji Zhang Li Xia De Zheng Dang Zhi Li* [Party-Building in Internet Firms: Party Governance under the Logics of Technology, Capital and Politics]. Beijing: Current Affairs Press.

Yang, Dali. 1997. *Beyond Beijing: Liberalization and the Regions in China*. New York: Routledge.
Yang, Guobin. 2003. "The Co-Evolution of the Internet and Civil Society in China." *Asian Survey* 43(3): 405–422.
Yang, Guobin. 2009a. *The Power of the Internet in China: Citizen Activism Online*. New York: Columbia University Press.
Yang, Guobin. 2009b. "China Since Tiananmen: Online Activism." *Journal of Democracy* 20(3): 33–36.
Yeo, Yukyung. 2020. *Varieties of State Regulation: How China Regulates Its Socialist Market Economy*. Cambridge, MA: Harvard University Press.
Zhao, Yuezhi. 1998. *Media, Market, and Democracy in China: Between the Party Line and the Bottom Line*. Urbana, IL: University of Illinois Press.
Zhao, Yuezhi. 2000. "From Commercialization to Conglomeration: The Transformation of the Chinese Press Within the Orbit of the Party State." *Journal of Communication* 50(2): 3–26.
Zheng, Yongnian. 2006. *De facto Federalism in China: Reforms and Dynamics of Central-Local Relations*. Singapore: World Scientific Publishing.

CHAPTER 3

Particularities of the Economic Regulation of the ISM Industry

Introduction

Communication scholar Negro advances two dialectic processes to analyse the Chinese regulation of the ISM sector: macro- and micro-dialectic processes (2017: 19–20). While the former involves three actors: the government, the market, and netizens, the latter only involves the government. The importance to study the micro-dialectic process is that "all ministries and departments involved in internet regulation tend to express a set of values that reflects the specific interests of lobbies that are often in war with each other" (Negro, 2017: 20). These two processes denote two parallel dynamics which are equally important for understanding Chinese internet development: (1) tensions between the government, the market, and netizens; (2) tensions within the government itself. This chapter aims to identify the Chinese particularities in the economic regulation of the ISM sector. The macro- and micro-dialectic processes are fully at play in shaping these particularities.

It is necessary to highlight the party-state's bifurcated approach to the ISM industry from the outset: economics-driven regulation (EDR) and politics-driven regulation (PDR). In this book, the EDR aims to achieve four principal economic objectives: economic growth and upscaling, job creation, technological innovation, and global competitiveness. For reasons of clarity, social benefits associated with the ISM, including the internet penetration, poverty alleviation, internet literacy, to name only a

A. Ma, *China as a Double-Bind Regulatory State*,
https://doi.org/10.1007/978-981-99-8857-0_3

few, are incorporated into economic objectives. Conceptually speaking, the party-state's economic objectives in this book involve all the aims unrelated to political objectives, and ergo, can be termed as non-political objectives. The PDR aims to realise the party-state's political objectives. This book highlights the following political objectives: regime security, leadership and legitimacy of the CCP in China, leadership and legitimacy of the party-state's ideology, and international prestige of China. Although economic objectives also serve the party-state's political objectives (Milhaupt and Zheng, 2015), the contradiction between the two exists in the ISM regulation and are difficultly reconcilable, as analysed in Chapter 5.

This chapter examines the particularities of the EDR and lays the historical foundations for in-depth analyses of the DBR in the chapters to follow. It is organised in 3 parts. The first part examines the issue of "motivations", highlighting the centrality of economic concerns behind the rapid growth of the Chinese ISM sector. The second and third parts turn attention to the "actors" in the Chinese ISM expansion. The second part analyses the party-state's spearheading role in building the key internet infrastructure and in nurturing large telecom operators. It also explains that from the outset the party-state reinforces the regime's security through "nudging" SOEs and private firms to respectively operate in the upstream and downstream sectors of the ISM. The third part introduces private firms on the stage and investigates the conditions enabling them to rise to international stardom.

ISM Industry: An Accelerator of Economic Development

Similar to the spearheading role of the American government in boosting the internet development in the USA, the Chinese government played the role of bulldozer for jumpstarting the internet development in China. The differences between the two countries is that while the internet development was motivated by military purposes in the USA, the Chinese government was motivated above all by the concern of economic modernisation. The Chinese government not only provided generous financial support, but also fixed the directions for the development of the internet sector. Bureaucratic competition existed and became an important factor for driving the initial strides of the Chinese internet industry. In this sense, the micro-dialectic process of Negro (2017) is of great help for explaining the internet development in the pre-2000 period, when major players of the Chinese market were public actors rather than private ones. The macro-dialectic process is most useful for explaining the regulation of the ISM industry in the post-2000 period, in which major privately-owned ISM firms came up and became quite competitive at both the domestic and the international markets.

In line with several leading scholars in Chinese internet (Tai, 2006; Lu et al., 2002; Qi, 2000, Negro, 2017), this book considers 1987 as the starting point of the internet development in China. On 20 September 1987, the first email signed by Professor Qian Tianbai was sent from Beijing to Karlsruhe University via a leased line from the Italian ITAPAC. The message contained in the email was "cross the Great Wall, join the world" and was sent at 300 bits per second, "an impossibly slow connection by today's typical consumer broadband speeds measured by the tens or hundreds of millions of bps" (Clark, 2016: 74). It was the first time that China reached foreign countries though the internet. Once this first stride was made, the ISM industry has been progressing at an astonishing pace in China. The results achieved are "the most impressive in the global history of the internet" (Negro, 2017: 19). In 1987, the first Chinese computer network CANET (China Academic Network) became operational. One year later, the CANET chose ".cn" as the domain name for the PRC. In 1989, China started building its first website, the NCFC (Networking and Computing Facility of China). In April 1994, the NCFC was connected to the NSFNET (National Science Foundation Network) of the USA, an event that marked China's entry into the global internet (Arsène, 2012: 100).

In the pre-1995 period, China started building the four backbone networks: the CERNET, CSTNET, ChinaNet, and China GBN. The interconnection of the four networks became a reality in June 1998, with the ChinaNet providing a high-speed line for connecting the other three networks (Qi, 2000). In addition, the construction of the CRNET (China Research Network), put under the supervision of the Chinese Academy of Sciences (CAS), was accomplished in 1990. To ensure a close follow-up of the Chinese internet sector, the CNNIC (China Internet Network Information Center) was created in 1997 and affiliated to the CAS. It has been producing biannual statistical reports on Chinese internet development since November 1997. These reports constitute widely authoritative sources for scholars on Chinese internet studies.

Imperative of Economic Modernisation Drives Internet Development

The imperative to accelerate economic modernisation drived the Chinese internet development. The tremendous support of the Chinese government was initially motivated by economic imperatives, rather than political or ideological reasons (Tai, 2006; Negro, 2017). The writings and speeches of Chinese top leaders in the late 1990s and early 2000s fully reflected the centrality of economic concerns for the party-state leadership. In his preface for the book *Explorations and Practices of Chinese Informatisation (zhongguo xinxihua tansuo yu shijian)*, book of Chinese scientist Hu Qili in 2002, Jiang Zemin emphasised the importance of information technologies for Chinese industry and national economy. For him, "industry and national economy should be equipped with information technologies to improve their international competitiveness and to leapfrog Chinese development" (Jiang, 2002). Considering the internet as a "new point of economic growth", Chinese leadership considerably invested into the ISM sector, hoping to change the latter into the engine for realising economic catch-up with advanced economies.

The writings of several influential scholars in the 1970s and the 1980s arguably aroused Chinese leaders' enthusiasm in developing information technologies, including among others Tadao Umesao (1963), Alvin Toffler (1980), and Daniel Bell (1976). A popular term frequently pronounced by Chinese leaders in the 1990s was "informatisation", which illustrates the party-state's central concern with the nascent internet technology: capitalising on technological innovation to realise economic growth. The concept of "informatisation" was coined by the Japanese ethnographer and anthropologist Tadao Umesao in his article "Information Industry Theory: Dawn of the Coming Era of the Ectodermal Industry" (1963). For him, informatisation included the modernisation of telecommunications, computerisation, and behavioural rationalisation (1963). He predicted that after the agricultural age and the industrial age, the information industry would create a new society where information would become an important economic factor. Popular in Western countries in the 1970s, "informatisation" gained popularity in China in the second half of the 1990s.

Two other scholars also played an important role in shaping the economy-oriented inclinations of Chinese decision-makers in the 1990s: Alvin Toffler with *The Third Wave* (1980) and Daniel Bell with *The Coming of Post-Industrial Society* (1976). In *The Third Wave*, Toffler declared that the first and the second waves, namely the agricultural and the industrial ages, became history. In contrast, the information age, coming along with information technology and co-driven by the tendency of individualisation, would lead to knowledge-demanding industries, including among others quantum electronics, molecular biology, and semiconductors. Within the third wave, knowledge-production and the processing of information would become the primary economic activity and the primary determinant of power and power distribution. Information technology was treated as one essential productive force in the third wave. This perspective described by Toffler is said to have enchanted Chinese leaders and academia alike. In his speech on 9 October 1983, Zhao Ziyang, then General Secretary of the CCP, quoted Toffler's *The Third Wave* as one of the valuable theories for guiding developing countries to become technological powerhouses (Zhao, 2016). Developing the ICTs to modernise China became the consensus of Chinese decision-makers. Chinese intelligentsia generally explained the country's backwardness from the 1840s to the 1940s by its failure to catch up with the two industrial revolutions. Since the internet stood at the heart of Toffler's "third wave", Chinese leaders betted on information technologies, so as not to see the country humiliated again by the West.

The book of Daniel Bell (1976) on the dawn of a post-industrial society also increased Chinese leaders' interest in the nascent internet sector. He wrote that "a post-industrial society is based on services. Life becomes a game between persons. What counts is not raw muscle, or energy; what counts is information" (1976: 576). For Bell, if raw muscle and energy were respectively the drivers of the agricultural and the industrial societies, information would be the driver of a post-industrial society. Having compared the modes of production in agricultural and industrial societies, Bell explained that a post-industrial society would undergo a shift from repetitive work, portrayed in the *Modern Times* of Charles Chaplin, to knowledge-demanding professions. Semi-skilled workers standing in front of assembly lines would be replaced by "professional, technical and managerial employees" (Bell, 1976: 577). The writings by Daniel Bell and Alvin Toffler provided the concepts which Chinese leaders frequently used in their speeches, including "knowledge economy" and "information society".

These futuristic writings based on the galloping development of information technologies in the Western world galvanised Chinese leaders. After the earliest internet connections were established in China in the late 1980s, Chinese policymakers quickly took measures to build the basic information infrastructure of the country. The special attention paid to the sector traced back to the Eighth Five-Year Plan (FYP) for the period 1991–1995. In fact, the economic importance of the internet sector was emphasised much earlier than its political importance. In April 1991, when the *Ten-Year Plan for the National Economic and Social Development (1991–2000) and the Eighth Five Year Plan (1991–1995)* was adopted, the internet industry barely kicked off in China. Although this document didn't explicitly mention the internet sector, it paid important attention to telecommunications and the electronic industry, two sectors underpinning the internet industry. One of the major tasks inscribed into the 8th FYP was to "reinforce the construction of infrastructure and fundamental industries" to which telecommunications belonged. To achieve this objective, "investment policies should lurch towards fundamental industries and infrastructure". In addition, the electronic industry was planned to occupy a "remarkable position" in the national economy and to become "the leading industry helping to upgrade the industrial structure and the whole national economy".

The importance of the electronic industry was further highlighted in the Nineth FYP (1996–2000), adopted on 17 March 1996. Different from its predecessor, the 9th FYP explicitly considered the electronic industry as one of the "pillar industries" (*zhizhu chanye*) of China. The emphasis on the infrastructure-building remained consistent throughout the 9th FYP. In addition, it pointed out the importance of manufacturing essential parts by national producers rather than foreign ones. According to this plan, "the priority is to develop integrated circuits, new components, computers and communications equipment", the objective being to provide essential equipment for realising the informatisation of China. Besides detailed stipulations on the types of integrated circuits that were to be developed from 1996 to 2000, supportive policies were included. For instance, China planned to "apply preferential policies to support the development of integrated circuits and continue to implement the project of integrated circuits".

The above discussion shows that in the 1980s, the Chinese leadership had already taken notice of the potential of the ICTs to drive up Chinese economic catch-up. In contrast, it was only in the 10th FYP (2001–2005) that central planners explicitly expressed their concerns on the political dimension of the internet sector. However, even at this stage, the FYP did not fully explain the political concern of Chinese policymakers, because the latter were not ready to impede the booming internet sector with overly restrictive measures. The reason behind the party-state's unequal emphasis on the political dimension of the internet development and its economic dimension is twofold. First, the economic impacts of the internet sector were felt much earlier than its political impacts. Except restrictive economic policies from 1989 to 1992 due to the Tiananmen crackdown, the 1990s in China was an epoch of economic liberalisation and modernisation during which the Chinese government significantly reduced state intervention into the market and encouraged the emergence of private actors (Huang, 2008; McGregor, 2010). The outcome of this move was the unleashing of market dynamism and a change of the social mentality from class struggles into profit-seeking. This state of mind reached as far as the ministries. For instance, the personnel reform of the Ministry of Post and Telecommunications (MPT) in the 1990s was motivated by economic gains rather than social or political concerns (Negro, 2017). In 1994, the MPT did not recruit any new staff to reduce its budgetary burden. The number of employees in the MPT decreased from 14 million at the beginning of 1990 to 1.13 million in 1996. In addition, the MPT's wage reform scheme in 1998 tied the employees' salary to the overall revenue growth (Negro, 2017: 31).

Second, due to the high price of the internet connection in the 1990s, the internet was used by a small number of well-to-do Chinese people. It was since the creation of the Ministry of Information Industry (MII) in 1998 that internet connection fees started going down. The average price of the internet connection dropped from $70 per month in 1997 to $7 by the end of 1999 (Clark, 2016: 95). However, this price was still high, especially for college students. Still in the end of 1999, in Beijing, internet connection fees were 10 yuan per hour for college students and 5.8 yuan per hour for families. Students at Peking University complained about the cost of internet connections by using the telephone card 201, issued by the Beijing Telegraph Bureau (Beijing dian bao ju).[1] High connection fees dissuaded Chinese young people from using the internet on a large scale. Therefore, talking about online freedom of speech in the 1990s was conceptually erroneous, because cyberspace was principally occupied by research institutions and big firms, not by ordinary netizens. Given that

[1] Lin, Da, "Will the fees of Internet connection through 201 decrease or not?," *Guang Ming Daily*, 24 November 1999, https://gmw.cn/01gmrb/1999-11/24/GB/GM^18250^16^GM16-209.htm.

netizens are the pillar for producing user-generated contents (UGCs) and that their access to the internet was limited in the 1990s, the UGCs were not quantitatively important enough to worry regulators.

Informatisation-Empowered Economy

Given the contributions of the ICTs to upscale economic growth models, informatisation was perceived as an important issue from the outset and continues to be highlighted by successive Chinese leaders. In January 1996, the SCIWSC (State Council Informatisation Work Steering Committee) was created. It was the first inter-ministerial agency to ensure coordinated actions of ministries and to counter vicious bureaucratic turf battles. On 18–21 April 1997, China held the first National Informatisation Work Conference in Shenzhen. This conference clarified on two terms: "informatisation" and "national informatisation". Informatisation was defined as a "historical process of nurturing and developing new productive forces, represented by intelligent tools, and mobilising these forces to create social wellbeing". "National informatisation" was defined as a process in which "under the unified planning and organisation of the state, modern information technologies are applied in agriculture, industry, science and technologies, defence, and all the aspects of social life. We thoroughly develop and widely use information resources to accelerate the process of national modernisation". In the same year, the National Information Infrastructure Steering Committee (NIISC) held a working meeting and adopted the *National Informatisation Plan of the 9th Five Year Plan and Long-Term Goals of 2000*. During this meeting, "Beijing officially recognised the economic importance of internet development and made network development a priority in the national economic policy" (Negro, 2017: 22). In May 2004, the MII created a bimonthly magazine *iChina (Zhong guo xin xi hua za zhi)*, which reported the latest applications of the ICTs in finance, insurance, education, real estate, energy, medicine, to name only a few. This magazine reflects well the Chinese *Zeitgeist* in that era: internet technology must serve economic upscaling.

In the following years, Chinese leaders continued to emphasise informatisation for shaping high-quality economic growth of China. On 19 March 2006, the General Office of the Central Committee of the CCP (CCCCP) and the General Office of the State Council (SC) jointly released a long-term informatisation plan, entitled *2006–2020 National*

Informatisation Development Strategy. This national strategy ranked informatisation of the national economy as the primordial priority that the party-state should strive for by 2020. It fixed four objectives to achieve. First, China should make use of informatisation to fundamentally transform the dynamics that drive Chinese economic growth. Second, big strides should be made in increasing indigenous innovation capacities and promoting the industrial development of information technologies. Third, China should improve the coverage of the internet penetration, better exploit information resources, and protect information security. Fourth, it should reinforce the capacity of the government to deliver public services through the internet, spread socialist advanced culture, "informatise" national defence, and improve internet literacy of the people. In a nutshell, in the mid-2000s, the party-state fixed the ambitious objective to use the ICTs to completely revamp the outlook of the country in economic, scientific, social, political, and cultural terms. The ordering of the four objectives reveals what the party-state values the most: while economic objectives were put in the first place, ideological concerns were put at the last place. The first objective is worth quoted in its entirety:

> Promote a fundamental change in the way in which Chinese economy grows. Widely use information technologies to revamp and upscale traditional industries. Develop information services to promote a strategic adjustment in the economic structure. Use information technologies in a thorough way. Mobilise them to reduce energy consumption and the material consumption per unit product, to reinforce the supervision and the reparation of environmental pollution, and to serve the development of circular economy. Use information technologies...to promote the shift of our economic growth model. The reliance on the capital and resources should be replaced by the reliance on scientific and technological progress as well as a more qualified labour force, so as to improve the quality and the efficiency of economic growth.

The quote above demonstrates the resolve of Chinese leaders to upgrade the economic growth models through informatisation. It also resumes pretty well the philosophy of the Chinese government in that epoch: although it can be an exaggeration to claim that informatisation was put up as the panacea for all economic and social ills, it is safe to say that Chinese leaders considered informatisation as an indispensable and probably the decisive ingredient composing that panacea.

Political leaders' official speeches in the 1990s also highlighted the role of information technologies in changing China into a world-class information society. President Jiang Zemin and his Prime Minister Zhu Rongji invoked the importance of the internet in their official speeches several times. In 2000, Jiang had mentioned the information society at least three times (Austin, 2014). These occasions include the opening speech at the National People's Congress (NPC) and the Chinese People's Political Consultative Conference (CPPCC) on 3 March, the 16th World Computer Congress on 21 August, and a meeting of the Central Military Commission (CMC) on 11 December (Austin, 2014; Negro, 2017: 43). One excerpt of his speech at the 16th World Computer Congress in August 2000 shows the leadership's determination in pushing forward an internet-empowered national economy:

> We should deeply recognise the tremendous power of the ICTs and vigorously promote their development. The speed and scope of their expansion have created a borderless information space around the world...The melting of the traditional economy and information technologies will provide the engine for the development of the economy and the society in the 21st century. (Quoted in Tai, 2006: 120).

Under Jiang Zemin, the combination between informatisation and economic growth stood at the heart of the national strategy of developing a "new industrialisation path". Jiang advanced this concept for the first time in November 2002 at the 16th Party Congress. It meant that China should "insist on promoting industrialisation with informatisation, promoting informatisation with industrialisation, so as to create a new industrialisation path characterised by high technological intensity, high economic benefits, low energy consumption, low environmental pollution, and full exploitation of the advantages in the human resources". The strategy of "new industrial path" took informatisation as a booster for upgrading Chinese industrialisation.

Under the two presidential terms of Hu Jintao from 2002 to 2012, China tried to extend the application of the ICTs from the industrial sector to a large variety of service sectors. The 10th and the 11th FYPs (2001–2005, 2006–2010) roughly corresponded to this period. The 10th FYP required to increase the percentage of the information sector in the national GDP, to promote a "new industrialisation path", and to maximise the comprehensive application of information technologies to economic sectors, including mobile payment, E-commerce, and fintech. The 10th FYP was the precursor of the national strategy of "Internet +" adopted in July 2015 by the State Council. It also paved the way for the booming of private ISM firms created at the turning point between the 20th and the 21st centuries (e.g., Sohu, Netease, Sina, Tencent, Baidu, Alibaba).

In the 11th FYP, central planners continued to reduce their direct intervention into the market. What distinguishes the 11th FYP from all its predecessors is that from the 11th FYP onwards, the title of the FYP in Chinese language changes from "five-year plan" (*wunian jihua*) to "five-year guidance plan" (*wunian guihua*). The change of wording implies important shifts in state-market relations because "guidance plan" means that the concerned actors—firms or regulators—are not bound to achieve the indicated objectives. To further demonstrate its commitment to liberalisation, from the 11th FYP onwards, the party-state introduced the distinction between binding indicators from predictive indicators. While binding indicators must be reached in the end of one FYP period, it is not the case for predicative indicators. Since informatisation of the service sector belongs to predictive indicators, firms obtain greater autonomy in operating in the E-commerce, Fintech, and other internet-related services.

The combination between industrialisation and informatisation is replaced by another buzzword when Xi came to power in 2012: "Internet +". At the same time, informatisation obtains greater importance. On the one hand, the party leadership directly takes in charge informatisation, and on the other hand, it recognises the importance of informatisation for realising high-quality economic growth. In China, state institutions (e.g., MII from 1998 to 2008, MIIT from 2008 to 2014) traditionally champion informatisation under the supervision of the State Council. This portfolio is transferred to party institutions in the era of Xi. The Central Cybersecurity and Informatisation Leading Small Group (CCI-LSG) is created on 28 February 2014, with Xi Jinping and his Prime Minister Li Keqiang respectively as the Director and the Deputy Director. The CCI-LSG is an interministerial coordination and decision-making institution in cybersecurity and informatisation. Its competences cover both the economic side and the political side of the ISM regulation. In March 2018, the *Plan on Deepening the Reform of Party and State Institutions* (*shenhua dang he guojia jigou gaige fang'an*) institutionalises the CCI-LSG by turning it from an informal institution into a formal institution and changes the name of CCI-LSG into Central Cybersecurity and Informatisation Commission (CCIC). Since 2014 onwards, the CAC takes charge of the routine work of the CCI-LSG, and since 2018, the CCIC. This move formally recognises the CCP as the chief regulator of informatisation, thus pushing the CCP from the backstage to the frontstage of the internet regulation.

The Chinese leadership recognises the potential of informatisation to improve the quality of economic growth. In the first meeting of the CCI-LSG on 28 February 2014, Xi Jinping emphasises that there is no national security without cybersecurity; there is no modernisation without informatisation. Under Xi, the new generation of information technologies figures among the "emergent industries with strategic importance" (EISI), a national project intended to pull up the Chinese position in the global value chain. During the second term of Xi from 2017 to 2022, economic growth slows down, at an average of 6.6% from 2016 to 2020 and 5.5% in 2021. The Chinese leadership does not seem to be excessively troubled by the slowing pace of economic growth, because "our national economy has shifted from a period of high-speed growth into an era of high-quality growth".[2] Despite the shifts of economic development models, the importance of informatisation remains unchanged. In his speech at the National Conference on Cybersecurity and Informatisation Work (*Quanguo wangluo anquan he xinxihua gongzuo huiyi*) on 20 April 2018, Xi states that informatisation should continue to drive Chinese social and economic development.

The earliest strides China made in the internet development are principally driven by the imperative of leapfrogging economic growth. Exercising political control in cyberspace is secondary to economic concerns in that era. This point echoes Negro (2017) who argues that the imperative of political control should not be overestimated in the establishment of censorship technologies like the Great Firewall. For him, "the strategy of blocking access to some foreign websites…cannot be seen as being solely for political reasons but must be set in a more complex environment that also takes into consideration the importance of national internet industries" (Negro, 2017: 47). The emphasis on the economic potential of the internet from the late 1980s to the 1990s, however, sows the seeds

[2] "Xi Jinping attended the National Conference on Cybersecurity and informatisation work and delivered an important speech," Xinhua, 21 April 2018, http://www.gov.cn/xinwen/2018-04/21/content_5284783.htm.

of the party-state's predicament because a wider internet use in subsequent years in China exposes the potential of the internet to destabilise the regime. The fundamental question facing the party-state becomes the following one: shall it continue to pursue the economy-centric approach in the internet sector, or, instead, concentrate on defusing the risks that the internet sector represents for the regime's survival?

Government-Piloted Internet Development in China

Internet development in China was a government-led process in the 1990s and the 2000s. Before private actors entered this sector in the mid-1990s (e.g., in 1995, Zhang Shuxin created one of the earliest private Internet Service Providers (ISP) in China, InfoHighWay), the Chinese government acted both as an operator and a regulator of the internet industry. The centrality of the Chinese government is investigated along two dimensions: (1) the creation and monopolisation of backbone networks by state-trusted agencies and (2) the nurturing of large public firms in telecommunications.

State-Trusted Agencies Spearhead the Creation of Backbone Networks

Ministries and government-patronaged institutions (e.g., research institutions) piloted the construction of the backbone networks in China. In 1993, funded by the Ministry of Posts and Telecommunications (MPT) and the State Development Planning Commission, the predecessor of the National Development and Reform Commission before 2003, the State Education Commission started building the China Education and Research Network (CERNET), the objective being to connect all the universities, secondary schools, and primary schools in China into one network. After the construction phase ended in 1994, the earliest backbone network in China came into being.

Among all ministries and government-patronaged institutions, the MPT demonstrated the highest zeal in building backbone networks, in large part for enlarging its turf. Created on 1 November 1949, one month after the creation of the PRC, the MPT obtained the monopoly in both postal services and telecommunications. After the internet appeared in China in 1987, the MPT deployed strong endeavours to incorporate the internet into its portfolio. In 1993, it began building its own packet-data network CHINAPAC. Wu Jichuan, then Minister of Posts and Telecommunications, advocated the MPT's monopolistic control of virtually all areas of voice and data communication. Despite the opposition and competition from other ministries, the MPT did not stop pushing its interests. In 1994, the MPT launched ChinaNet in Beijing and Shanghai to provide public commercial services. ChinaNet was licensed as one of the government's major international gateways connecting China to foreign countries. ChinaNet was a wholesale provider of the internet bandwidth. Therefore, provincial telecommunications administrations (PTAs), namely local outposts of the MPT, relied on it to offer retail services in their respective jurisdiction. It was the first commercial network in China, connecting Beijing and Shanghai to the USA (Clark, 2016). It was also the first to offer internet services to ordinary citizens (Negro, 2017: 29). In 1996, the State Council recognised ChinaNet as the only access channel by which domestic networks could be connected to the global internet.

Other ministries did not wait long before showing their interest in the internet. Attracted by the benefits coming from the management of internet networks, they started building their own networks. In the mid-1990s, the MPT faced strong competition from two ministries: the Ministry of the Electronic Industry (MEI) and the Ministry of Broadcasting, Films and Television (MBFT). The leverage of the MEI in comparison with the MPT was that it was both the regulator and the producer of the essential parts for information technologies. Since the "profit motive" became "endemic" inside Chinese bureaucracies in the period of economic transition of the 1990s (Negro, 2017: 31), the MEI tried hard to split the dominant position of the MPT.

In late 1993, the MEI, together with two other SOEs, created a new corporation called Jitong. It was meant to be a satellite-based telecommunications network that used the MPT's land-based telephone lines for customer local access. In parallel, the ChinaGBN (Golden Bridge Network) was launched under the auspices of the vice-Prime Minister Zhu Rongji, who charged the MEI of building this new network. Golden Card and Golden Customs project were launched shortly after to create the so-called Three Golden Projects (Tai, 2006: 124). The first-phase construction of China GBN started on 8 June 1994 and was terminated in August 1995. 24 provinces and cities were covered by ChinaGBN, which was also connected to the internet network outside the country.

Besides the MEI, another powerful competitor of the MPT was the aforementioned MBFT. While the MEI's competitive advantage was that it produced essential parts of the ICTs that the MPT relied upon, the MBFT's trump was that it participated in guarding the party-state's ideological leadership, an advantage that the MPT did not have. In practice, the MBFT's ambitions went beyond the defence of the party-state's ideology and it endeavoured to build its own nationwide network to extract commercial benefits. Under the pressure of the SCIWSC, the MPT was obliged to reduce the charges of the lines it leased to the MBFT. The latter seized the opportunity and engaged itself in building a nationwide network for radio and television information. The MBFT

> expressed its interests in gaining benefits from the internet, and for this reason in 1997 it urged 14 provincial broadcasting authorities to undertake an all-out effort to upgrade the existing cable network, interconnect dispersed networks, deploy fiber-optic trunk lines, and promote intra- and inter-provincial interconnectivity. (Its) ambitious plan changed again in 1997 with the creation of a quasi-business entity, the Radio and Television Information Network Centre, which was in charge, among others, of the interconnection and the provision of internet and other value-added services to society. Although the Radio and Television Information Network Centre's mission was mainly based on broadcasting activities, its project was clear: to engage in a full range of broadcasting, information and telecommunications services. (Negro, 2017: 25)

Facing fierce inter-ministerial turf battles, the SCIWSC failed to ensure coordinated development of the internet industry. The reason was that as an interim organisation, the SCIWSC could not take over the regulatory responsibilities of the MPT, the MEI, and the MBFT. To clarify on the competences of each agency and to mitigate competition, on 1 February 1996, the State Council issued *Provisory Regulations of the PRC on International Internet Connections of Computer Information Networks (jisuanji xinxi wangluo guoji lianwang guanli zanxing guiding)*. This document recognised the MPT's preeminent position in regulating international internet connections. Article 6 stipulated that "it is mandatory to use the international access channels provided by ChinaNet [controlled by the MPT, added by the author] for making international connections… It is forbidden for any organisation and individual to set up on their own or use alternative access channels to make international connections". However, this document *de facto* ended the MPT's monopoly by recognising that each ministry or government-patronaged institution could regulate its own network. The power of deciding the establishment of new networks was centralised to the State Council, and ministries could no longer set up new networks as they saw fit. Article 7 added that "after relevant adjustments by the State Council, internet networks that have already been established are respectively managed by the MPT, the MEI, the State Education Commission and the CAS. It is mandatory to inform the State Council and to obtain its approval for establishing any new network".

The SCIWSC also launched an institutional reform to reduce overlapping responsibilities among different ministries and to curb the negative impacts of bureaucratic turf battles, including the waste of resources. This reform led to the merger between the MEI and the MPT into the Ministry of the Information Industry (MII) in March 1998. Since the creation of the MII was intended to end bureaucratic fragmentation in the operation and regulation of internet networks, the MII inherited all the competences of the MPT and the MEI, as well as the two key backbone networks under their respective control: ChinaNet and ChinaGBN. It also "gathered within it all the functions previously performed by the MBFT as well as the China Aerospace Industry Corporation and the China Aviation Industry" (Negro, 2017: 26).

From 1987 to 1995, the earliest four backbone networks were all set up by government-trusted agencies, including ministries (e.g., the MPT, the MEI, and the State Education Commission) and government-patronaged institutions (e.g., the CAS). No private actor was involved. The four agencies were tasked with granting operating licenses to downstream ISPs and overseeing them (Negro, 2017: 30). The government-piloted design for the construction and management of the internet infrastructure in China ensured that the provision and operation of the internet was managed by trusted government agencies (Negro, 2017: 30).

Inter-ministerial competition held two implications on the incipient stage of the Chinese internet development. First, it contributes to shape the institutional framework of the internet regulation. For instance, the creation of the centralised regulatory agency, the MII, was partly intended to end the turf battles between the MPT and the MEI. Second, inter-ministerial competition accelerated the unfolding of the internet industry in China in the second half of the 1990s. In the absence of coherent rules and a unified regulatory agency, ministerial infighting over the control of networks provided stimulus for a quick deployment of backbone networks. The "process of institutional bargaining and 'coordinated competition' has led to the most rapid network development in the history of world telecommunications" (Negro, 2017: 31).

Rise of Big SOEs: State-Orchestrated Competition in the Telecommunications Sector

That Chinese leaders supported the ICTs to accelerate economic modernisation stimulated the mushrooming of internet firms at the turning of the 20th and the 21st centuries, including both SOEs and private firms. However, the two types of firms did not operate in the same area. While SOEs controlled key internet resources and backbone networks, their private counterparts mostly operated in the downstream sectors of the internet industry and acted as ISPs and ICPs (Internet Content Providers). This distinction originated in the 1996-released *Provisory Regulations of the PRC on the International Internet Connections of Computer Information Networks*. Its article 3 introduced the distinction between two types of networks: interconnecting networks (*hulian wangluo*) and access networks (*jieru wangluo*). Interconnecting networks were defined as "the computer information networks that conduct direct connection with the international internet". Access networks were defined as "the computer information networks that connect to the international internet via the interconnecting networks". In other words, interconnecting networks were located at a more upstream position than access networks, and the latter could not reach the international internet without using the former. Given the strategic importance of both types of networks, SOEs, rather than private firms, dominated the provision of interconnecting networks and access networks. This situation persists today in China.

Private firms made attempts to step in this profitable area. However, these attempts failed, due to lack of governmental support. In May 1995, Zhang Shuxin (alias Jasmine Zhang) created Yinghaiwei, a phonetic equivalent to the English term "information highway". It was the first privately owned internet service provider in China, and enjoyed high popularity among college students at Peking University and Tsinghua University. In June 1998, "Jasmine Zhang had been forced out of Yinghaiwei by her largest shareholder, rumoured to be connected with China's Ministry of State Security. Other entrepreneurs, especially those who had set up ISPs to roll out dial-up services to consumers, found themselves squeezed out by large SOEs, like China Telecom" (Clark, 2016: 85–86). Similar to private entrepreneurs' failure to operate interconnecting networks and access networks, the SOEs failed to become mammoth ICPs. In Clark's words, "while the telecom SOEs were actively protecting their turf from encroachment by the private sector, China's state-owned media companies proved to be surprisingly incapable of competing with entrepreneurs building out internet content businesses" (2016: 86). Over time, an informal "division of labour" between SOEs and private firms has taken form: while SOEs operate as ISPs, private firms boom in content provisions and other internet-based services like E-commerce.

Competition Between SOEs

Competition also existed among SOEs, and, as will be shown later, the party-state voluntarily encouraged this competition to improve market efficiency and drive down internet connection fees. In addition, before the creation of the SASAC in 2003, year after which the SASAC became the supervisory agency of all SOEs, since telecom firms were attached to and managed by separate ministries, ministerial competition over competences and resources inevitably translated into the competition between SOEs.

From 1994 to 2008, 2008 being the year of the last restructuring of telecom operators, a great number of SOEs were breathed to life against the backdrop of intense institutional competition over the control of the internet sector. Jitong was the earliest telecom operator in China. In July 1994, the MEI joined efforts with the Ministry of Electricity and the Ministry of Railroads to create the second earliest telecom operator, China Unicom. China Telecom, initially called "General Administration for Posts and Telecommunications" (GAPT, *youdian dianxin zongju*, 1995–1997), was registered in 1995. In January 2009, with the merger of China Netcom into China Unicom, the latter became one of the three major SOEs to provide internet access services, the other two being China Mobile and China Telecom. The monopolistic position of the three SOEs was established after three restructuring waves in 1999, 2002, and 2008. However, despite the tortuous trajectory of the three operators, enterprises that came and went in this period were all SOEs supervised by different governmental institutions put under the direct supervision of.

Shanghai municipal government, for instance, was one of the four stakeholders in the creation of the Little China Netcom on 6 August 1999. The other three co-founders were the CAS, the Ministry of Railroads, and the MBFT. Figures 3.1, 3.2, and 3.3 resume the mergers and acquisitions of state-owned telecom operators within these three restructuring waves.

- **First Restructuring**

In February 1999, the State Council adopted the restructuring plan of the telecommunication sector, the objective being to reduce the monopoly position of China Telecom and to introduce competition in the state-controlled telecommunications sector. In fact, as early as

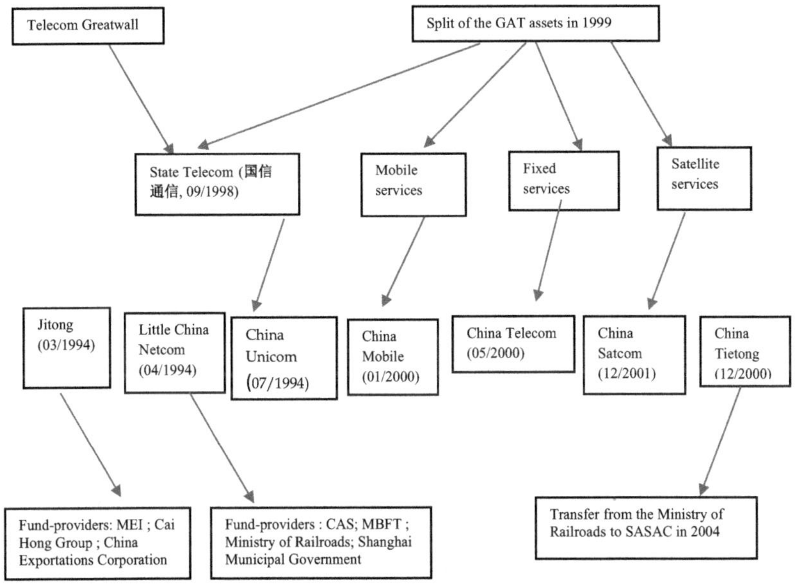

Fig. 3.1 First wave of telecom restructuring in 1999 (*Source* Compiled by the author based on interviews and the information collected from Chinese-language sources)

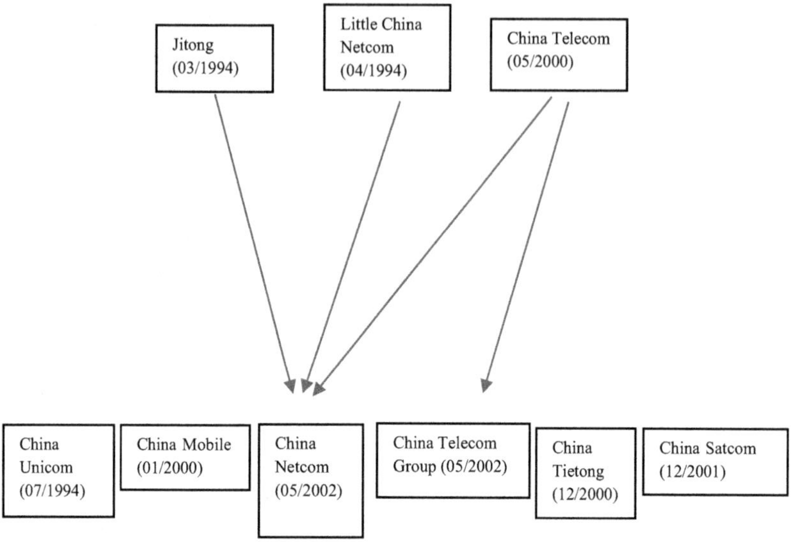

Fig. 3.2 Second wave of telecom restructuring in 2002 (*Source* Compiled by the author based on interviews and other information collected from Chinese-language sources)

in 1998, the General Administration of Posts and Telecommunications (GAPT), the first predecessor of China Telecom, was split in two institutions: profit-making General Administration of Telecommunication (GAT, second predecessor of China Telecom) and the non-profit-making General Administration of Posts (GAP). The restructuring started with the splitting of the GAT's responsibilities. Its paging service was transferred to China Unicom. Its satellite services and mobile services respectively became the prototype of China Satcom and China Mobile, both of which were created soon after the restructuring. The only competence left to the GAT was fixed network. It became the principal activity of China Telecom which was officially registered in 2000. After the 1999 restructuring, China counted 7 state-owned telecom operators: China Telecom, China Mobile, China Unicom, China Satcom, Jitong, Little China Netcom, and China Tietong (Fig. 3.1).

- **Second Restructuring**

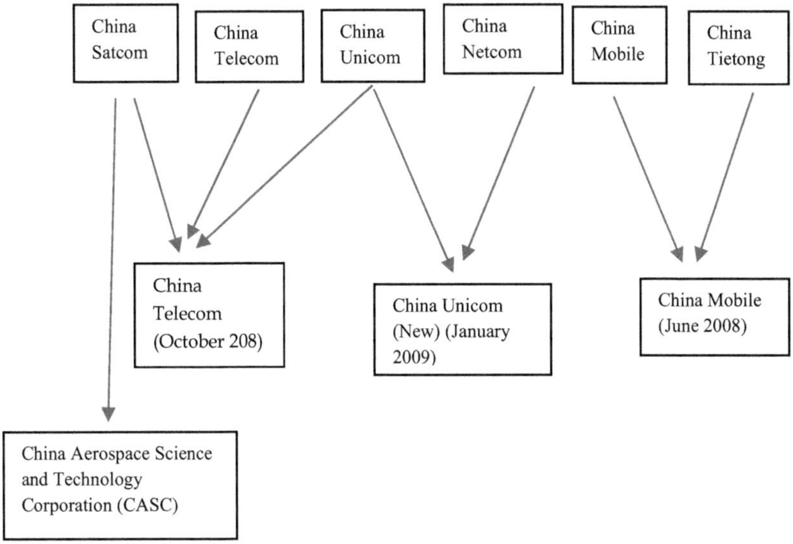

Fig. 3.3 Third wave of telecom restructuring in 2008 (*Source* Compiled by the author based on interviews and other information collected from Chinese-language sources)

...happened on 16 May 2002. It was guided by the same spirit as the 1999 restructuring: introducing greater competition in the state-controlled telecommunication market. The difference was that the 2002 restructuring aimed to split the China Telecom's monopoly in the fixed network service. 10 subsidiaries of China Telecom, based in 10 provinces and provincial-level municipalities in the North of China, left their parent company to join the newly created China Netcom. Jitong and Little China Netcom were merged into China Netcom as well. The rest of China Telecom became China Telecom Group Corporation. The new situation in the fixed network sector was called "China Telecom in the South, China Netcom in the North" (*nan dianxin bei wangtong*). The monopoly position of China Telecom continued to be torn apart after the second restructuring. At the same time, the telecommunication market became more concentrated with the number of telecom operators decreasing from 7 to 6 (Fig. 3.2).

- **Small-Scale Restructuring in 2006**

Before the third restructuring in 2008, a small-scale restructuring happened on 29 April 2006: merger of the China Information Consulting and Design Institute (CICDI) into China Unicom. Created in 1952, the CICDI was initially called the Design Institute of the MPT. The MII replaced the MPT to become the supervisory body of the CICDI in 1999. The separation between governmental institutions and their assets in the late 1990s, called "separation between government and capital" (*zheng zi fen kai*), led to the creation of the State-Owned Assets Supervision and Administration Commission (SASAC) on 24 March 2003. This sweeping reform changed the ownership of the Design Institute again. After changing its name from the Design Institute of the MII into the CICDI in 2002, it became part of the state assets managed by the SASAC since 2003. From 2006 onwards, it was affiliated to China Unicom. The merger in 2006 benefitted China Unicom because the CICDI was known for its capacities in designing backbone networks and large-scale telecommunications projects. However, this restructuring didn't encroach on the competences of the SASAC, because China Unicom as a SOE was under the supervision of the SASAC as well. Since the merger of the CICDI into China Unicom in 2006 did not change the control by the SASAC, it did not count as a major restructuring.

As Fig. 3.3 shows, the restructuring of 2008 reinforced the market concentration of the telecommunication sector by halving the number of telecom operators from 6 to 3. The *Notice on Deepening the Reform of the Telecommunication System*, co-released on 24 May 2008 by the MIIT, the National Development and Reform Commission, and the Ministry of Finance, explained the motivations, principles, and objectives of the third wave of restructuring:

> Taking the development of the third-generation mobile telecommunications (thereafter 3G) as an opportunity, we should rationally allocate the existing telecommunication resources, realise full-service operations, create moderate and healthy market competition, so as to avoid monopoly on the one hand, and excessive competition and redundant construction on the other.

The spirit of the 2008 restructuring is different. While the restructuring in 1999 and 2002 aimed to introduce greater competition, the restructuring in 2008 was intended to reduce competition between operators and to create large operators. Full-service operations in the telecommunication sector meant that an operator could provide a whole range of services to its consumers, ranging from fixed network, mobile network, data network, to internet access service, telecommunications, and other services. During this restructuring, China Telecom acquired the CDMA networks of China Unicom in October 2008. It also obtained the basic telecommunication service of China Satcom. The other services of China Satcom were transferred to China Aerospace Science and Technology Corporation (CASC, a SOE in China). With the approval of the State Council, the SASAC announced on 10 June 2008 that China Tietong became one subsidiary of China Mobile. The merger of China Netcom into China Unicom led to the creation of an aggrandised China Unicom (new) on 7 January 2009. On the same day, the MIIT granted 3G licences to the three revamped telecom operators, marking officially the Chinese entry into the 3G era. China Telecom, China Unicom, and China Mobile, all of three being SOEs, have been controlling the backbone networks in China without being competed by private or foreign firms. By January 2009, the SOEs in the internet industry were formed and have stabilised since then. The Chinese government was the big winner of the three waves of restructuring, because all access channels linking China to the world have been controlled by state-trusted firms since then.

The development of large telecom operators in China, characterised by the shift between more competition (restructuring in 1999 and 2002) and less competition (restructuring in 2008), reflects the party-state's hesitation. Out of suspicion towards private operators, the party-state tasked SOEs with running the backbone networks. However, worried about SOEs' lack of motivations, it introduced some competition to improve their efficiency. At the same time, the party-state brought the competition between SOEs under control and did not put them into genuine market competition, the objective being to avoid redundant construction and vicious competition. Clark called the predicament between more competition and less competition a "uniquely Chinese approach to telecom deregulation" (2016: 76). Striking a right balance between these two concerns is already difficult for the party-state. However, when a new type of actors—private ISM firms—enters the equation, the situation can only be more delicate.

National Champions are at First Local Champions

Private ISM firms forcefully emerged in the late 1990s in China. Since SOEs occupied a quasi-monopolistic position in internet access services, a position nurtured by the party-state for both commercial and security reasons, private firms mainly operated in the downstream sectors and have prospered in content provision. The booming of private ISM firms in the late 1990s has nothing natural if examined with today's eyes, because in a similar way to propaganda, whoever is able to spread contents influences the public opinion. Through allowing private firms to provide contents, the party-state creates its own rival because it allows private businesses to influence people' minds. However, when associating the rise of private ISM firms with the social, political, and economic contexts of the 1990s, the booming of private firms becomes less unusual, for it fits into the economic-centric Zeitgeist of that epoch. This section investigates the background of emergence of the earliest private ISM firms in China. It also analyses how private ISM firms are tied into a different configuration of opportunities and constraints than the SOEs.

Emergence of Private ISM Firms

In the post-1978 period, the priorities of the newly established Chinese leadership changed from class struggle to economic development. The widely quoted dictum of Deng Xiaoping in the 1980s vividly illustrated a mentality shift in China: "as long as a cat can catch mice, it should be considered as a good cat, no matter it is white or black". In other words, even though capitalism ran against the core principles of communism (collectivisation, price controls, suppression of private property), it should be positively perceived if it helped economic recovery. The shift of mindset among the party-state leadership motivated the progressive market liberalisation in the 1980s.

The evolution of the international environment also increased the salience of economic liberalisation for Chinese leaders. To benefit from the integration into global economy in the respects of investment, technologies, and export–import, China had been endeavouring to join the WTO since 1986. On 10 July 1986, Qian Jiadong, then Permanent Representative of China to the United Nations' Office and other International Organisations in Switzerland, officially advanced the Chinese request to resume the status of China as the signatory country of the

GATT. The existence or not of a market economy was an essential criterion for obtaining the membership. This endeavour continued when the GATT changed its name into WTO in 1999. Therefore, that the Chinese government progressively allowed private entrepreneurship since the 1980s was an inevitable choice for complying with the accession rules of the WTO.

The approach to market liberalisation for SOEs and private firms differed, but both reforms demonstrated the resolve of Chinese leaders to dynamise the Chinese economy by granting a higher level of autonomy to market actors and reducing governmental intervention into the market. The liberalisation of the SOEs was effectuated according to what the Prime Minister Zhu Rongji called "catch the big and free the small" (*zhua da fang xiao*), a strategy advanced in 1995. The central government concentrated its energy in nurturing cross-regional and cross-sectional large corporation groups in upstream economic sectors related to national economic lifelines (e.g., energy, resources, chemistry, telecom), so as to help them acquire global competitiveness. The last wave of restructuring of state-owned telecom operators in 2008 epitomised this approach because it allowed China Mobile, China Telecom, and China Unicom to increase their size and global competitiveness through annexing small operators and to develop full-service operations. The imperative needs for improving efficiency and competitiveness pushed the SOEs to lay off their workers. McGregor reported that "workers at the centrally controlled urban state enterprises dropped from a peak of 76 million to 28 million in just ten years from 1993" (2010: 43).

In the case of small and medium SOEs operating in downstream economic sectors (e.g., textile, food, home appliances, etc.), the central government granted them full autonomy and let them ensure their survival on their own. Officials previously dispatched by central and local governments into SOEs were revoked, enabling firms to choose their leaders in full autonomy. High-level managers could acquire the shares of firms, which ended up with transferring the ownership from public hands to private hands.

From "Three Irons" to Entrepreneurship Frenzy

With liberalisation, private entrepreneurship acquired greater legal protection. After the rollout of the corporation law in China in December 1993, it became possible for entrepreneurs to register their business as private firms, and the attitude of public authorities towards private entrepreneurs changed from suspicious to supportive. Private entrepreneurship was not only legally authorised by the Chinese government, but also enthusiastically sought after by Chinese people. This trend even influenced Chinese civil servants supposed to stick to communist ideological lines rather than money-seeking ethos. At the heyday of economic liberalisation in the 1990s, the "three irons" of civil servants, namely the "iron chair" (lifelong job), "iron rice bowl" (permanent job without risk of being fired), and "iron wage" (guaranteed income and pension), were replaced by a new phenomenon called "xia hai" ("diving into the sea"). In other words, the employment of civil servants was losing attractivity in comparison with making money in the private sector. Some municipalities, like Yushu in Jilin province, even accepted that their employees engaged in the private sector without having to resign from their public functions. The data of the Ministry of Human Resources showed that in 1992, 120,000 civil servants resigned to fully operate in private business. Another 10 million civil servants engaged in private business without resigning from their public functions.

Private businesspersons were keener than government officials to embrace market liberalisation. In the internet sector, although private firms were discouraged from providing internet access services, they were encouraged to operate as ICPs thanks to their greater creativity and competitiveness than SOEs (Jiang, 2012). The outgrowth of a booming ISM industry in China, sustained mainly by private ISM firms, demonstrates that it is not because private firms are not encouraged to become ISPs that they wither to disappearance. Instead, this constraint is insignificant when compared with the opportunity generated by the exploitation of access networks.

From 2013 onwards, the MIIT and the Internet Society of China (ISC) have been releasing in July or August of each year a ranking of the top 100 internet companies in China. Although the criteria of the ranking vary year on year, four indicators remain consistent: the size of the firm, its social influence, development potential, and corporate social responsibility (CSR). Both SOEs and private firms are taken into consideration for the ranking, although no SOE has ever managed to enter the top 10 from 2013 to 2022. According to the ranking of 2022, the top 10 included Tencent, Alibaba, Meituan, Ant Financial, ByteDance, JD.com, Baidu, Pinduoduo, Kuaishou, and Ctrip. Although the list of the top 10 firms changes every year, firms which frequently appear in the top 10 are mostly founded during the first internet boom in China at the turning point between the twentieth and the twenty-first centuries (Table 3.1). Despite the rising economic importance of latecomers—some of which overtake the pioneers (for instance Meituan ranks before Baidu in 2020)—the business models and the scope of activities were well in place in the 2000s.

Table 3.1 shows that from 2013 to 2022, a total of 23 ISM firms are listed among the top 10. Since Ant Financial and Tencent Music, marked in green, are respectively the subsidiaries of Alibaba and Tencent, the number of top 10 firms in the covered period is actually 21. The pink zone includes the 15 firms created during the first dot-com boom and the burst of the boom from mid-1990s to mid-2000s. The public flotation of Netease, Sina, and Sohu in Nasdaq in 2000 marked the dot-com boom in this decade. After the puncturing of the bubble in 2003, two other Chinese ISM firms, Tencent and Baidu, were respectively listed in Hong Kong and Nasdaq in 2004 and 2005.

The white zone covers the other 6 firms founded after the internet boom. The firms listed among the top 10 more than 5 times are all created before and during the 2000s. Except the fintech and the online car hailing, firms created in that incipient period have already developed a relatively wide range of activities, including search, E-commerce, news, social media, online gaming, online reservations, and online house renting. Readers may raise objections about the case of Xiaomi, arguing that none of the top 10 ISM firms fabricated smartphones before Xiaomi and that it was created after the internet boom. However, the fabrication of smartphones was not a new activity because Huawei started manufacturing smartphones earlier than Xiaomi. That Huawei does not appear in the ranking is plausibly explained by the fact that the company's major activity is the installation of optic fibres and the production of other telecommunication equipment. The manufacturing of smartphones is not Huawei's principal activity.

In the first decade of internet development from the mid-1990s to the mid-2000s, private ISM firms grew quickly without attracting special attention from the central government, firms being too young and too small to draw special attention from the decision-makers. It was a period in which the party-state put emphasis on building internet infrastructure and combining information technologies with the real economy. As a result, the central government paid special attention to large SOEs that

Table 3.1 List of top 10 ISM firms from 2013 to 2020

Firms	Activity	Times of being listed in top 10	Years	Year of creation
Alibaba	E-commerce	10	2013-2022	1999
Tencent	Social media	10	2013-2022	1998
Baidu	Search	10	2013-2022	2000
Netease	News	9	2013-2021	1997
JD.com	E-commerce	9	2014-2022	2004
Qihoo 360	Search	9	2013-2021	2005
Meituan	Booking	9	2013-2015; 2017-2022	2007
Sohu	News	6	2013-2018	1996
Sina	News	6	2013-2015; 2018; 2019	1998
Ctrip	Booking	4	2015-2017; 2022	1999
Bytedance	Social media	4	2019; 2020-2022	2012
Pinduoduo	E-commerce	3	2020-2022	2015
Suning	E-commerce	2	2014; 2016	1996
Xiaomi	Smartphone	2	2014; 2018	2010
Kuaishou	Video-sharing	2	2021; 2022	2011
Ant Financial (Alibaba)	Fintech	2	2019; 2022	2014
Tencent Music	Social media	1	2020	2016
Didi Chuxing	Car hailing	1	2020	2012
VIPS (wei pin hui)	E-commerce	1	2016	2008
Shengda Wangluo	Games	1	2013	1999
Giant Network	Games	1	2013	1997
Wanmei Shijie	Games	1	2013	2004
Sofang.com	House renting	1	2015	1995

Source Compiled by the author based on the ranking of the MIIT and the ISC of the top 100 ISM firms in China from 2013 to 2022

played a key role in building essential equipment of information technologies. Private firms were neither perceived as the mainstay of national economy nor as a threat to the party-state's grip on power. In this sense, Beijing did not bother to make special arrangements to boost young ISM firms. The trajectory traversed by most Chinese POEs (privately-owned enterprises) in the ISM industry is similar to the transformation of Cinderella into queenship, rather than the ascension of a well-born princess into queenship, "well-born" in the sense of being mollycoddled by resourceful governmental institutions in terms of access to bank loans and public procurement contracts. In contrast, private firms had to trudge in unchartered zones and develop new business models for building their competitive advantages. These difficulties, reflected in lack of existing experiences to draw upon, turned out to be an advantage afterwards because they pushed private firms to explore new areas of the internet economy and to work out new models.

Although young ISM firms were not taken as seriously as they are today by the party-state, since informatisation was a priority concern of the Chinese leadership from the mid-1990s onwards, the party-state encouraged the involvement of private capital into informatisation. At the 5th Plenum of the 15th Central Committee of the Chinese Communist Party (CCCCP), held on 9–11 October 2000, the party leadership recognised the strategic importance of informatisation for national economy. The importance of informatisation continued to grow in the following years. On 19 March 2006, the General Office of the CCCCP and the General Office of the State Council released the *2006–2020 National Informatisation Development Strategy (2006–2020 nian guojia xinxihua fazhan zhanlue)*. In this long-term plan, Beijing emphasised the role of medium and small firms in contributing to the national enterprise of informatisation. Among the six strategic action plans in this strategy, the party-state planned to involve medium and small firms in the development of E-commerce. The other five sectors solicited public actors, including governmental institutions, government-patronaged social associations, and SOEs.

In the course of building the E-commerce sector in China, Beijing granted greater attention to big firms than to medium and small firms. The *2006–2020 National Informatisation Development Strategy* stated that the construction of a complete value chain in E-commerce relied above all on "large companies with strategic importance". Given lack of large private firms in E-commerce in the mid-2000s, "large companies with strategic importance" meant above all SOEs and government-patronaged companies, including Suning, the largest retailer in China in that epoch. Created in December 1990, Suning figured among the "15 big business corporation groups" to which the Ministry of Commerce provided targeted support.

The central government adopted a highly selective approach to dealing with firms based on their size and economic performance. In this context, large firms were favoured compared with small firms, and performing firms were favoured compared with less performing firms. In the downstream economic sectors perceived as being less strategic than upstream ones, performance and the growth potential of a firm were more decisive than its ownership structure for influencing the way the party-state treated it. As Milhaupt and Zheng put it, "where POEs demonstrate growth potential and technological innovation in markets valued by the state, their rise had not been blocked; indeed, it has been nurtured, particularly at the local level" (2015: 699). As shown by the national strategy on informatisation, the general policy environment for internet-based economic sectors was favourable for the creation of large performing firms in the E-commerce sector.

Growing Contributions of Private ISM Firms to National Economy

The relationship between the central government and private ISM firms became closer when the latter proved their potential to be a mainstay for Chinese economic growth. There is no precise moment at which the party-state suddenly abandoned their aloofness towards private enterprises and developed comradery with them. The change in the party-state's attitude has been gradualist and co-variates with the growing economic importance of firms. The more private ISM firms acquire economic importance, the more the party-state takes interest in them. One useful indicator to measure the contributions of private ISM firms to the Chinese economy is the GDP shares created by the ISM sector in each year. The *Digital China Development Report* (*shuzi zhongguo fazhan baogao*), annually released by the CAC since 2017, provides authoritative data on the contributions of the digital economy to the overall GDP growth in China. However, since the data of the CAC cover a short period of time (2017–2020), I complement them with the data of the China Academy of Information and Communication Technology (CAICT) which inform of the contributions of the digital economy to national economy from

2005 to 2017. Figure 3.4 plots the contributions of the ISM sector to Chinese economic growth from 2005 to 2020. ‘

Figure 3.4 demonstrates that from the mid-2000s onwards, the importance of the ISM industry for Chinese economy has been growing steadily. Since 2014, the digital economy has been creating more than 25% of Chinese GDP each year. The percentage reached 36.2% in 2019. The *New Perspectives of the Global Digital Economy in 2020*, a report annually released by the CAICT since 2017, revealed that the contributions of the digital economy in developed countries achieved 51.3% (2020: 3). The percentage in Germany, the UK, and the USA even exceeded 60%. In 2019, it was only in Beijing and Shanghai that the digital economy contributed to more than 50% of the overall economic growth. The percentage variated between 40 and 50% in Guangdong, Zhejiang,

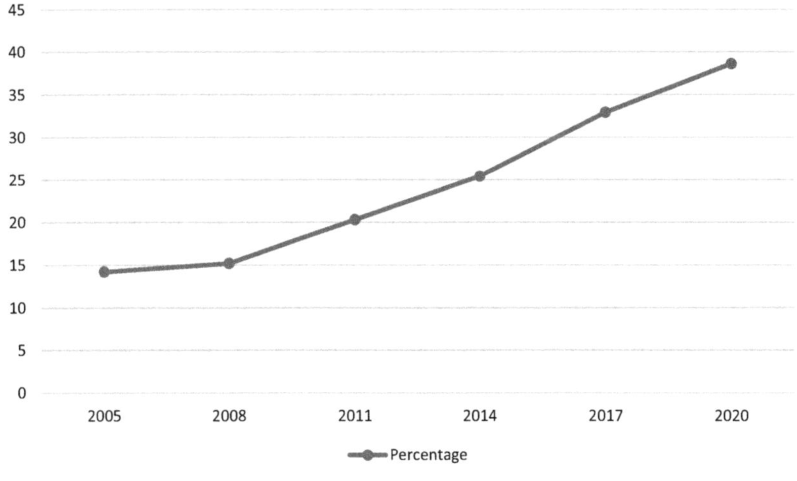

Fig. 3.4 Contributions of the digital economy to Chinese GDP growth 2005–2020 (%) (*Source* Compiled by the author in March 2021. The data on Chinese GDP size come from the National Bureau of Statistics. The data on the size of the Chinese digital economy come from the Development Report on the Construction of a Digital China [2017, 2018, and 2019]. The data for 2020 come from the 2021 Report on Chinese Internet Development released by China Cyberspace Institute in September 2021. The data for 2005, 2008, 2011, 2014, and 2015 came from the China Academy of Information and Communication Technology [CAICT])

Jiangsu, and Fujian provinces. Although China still falls behind developed countries in maximising the economic contributions of the digital economy, the growth of the digital economy in China is the fastest. Chinese digital economy registered a growth of 15.6% in 2019, much higher than the average growth rate of 5.4% in developed countries and 7.9% in developing countries (CAICT, 2020: 4).

Private ISM firms demonstrate strong capacities in technological innovation, which interests central-level decision-makers. In the post-Mao China, the emphasis on science and technology has remained consistent for successive Chinese leaders. Xi Jinping is no exception. In his article published in *Qiushi*, an official magazine of the CCP, Xi Jinping wrote: "practices tell us over and over again that we cannot obtain core technologies through coercion, purchase, or flattering (from the West). It is only through mastering the core technologies by ourselves that we can fundamentally guarantee economic security and national security…" (Xi, 2021). In China, the SOEs, their affiliated research institutions, and government-patronaged research institutes serve as the principal cradle of innovation in the fields of biology, aerospace technology, and energy. However, when it comes to futuristic technologies which increasingly become a mainstay of national economy, the party-state turns to private ISM firms, which have stronger innovation capacities than SOEs. The reliance on private firms in this field is expected to grow further in the future. In artificial intelligence for instance, private ISM firms occupy a privileged position and have largely dwarfed SOEs thus far (Ma, 2018). On 15 November 2017, the Ministry of Science and Technology released the list of the first four open innovation platforms in artificial intelligence: Baidu for autonomous driving, Alibaba for city brains, Tencent for medical imaging, and iFlytek for voice recognition. All these four firms are private ones. Except iFlytek, the other three are also important players in the ISM sector and control huge amounts of data.

With the increasing importance of private ISM firms for national economy and security, the relationship between private firms and the party-state becomes closer, such that the dividing line of the public–private tends to blur since the late 2000s. For Beijing, that a firm has a public, private, or mixed ownership matters less than its growth potential. Milhaupt and Zheng explain the weakening importance of firms' ownership structure for influencing state-firm relations (2015: 694). Therefore, it is no surprise that "the overriding primacy placed on sustained economic growth has enabled some private firms to obtain special benefits from the state by demonstrating the potential to deliver that growth" (Milhaupt and Zheng, 2015: 694). They take the example of Huawei, a Shenzhen-based private firm specialised in telecommunications equipment. In the 1990s, the digital telephone switch of Huawei had greater capacity than all its competitors, including Shanghai Bell, backed among others by the MPT. Technological advantages of Huawei won the firm particular attention from Beijing, attention reflected by generous governmental subsidies, privileged access to bank loans, and public procurement contracts.

China is not the only country where firms' growth potential matters more than their ownership structure in influencing state-firm relations. This pattern exists in several Asian economies like South Korea and Japan. During the reign of Park Chung-Hee in South Korea from 1961 to 1979, the military junta of Park allied with the Chaebol (family-centric conglomerates), which enabled the economic take-off of the country. In 1965, the income per capital in South Korea was merely $100. This number jumped to $10,000 in 1995 and 16,000 in 2005 (Gilson and Milhaupt, 2011: 243). Therefore, similar to China, the ownership structure of private firms in South Korea was not that relevant for explaining the governmental support they obtained. In interpreting both Chinese and Korean cases, a focus on ownership alone is likely to miss the mark (Milhaupt and Zheng, 2015: 669). With the expansion of the digital economy and its rising contributions to national economic growth since the 2000s onwards, the party-state has been rolling out supportive policies to capitalise even further on the economic potential of the internet.

As analysed above, in private ISM firms' lead-up to national stardom, two factors matter: (1) money-making *zeitgeist* in China in the 1980s and the 1990s, and (2) growing contributions of private ISM firms to national economy. Besides these two factors, the role of local governments as incubators of national digital champions is no less important. This is the point we now turn to.

Local Governments as Incubators of Private ISM Firms

The economy-driven regulation (EDR) and the politics-driven regulation (PDR) of the ISM industry differ from each other. While the PDR is characterised by a centralised structure, the EDR is effectuated in a decentralised manner. The decentralisation in Chinese economic governance has been agreed upon in an extensive literature discussion (Oi, 1992; Birney, 2013; Landry 2008; O'Brien and Li, 2006; Lieberthal and Lampton, 1992). Decentralisation, also called "mark-preserving federalism, Chinese style", by Qian and Weingast (1996), matters crucially for Chinese economic success in the 1990s. For Milhaupt and Zheng, "a unique attribute of state capitalism in China is the large role played by local governments. Contrary to popular perception, China's model of economic decision-making is highly decentralised" (2015: 695).

When analyzing the features of China's regulatory state, Hsueh emphasises the Chinese reconciliation between "extensive market liberalisation and decentralisation of economic decision-making with regulatory centralisation and enhanced state control" (2011: 2). Her central argument is that regulation intensity depends on the strategic importance of the concerned sector: centralised regulation for strategic sectors and decentralised regulation for non-strategic ones. To complement Hsueh's argument, I argue that strategic importance alone is not always sufficient to explain observed regulatory patterns. Instead, the actors who perceive this strategic importance equally influence how a regulation unfolds. For instance, since central regulators perceive SOEs as having "strategic value", they support SOEs. Likewise, since local regulators perceive private ISM firms as having "strategic value", they buttress private ISM firms in their jurisdiction. This section examines the role of local regulators in boosting private ISM firms by taking the example of Alibaba. Since Alibaba has shown its potential to drive economic development of Hangzhou in Zhejiang province since the mid-2000s, municipal and provincial regulators reinforced their regulation of the firm. However, in this case, more regulation means more support, as will be shown shortly. At the local level, more regulation also means more autonomy for firms because the latter tend to have greater bargaining power vis-à-vis local regulators.

Hardship of Private ISM Firms: Case of Alibaba

Among the factors that have allegedly contributed to form decentralised economic governance in China, fiscal reforms have attracted wide attention (Zheng, 2010; Milhaupt and Zheng, 2015; Xiao, 2008; Oi, 1992). Libman and Rochlitz (2019) consider that "fiscal federalism" in China provides reliable explanations for the activism of local authorities in boosting local economic development, fiscal federalism being understood as a situation in which local governments are autonomous or quasi-autonomous in establishing their budgets and expenditures. After the fiscal reform in 1994, a higher percentage of taxations collected at the local level go to Beijing, leaving local governments a smaller share (Milhaupt and Zheng, 2015). For instance, Beijing and local governments share 2 of the 10-odd taxations paid by firms: the valued-added taxations (VAT) and the income taxations. For the VAT, while Beijing takes 75%, local governments take only 25%. The inequality also exists for the sharing of the income taxations: the percentage for Beijing and local governments is respectively 60% and 40%. On the contrary, local governments take up greater expenditure responsibilities, a highly uncomfortable situation pushing local governments to seek new sources of income. Since private firms provide extra sources of income to local governments and that the latter can collect them as non-fiscal income, local governments are enthusiastic in supporting their private firms to grow bigger and stronger.

The importance of local regulators for the success of private ISM firms can be illustrated by the rise of Alibaba. The trajectory trekked by Alibaba from its creation to its rise to global stardom is representative of the path borrowed by most private ISM firms in China. The representativity of Alibaba's experiences is twofold: hardship at the initial stages on the one hand and caring from local authorities on the other hand. The Alibaba case illustrates the hardship experienced by most private ISM firms, hardship reflected above all by lack of access to bank loans. Before launching Alibaba in 1999, Jack Ma created Hangzhou Haibo Network Consulting (HHNC) in April 1995, more widely known as China Pages, one of the earliest internet companies in China. In the late 1990s, since governmental attention was focused on the SOEs and large private firms, bank loans were made available for them. On the contrary, private entrepreneurs had to rely on themselves to secure financing. Jack Ma borrowed money everywhere: his sister, his brother-in-law, and his parents. The office of the HHNC, situated at 38, Wen'er Road, Hangzhou, was only 12-square-metre large. His wife, Ying Zhang, was his first employee. Other employees of China Pages were former students of Jack Ma when he taught English language at Hangzhou Institute of Electronic Engineering. Cui Luhai, now a lecturer at China Academy of Art, was one of the earliest visitors to China Pages of Jack Ma. He recalled this experience as follows:

> I can still remember the first scene I saw when I walked into his office...It was a pretty empty space with only one desk set up in the middle of the room. There was only one very old PC desktop surrounded by a lot of people. (Quoted in Clark, 2016: 71)

The business of Jack Ma thus started as a "family affair" (Clark, 2016: 71) and remained so until the growth potential of his business attracted the attention of public authorities. The opportunity initially came from local regulators, rather than national ones. Within the framework of the Three Golden Projects in March 1993, Beijing called on local governments to build their own websites to improve the provision of public services. Zhejiang provincial government invited China Pages to help it build the website. In collaboration with a unit of the Zhejiang Telecom, the team of Jack Ma managed to create the website and hosted it on the platform of China Pages. That Zhejiang succeeded in getting online won it good publicity at home and abroad. Yang Jianxing, the governmental official in charge of the project, received the congratulations from the American Congress. As a sign of gratitude, local official newspaper covered the story of Jack Ma and his company. However, since private entrepreneurship was perceived with suspicion at that epoch, let alone the cooperation between public authorities and private companies, one colleague of Yang wrote a report to criticise him of "hobnobbing with a *getihu*" (a getihu means a self-employed person). For this colleague, "government information dissemination was a serious issue, how could it be handled and published via a *getihu*?" (Clark, 2016: 79). Since the relations with the Zhejiang government soured, Jack Ma started spending more time in Beijing, hoping to obtain the attention directly from central-level decision-makers, but this failed and he was mocked in the Central Television broadcasting in July 1996 (Clark, 2016).

Ma had to return back to Hangzhou in 1998 after spending two unhappy years as the General Manager of Infoshare Technology, an SOE created by China International Electronic Commerce Centre (CIECC), itself affiliated to a department of the Ministry of Foreign Trade and Economic Cooperation (MOFTEC) (Clark, 2016: 88–89). In 1999, he created Alibaba, then a B2B platform. Before the company demonstrated its growth potential, Ma had to seek investment on his own and managed to get the investment from two foreign investors, Goldman Sachs and SoftBank. The American and Japanese investors respectively agreed to pour $5 million in October 1999 and $20 million in October 2000.

The other initial investors of Alibaba were all foreign investors (Transpac Capital Investor AB of Sweden, Technology Development Fund of Singapore, Fidelity International of the USA),[3] and Chinese authorities were not interested. There was only one Chinese investor, TDF Capital, a Shanghai-based private investor. In this sense, despite the heavy hand of the Chinese state, "Chinese internet companies are tied tightly into global capitalism" (Jia and Winseck, 2018: 55), like their American counterparts. In both countries, the logic of commerce, capitalism, and advertising have dominated the growth of internet firms (Fuchs, 2016). For other earliest ISM firms in China (Sina, Netease and Sohu), due to lack of governmental support, they must seek investment abroad, too: Sina from Goldman Sachs and SoftBank; Netease from Goldman Sachs. In 2000, all of the three tech firms were listed in Nasdaq, bringing them fresh money. In a nutshell, before private ISM firms demonstrated their potential to boost Chinese economic growth, they seldom attracted governmental actors' attention.

Economic Potential of Alibaba Attracted Local Governments

Although the first collaboration between Jack Ma and the Zhejiang government failed, the fame of Alibaba as a rising E-commerce star attracted the attention of Zhejiang officials in the 2000s. By the end of 2002, Alibaba edged towards profitability (Clark, 2016: 146). On 10 May 2003, Alibaba launched Taobao, a B2B and C2B platform. The business model of Taobao seriously challenged the American E-commerce firm eBay which then held 80% of the market shares in China. Taobao's popularity among Chinese consumers rose quickly, helping it to gnaw eBay's shares progressively. In the first half of 2007, Ebay's market shares in China decreased to 7.2%.[4] In the 2000s, that Alibaba defeated eBay within a record time was widely mediatised in China. Another eyeball-catching prowess of Alibaba in the 2000s was the payment platform Alipay. Launched in October 2003, it was awarded the "most innovative company" of China in 2005 by 21st Century Business Herald and

[3] "SoftBank invested $80 million to Alibaba and the rate of return is 71 times high," Sina.com, 8 November 2007, http://tech.sina.com.cn/i/2007-11-08/00341839107.shtml.

[4] "Ebay comes back and accepts the Chinese free model", Caixin.com, 31 August 2007, https://china.caixin.com/2007-08-31/100058300.html.

21st Century Business Review, two leading Chinese business magazines. In 2006, around 48.2% of online payment in China were effectuated via Alipay. Realising the huge potential of Alibaba in driving up local economy, Zhejiang and Hangzhou officials observed Alibaba with great interest.

Since the mid-2000s, the name of Alibaba started appearing in the official policy documents of Hangzhou and Zhejiang. Local regulators' attention to Alibaba was justified by their pressure to survive the fierce interjurisdictional competition. In the *Several Opinions on Promoting the E-Commerce Development in Hangzhou*, released on 28 December 2005 by the Hangzhou government, the latter expressed the ambition to build Hangzhou into the "E-commerce Hub of China". Firms, private and public confounded, were encouraged to develop innovative products and technologies "made in Hangzhou" to improve the city's position in the E-commerce sector. Hangzhou government chose three local private companies to spearhead this move: Alibaba, ChemNet (belonging to NetSun) and the ccf.com.cn (belonging to Huarui Information). In exchange for generous financial support (e.g., deduction of the investment in equipment from taxations), the three firms must help Hangzhou to build E-commerce platforms with national and international clout.

Zhejiang government started bringing targeted caring to Alibaba during the second half of the 2000s. Aiming to change Zhejiang into a national leader in E-commerce, provincial regulators emphasised the importance of Alibaba in their official plans. For instance, the name of Alibaba appeared 5 times in the 12,069-word plan on the E-commerce development of the province. This document, entitled *E-commerce Industry Development Plan of Zhejiang Province during the 12th FYP Period*, was the first provincial planning for the E-commerce in China. Taobao, Alibaba's C2C platform launched in 2003, was mentioned 9 times. Alipay, Alibaba's third-party payment platform to support the transactions on Taobao, was mentioned 3 times. That Zhejiang policy-makers mentioned Alibaba and its core products 17 times demonstrated Alibaba's importance to the provincial government.

This plan treated Alibaba as a star company of Zhejiang and praised it as the largest B2B platform in the world. Likewise, provincial regulators did not shy away from highlighting that Taobao and Alipay respectively had become the largest online retail platform and the largest third-party payment platform in the world by 2010. Alibaba and its platforms were involved in 5 of the 8 strategic projects contained in the plan. Provincial regulators declared that they would "support Alibaba to enlarge its market shares, to enter into the international market and to consolidate its position of the largest B2B platform of the world". In addition, Zhejiang government promised to support Alibaba in its development of cloud computing service and to support Alipay in its expansion of services, such that Alipay could keep its position of the largest third-party payment platform in the world. The same emphasis can be found in the *E-commerce Industry Developmental Plan of Zhejiang Province during the 13th FYP Period*. In this document, provincial regulators mentioned Taobao and Tmall 3 times and emphasised their spearheading role in increasing Zhejiang's influence in the E-commerce.

Conclusion

This chapter has identified three key features of the economic regulation of the Chinese ISM industry. First, Chinese ISM development is a government-piloted process. Different from the USA where private actors quickly replaced the state to spearhead internet development, the Chinese party-state dominated the initial strides of internet development and nurtured an informal division of labour between SOEs and private firms, ensuring that the management of internet infrastructure remained in the hands of state-trusted agencies. Second, economic concerns are central to the rise of the Chinese ISM industry. Since content regulation was likely to prevent the economic potential of the ISM from being unleashed, it was not a priority concern of Chinese leaders in the 1990s (Miao et al., 2021: 2005–2006). Third, local governments have shown much activism in boosting local digital champions. Without their activism, local ISM firms would hardly turn into national champions, let alone international stars.

The rise of private firms is an important event against the backdrop of government-piloted internet development in China. Their trajectory differs significantly from SOEs because in the initial stage of their development, they were not blessed with governmental support. Nevertheless, three factors have played to their favour: frenzy of private entrepreneurship since the 1980s, the Chinese government's hands-off regulatory approach to private business, and local governments' support. Our analysis of local governmental support to private firms enriches Hsueh's research (2011) on the correlation between the strategic value of an economic sector and the regulatory approach applied to it. Besides "strategic value", this chapter has demonstrated the importance of investigating who are the perceivers of that "strategic value". It is not because private firms operate in less strategic sectors—downstream sectors of the ISM industry—that they receive less attention. Instead, since they have "strategic value" to local regulators, they receive close attention from local governments and acquire important bargaining power vis-à-vis them.

This chapter has developed an eclectic approach to explaining the growth path of private ISM firms in the Chinese authoritarian context and provides an alternative explanation to the bifurcated interpretations dominating the existing literature. Thus far, the growth of Chinese internet companies has been explained in two ways: while the first view argues that Chinese tech firms have undergone a distinctive growth path from the American one (Jiang, 2012; Jia and Kenney, 2022; McNight et al., 2021), the second view considers that Chinese and American firms have borrowed similar growth models, namely reliance on global capitalism and financialisation (Fuchs, 2016; Jia and Winseck, 2018).

This chapter has shown that, in China, the development stages at which internet firms find themselves strongly influence their growth patterns and the regulator-firm interaction patterns. Through Alibaba's example, I have argued that before internet firms demonstrate their growth potential, they seldom attract attention from public authorities and grow in a similar way as their American counterparts, namely integration into global capitalism. After becoming influential players, the growth patterns of Chinese firms distinguish themselves from American firms. Thanks to their potential to boost economic upscaling and technological innovation, the private ownership structure of internet firms does not prevent them from benefitting from privileged treatment from local and national policymakers. As Milhaupt and Zheng (2015) argue, promising private firms enjoy similar treatments as SOEs. Therefore, the growth path of Chinese internet firms is contingent on their development stages and their "strategic value" perceived by regulators at the right governance level.

References

Arsène, Séverine. 2012. "Protester sur le web chinois (1994-2011)." *Le temps des médias* 18(1): 99–110.

Austin, Greg. 2014. *Cyber Policy in China*. Cambridge, UK: Polity.

Bell, Daniel. 1976. *The Coming of Post-Industrial Society*. New York: Basic Books

Birney, Mayling. 2013. "Decentralisation and Veiled Corruption Under China's 'Rule of Mandates'." *World Development* 53: 55–67.

CAICT. 2020. "Quan Qiu Shu Zi Jing Ji Xin Tu Jing 2020 Nian" [New Perspectives of the Global Digital Economy in 2020]. http://www.caict.ac.cn/kxyj/qwfb/bps/202010/P020201014373499777701.pdf

.Clark, Duncan. 2016. *Alibaba: The House That Jack Ma Built*. New York: HarperCollins Publishers.

Fuchs, Christian. 2016. "Baidu, Weibo and Renren: The Global Political Economy of Social Media in China." *Asian Journal of Communication* 26(1): 14–41.

Gilson, Ronald, and Curtis Milhaupt. 2011. "Economically Benevolent Dictators: Lessons for Developing Democracies." *The American Journal of Comparative Law* 59(1): 227–288.

Hsueh, Roselyn. 2011. *China's Regulatory State: A New Strategy for Globalisation*. Ithaca, NY: Cornell University.

Huang, Yasheng. 2008. *Capitalism with Chinese Characteristics: Entrepreneurship and the State*. Cambridge: Cambridge University Press.

Jia, Kai and Martin Kenney. 2022. "The Chinese Platform Business Group: An Alternative to the Silicon Valley Model." *Journal of Chinese Governance* 7(1): 58–80.

Jia, Lianrui, and Dwayne Winseck. 2018. "The Political Economy of Chinese Internet Companies: Financialization, Concentration, and Capitalization." *International Communication Gazette* 80(1): 30–59.

Jiang, Min. 2012. "Internet Companies in China: Dancing Between the Party Line and the Bottom Line." *Asie Vision* 47: 1–50.

Jiang, Zemin. 2002. "Jia Kuai Wo Guo De Xin Xi Hua Jian She" [To Rapidly Informatise Our Country]. Preface in Qili Hu. Zhong Guo Xin Xi Hua Tan Suo Yu Shi Jian [Explorations and Practices of Chinese Informatization]. Beijing: Dian Zi Gong Ye Chu Ban She [Publishing House of Electronics Industry].

Landry, Pierre F. 2008. *Decentralised Authoritarianism in China: The Communist Party's Control of Local Elites in the Post-Mao Era*. Cambridge: Cambridge University Press.

Libman, Alexander, and Michael Rochlitz. 2019. *Federalism in China and Russia*. Cheltenham: Edward Elgar Publishing.

Lieberthal, Kenneth, and David M. Lampton. 1992. *Bureaucracy, Politics, and Decision Making in Post-Mao China*. Berkeley, CA: University of California Press.

Lu, Wei, Jia Du, Jin Zhang, Feicheng Ma, and Taowen Le. 2002. "Internet Development in China". *Journal of Information Science* 28(3): 207–233.

Ma, Aifang. 2018. *L'intelligence artificielle en Chine: Un état des lieux*. Paris: Fondation pour l'Innovation Politique.

McGregor, Richard. 2010. *The Party: The Secret World of China's Communist Rulers*. New York: HarperCollins Publishers.

McNight, Scott, Martin Kenney, and Dan Breznitz. 2021. "Platformizing the Economy? Building and Regulating Chinese Digital Platforms." Available at SSRN https://papers.ssrn.com/sol3/papers.cfm?abstract_id=3885190

Miao, Weishan, Min Jiang, and Yunxia Pang. 2021. "Historicizing Internet Regulation in China: A Meta-Analysis of Chinese Internet Policies (1994–2017)". *International Journal of Communication* 15: 2003–2026.

Milhaupt, Curtis, and Wentong Zheng. 2015. "Beyond Ownership: State Capitalism and the Chinese Firm." *The Georgetown Law Journal* 103: 665–722.

Negro, Gianluigi. 2017. *The Internet in China: From Infrastructure to a Nascent Civil Society*. Cham: Palgrave Macmillan.

Oi, Jean C. 1992. "Fiscal Reform and the Economic Foundations of Local State Corporatism in China." *World Politics* 45: 99–126.

O'Brien, Kevin, and Lianjiang Li. 2006. *Rightful Resistance in the Chinese Countryside*. New York: Cambridge University Press.

Qi, Yanli. 2000. "The Current Situation and Prospect of Chinese Information Resources on the Web." *Social Science Computer Review* 18(4): 484–489.

Qian, Yingyi, and Barry Weingast. 1996. "China's Transition to Markets: Market-Preserving Federalism, Chinese Style." *Journal of Economic Policy Reform* 1(2): 149–185.

Tai, Zixue. 2006. *The Internet in China: Cyberspace and Civil Society*. New York: Routledge.

Toffler, Alvin. 1980. *The Third Wave*. New York: William Morrow Publishing House.

Umesao, Tadao. 1963. "Information Industry Theory: Dawn of the Coming Era of the Ectodermal Industry." Hoso Asahi, January: 4–17.

Xi, Jinping. 2021. "Nu Li Cheng Wei Shi Jie Zhu Yao Ke Xue Zhong Xin He Chuang Xin Gao Di" [Try Our Best to Become a Principle Global Science and Technology Centre and a Global Innovation Highland]. *Fen Dou* [Hard-Working] 6: 4–11.

Xiao, Lihui. 2008. "Xian Wei Shu Ji Yan Zhong De Zhong Yang Yu Di Fang Guan Xi" [Central-Local Relationship in the Eyes of the Party-Secretaries of the Counties]. *Jing Ji She Hui Ti Zhi Bi Jiao* [Comparison of Economic and Social Systems] 4: 145–150.

Zhao, Ziyang. 2016. *Zhao Zi Yang Wen Ji 1980–1989* [Collected Works of Zhao Ziyang 1980–1989]. Hong Kong: Chinese University Press.

Zheng, Wentong. 2010. "Transplanting Antitrust in China: Economic Transition, Market Structure, and State Control." *University of Pennsylvania Journal of International Law* 32(2): 643–721.

CHAPTER 4

Particularities of the Political Regulation of the ISM Industry

Introduction

The party-state's massive investment in building backbone networks and creating state-owned ISPs since the late 1980s was motivated above all by the concern of economic modernisation. The perspectives presented by Alvin Toffler (1980) concerning information societies enchanted Chinese leaders like Zhao Ziyang and Ma Hong and became a "central part of the worldview" of Chinese policymakers (Gewirtz, 2019: 116). The potential of the ICTs to leapfrog the backward national economy in a record period of time was particularly attractive. The sweeping informatisation wave, represented by the massive expansion of information infrastructure in China in the 1990s and the 2000s, illustrated Chinese leaders' determination to make Toffler's predictions come true in China.

Economic and political concerns are inseparable in authoritarian China. When explaining the Chinese government's early support for internet development, Clark highlighted that "China's Communist rulers had watched the collapse of the Soviet Union in 1990 with alarm, attributing it in part to the yawning technological gap that had opened up with the United States" (2016: 75). Therefore, governmental investment into the internet sector was motivated by the concern of preserving the regime's survival, the utmost priority of the Chinese party-state. Since the ISM challenged the party-state's monopoly in the thought work, the party-state established a streamlined structure to facilitate information control.

A. Ma, *China as a Double-Bind Regulatory State*,
https://doi.org/10.1007/978-981-99-8857-0_4

However, this centralised institutional design for the political regulation of the ISM was not put in place immediately after the internet turned up in China. Instead, it was a "fragmented, sector-oriented regulatory regime" (Negro, 2017: 20). Likewise, although political concerns were present in the minds of decision-makers from the outset, they were not inscribed into the nationwide planning until the 10th Five-Year Plan (2001–2005).

This chapter investigates the political regulation of the ISM industry and discusses two interrelated issues. First, it explains how political and ideological concerns have risen to eminence while they were secondary to economic considerations before the first half of the 2000s. Second, it discusses regulators' approaches to ensuring "political correctness" in cyberspace. The second part analyses in priority two pairs of asymmetric power relations as central features of the political regulation of the ISM sector: power asymmetry between party institutions and state institutions, on the one hand, and power asymmetry between central regulators and local regulators, on the other hand. The conclusion discusses the contributions of this chapter to existing Chinese internet studies.

Political Concerns Rise to Eminence

That the Chinese internet development was initially driven by economic imperatives does not mean that political concerns were absent in the minds of decision-makers. The party-state grasped the political risks of the internet from the beginning. However, the rudimentary level of the ICTs and the limited number of netizens prevented the party-state from fully figuring out the scope of the ISM-related political and ideological challenges in China. With the rapid generalisation of the ISM and the unfolding of a large variety of online mass incidents, the political regulation has progressively risen to an important position. The Chinese government established a streamlined institutional design and ensured that political regulators could easily report to the top party-state leadership.

Political Concerns were Present at the Outset

The party-state's political concerns in the ISM sector cover a large range of issues, including (1) the regime's security (information security and infrastructure security are included in the regime security), (2) the leadership and legitimacy of the CCP, (3) the leadership and legitimacy of the party-state's ideology, (4) and the international prestige of China. Security is the party-state's highest concern because without security, the discussion of leadership, legitimacy, and international prestige would be void of meaning. The name of the meta-regulator of the ISM industry, the Central Cybersecurity and Informatisation Commission (CCIC), presided by Xi Jinping, shows the importance of the security in the ISM regulation. That "cybersecurity" is put before "informatisation" in the title of the CCIC illustrates how security concerns weigh high. Construed *lato sensu*, the pursuit of goals (2) (3) and (4) also serve to maintain the regime's security. In this sense, the regime's security in an authoritarian country becomes a highly encompassing term. Nevertheless, I will discuss the regime's security separately, because this objective is realized differently from the other three goals. More precisely, to achieve the first goal, the party-state must keep its upper hand over the fundamental internet infrastructure. In contrast, the realization of the other three goals goes through the implementation of content regulation.

Although central decision-makers took into consideration all the four goals at the outset, not all of them translated into concrete regulation practices. The regime's security and its two interrelated components (information security and infrastructure security) were spotted in the first place. The earliest statute on the internet regulation in China was rolled out on 18 February 1994 by the Order 147 of the State Council. It emphasised the security protection of computer information systems, in particular the information systems related to politics, economy, national defence, science, and technology. The Ministry of Public Security (MPS), earliest political regulator of the ISM industry, was charged with the portfolio (article 6) and spearheaded the elaboration of technical norms and implementation methods for accomplishing this task. The ministry of the State Security (MSS) and National Administration of State Secrets Protection (NASSP) assisted the MPS. The computers connected with the WWW must be registered with the MPS by the work unit of the concerned user. Violation cases were handled by the MPS and its local outposts at the prefecture-level municipalities or above. That public security organs were the earliest political regulator of the ISM shows the importance of ensuring the regime's security for the Chinese leadership.

Securing the regime's security is the principal theme of the earliest laws and statutes pertaining to internet networks. Apart from the 1994-released *Regulations for the Protection of Computer Information System Security in the PRC*, the State Council released the *Provisory Regulations of the Administration of International Networking of Computer Information in the PRC* (thereafter Provisory Regulations, *jisuanji xinxi wangluo guoji lianwang guanli zanxing guiding*) on 1 February 1996, which was revised on 20 May 1997. This regulation is the "most comprehensive and perhaps most important official ordinance" in the 1990s (Tai, 2006: 98). Article 13 demonstrated the party-state' high attention to the security issue:

> The institutions and individuals working on international internet connections should respect the relevant laws and statutes of the country, and strictly protect state secrets. It is forbidden to use international internet connections to undertake illegal and criminal activities, including the damage to the national security or the divulgation of state secrets. It is forbidden to produce, consult, replicate and spread the information that disturbs public security, or relates to pornography.

The *Provisory Guidelines on the Secrets Protection of the Computer Systems (jisuanji xitong baomi guanli zanxing guiding)*, released by the NASSP on 26 February 1998, aimed to protect national security in the internet era. The level of secret protection of computer networks should correspond with the sensitiveness of the information they delt with (article 6). The level of protection of the computers having stocked state secrets should not be lowered after the removal of secrets. The databases without security protection measures could not be connected to the internet (article 13).

The party-state's approach to handling online mass incidents in the 1990s also illustrated the high importance attached to the regime's security. A highly mediatised incident in the 1990s was the alleged divulgation of state secrets by Lin Hai, arrested in March 1998. Running a software company in Shanghai, Lin sent 30 000 email addresses to VIP Reference, an American online publication spreading anti-Chinese news. Since the court found that the names provided by Lin had been used to distribute large numbers of articles to incite subversion of state power and the socialist system, Lin was sentenced to 2 years prison by Shanghai Number 1 Intermediate Court in January 1999. In the 1990s, compared with the preservation of the regime's security, the other three goals attracted less attention from the party-state, which explains lack of content regulation in that epoch (Table 4.1). Likewise, the low level of internet penetration in that period explains the irrelevance of defending the national prestige in cyberspace because propaganda battles inside and outside China were principally waged offline, not online.

Lack of nationwide laws and statutes on content regulation in the 1990s did not mean that the party-state ignored the challenges that the ICTs posed to its ideological leadership. It was quick to realise the underlying risks associated with the unprecedented scale of technological transformation (Tai, 2006: 81). *Financial Times* reported Chinese leaders' aversion to the "flood of trash" coming along with the internet,[1] "flood of trash" being defined as "anti-science, false science, and information that is unhealthy to the point of being downright harmful" (Tai, 2006: 120). Nevertheless, content regulation remained primitive in the 1990s, in stark contrast with the sophistication of deployed means to defend the regime's security. The primitiveness of the content regulation in the 1990s was reflected in the resistance to foreign cultural products conveying a different set of values and norms, instead of the control of domestic speeches. Some foreign sources of information like *The Economist* and Amnesty International were blocked in China as early as in 1995 (Negro, 2017: 48).

Even so, the vigilance towards foreign cultural products did not lead to drastic regulatory moves. "The high degree of government control over foreign web access foreseen by Mueller and Tan has yet to materialize and...there is actually no intention to develop or deploy omniscient monitoring tools" (Harwit and Clark, 2001: 398). Strictly speaking, the filtering technologies against foreign contents, epitomised by the Great Firewall (GFW), were deployed to preserve the regime's security. The reason was that Chinese leaders feared that "peaceful evolution" would happen in China. Therefore, content regulation in the 1990s had a quite different meaning from today because it served to preserve the regime's security against foreign threats, and it was not an objective in its own rights. In terms of target, while today's content regulation mainly targets the speeches of Chinese netizens, with foreign sources already cut off

[1] "Jiang Calls for Vigorous Promotion of the IT," *Financial Times*, 22 August 2000, p. 8.

Table 4.1 Topics of the ISM regulations released in the 1990s

Regulations (Chinese)	*Translation*	*Release date*	*Topic*
《中华人民共和国计算机信息系统安全保护条例》	• Regulations of the People's Republic of China for **Security Protection** of Computer Information Systems	18 Feb 1994	Security protection
《中华人民共和国计算机信息网络国际联网管理暂行规定》	• Provisory Regulations of the International Connection of Computer Information Networks in the PRC	1 Feb 1996	Security protection (major) Pornography (minor)
《中国公用计算机互联网国际联网管理办法》	• Administrative Measures of the International Internet Connection through ChinaNet	9 Apr 1996	Security issues for international connection via ChinaNet
《中国金桥信息网国际联网管理办法》	• Administrative Measures of the International Internet Connection through China GBN	7 Nov 1996	Security issues for international connection via ChinaGBN
《计算机信息网络国际联网安全保护管理办法》	• Administrative Measures for the **Security Protection** of Computer Information Networks Linked to the Internet	11 Dec 1997	Security protection
《计算机系统保密管理暂行规定》	• Provisory Guidelines on the **Secrets Protection** of the Computer Systems	26 Feb 1998	Secrets protection
《计算机信息系统国际联网保密管理规定》	• Administrative Provisions on the **Secrets Protection** of the International Networking of Computer Information Networks	27 Dec 1999	Secrets protection

Source Compiled by the author based on R. Han (2018: 35–36) and D. Han (2016)

from China, content regulation in the 1990s targeted in priority foreign sources of information.

Several factors explain lack of systematic content regulation in the 1990s: regulators' unfamiliarity with filtering technologies, low level of internet penetration, low quantity of UGCs in cyberspace and, consequently, insufficient unfolding of political risks for political leaders. For instance, on the pouring of Western cultural products, the Chinese government started building the Great Firewall (GFW) in 1998. However, lack of technological expertise obliged it to seek help from foreign firms, most of which were based in the USA, so as to prevent Chinese netizens from being "polluted" by foreign liberal-leaning thoughts. The companies listed in the 2002 report of Amnesty International, which were accused of helping to build Chinese internet control, included Sun Microsystems, Cisco Systems, Microsoft, Nortel Networks, and the filtering software supplier Websense. Even with the assistance of technologically advanced American companies, it was only in 2003 that the GFW became fully operationalised (Negro, 2017).

Low level of internet penetration in China in the 1990s also explained why the party-state did not pay great attention to content regulations. Regulators cannot regulate what does not exist or what barely exists, which is obviously the case of the UGCs in the 1990s. One survey released by Xinhua News Agency in January 1999, a state-owned media in China, revealed the stark contrast between Chinese young people's enthusiasm with the internet, and their inaccessibility to it. The survey covered Chinese youth aged between 14 to 28 years old. While only 7% said they had no interest in the internet, merely 3.4% declared that they regularly surfed in cyberspace.[2] In addition, 69% of the respondents admitted that they had no way to get online. At the national level, by the end of 1999, China counted 8.9 million internet users, with an internet penetration rate at 7%.[3] Moreover, although monthly internet connection fees were slashed by 90% from 70 dollars in 1997 to 7 dollars in the end of 1999 (Clark, 2016: 95), the price remained high for average Chinese families. Therefore, it can be inferred that in the 1990s the central concern of

[2] Faison, Seth, "E-mail to US Lands Chinese Internet Entrepreneur in Jail," *The New York Times*, 21 January 1999, section A, page 10.

[3] CNNIC, *The 5th Statistical Report on Chinese Internet Development*, January 2000, sina.com, https://tech.sina.com.cn/news/internet/2000-01-18/15795.shtml.

political regulators was to ensure the regime's security, rather than the content regulation.

The above discussion shows that the party-state was aware of the political risks related to the generalisation of the internet from the beginning. However, faced with technological constraints and insufficient unfolding of subversive UGCs in cyberspace, the political regulation of the ISM was under-developed when compared with the economic regulation. For instance, while the State Council Informatisation Work Leading Small Group was set up in January 1996 to informatise the national economy, no equivalent organisation was created in the 1990s to take up the political regulation. *Stricto sensu*, no content regulator was created in this period. In the late 1990s, with the rise of the Falun Gong movement, perceived as a destabilising force by the regime, the political regulation started becoming a priority concern. However, as it will be shown later, the way in which the party-state notices the importance of the regime's security and the content regulation varies. While it pre-emptively engages in maintaining the regime's security through the ISM regulation, it has been, more often than not, reactionary in regulating the UGCs.

Increase of Netizens Drives the Increase of ISM Platforms and UGCs

In the 2000s, the party-state was less preoccupied with preserving the regime's security through technologically cutting China from foreign countries, because this mission was accomplished by November 2003 with the operationalisation of the GFW under the close supervision of the Public Information and Network Security Supervision Bureau, a work unit affiliated to the MPS. As the infrastructure of information control, the GFW has not changed too much over time. What vastly changed was the intensity of information flows circulating on both sides of the GFW: inside and outside China. With the generalisation of internet use, Chinese netizens have learnt to climb up the GFW by using VPNs (Virtual Private Networks). In addition, a laxist implementation of the VPN regulation in the 2000s explained the prevalence of netizens' climbing the GFW without getting in trouble. It was not a taboo among college students to talk about climbing the GFW to obtain information. Classmates exchanged experiences on the choice of appropriate VPNs products. Netizens who were not nimble enough to climb the GFW also caused trouble to the party-state with their newly found enthusiasm in expressing diverse opinions in cyberspace. In other words, in the 2000s, what the internet carried rather than what the internet infrastructure itself became a prominent concern of ISM regulators.

The rising importance of content regulation is closely related to the rising number of Chinese netizens. The personal experiences of Jack Ma, brilliantly reported by Duncan Clark (2016), illustrate the great strides that China has made in improving internet connection conditions and in increasing the size of Chinese netizens. On 10 May 1995, Jack Ma registered his first website in Seattle, entitled "China Business Pages: The Online China Business Directory", shortened as "China Pages". As a B2B platform, China Pages gathered the information of the Chinese companies that Jack Ma collaborated with to help them find clients overseas. However, most of these firms had never heard about the internet then and suspected Jack Ma of fraud. Jack Ma had to ask his American partners to take snapshots. He then sent the snapshots to his Chinese partners to convince them that their business really went online. The conditions of internet connections had improved in subsequent years. With the establishment of ChinaNet connecting Beijing and Shanghai to the United States in 1995 and with the significant reduction of the internet connection fees since 1998, internet use was extended from government officials and wealthy proprietors of small businesses to ordinary families and college students. The number of Chinese internet users increased exponentially at the turning point between the 20th and the 21st centuries. Figure 4.1 shows the growth of Chinese internet users during the decade from 1995 to 2005. It shows that 1998, 1999 and 2000 mark the strongest internet development, with the growth rate, respectively, achieving 239%, 324%, and 153%.

The increase of netizens explains the increase of online content in the 2000s. The earliest UGCs that appeared in China at the beginning of the 1990s were the posts on the BBS (bulletin board system). In 1984, an American called Tom Jennings invented FidoNet by using dial-up connections. In 1991, inspired by the FidoNet, the Beijing-based Luo Yi created the first BBS in China: Great Wall Station (*changcheng zhan*). However, the latter did not attract many visitors due to the weak internet coverage in China in that epoch. In 1993, it joined the Shantou-based Hand in Hand Station (*shoulashou zhan*). This merger led to the creation of CFido (*huiduo wangluo*).[4] Luo became the coordinator of FidoNet in

[4] See https://www.cfido.com/. The last update of the website is on 17 October 2001. There are no more updates after this date.

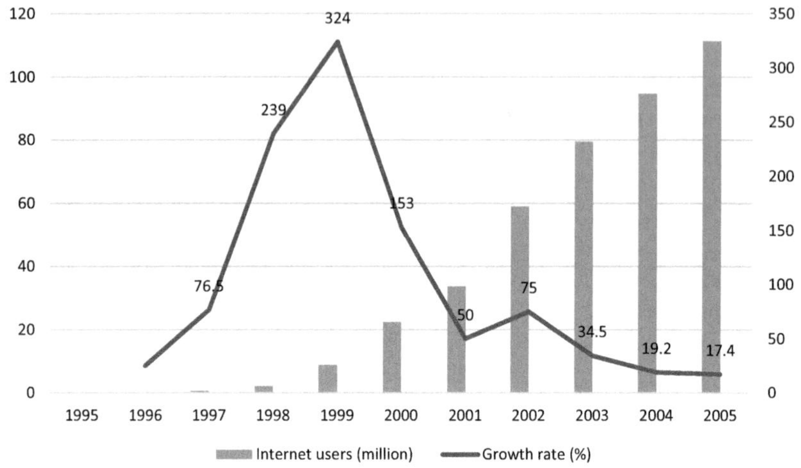

Fig. 4.1 The growth of Chinese internet users 1995–2005 (*Source* Compiled by the author, based on Tai [2006]. In 1995 and 1996, China counted, respectively, 0.0064 million and 0.008 million internet users)

China after the merger. According to the last update posted on CFido on 17 October 2001, CFido was expanded into 13 provinces and cities, set up more than 100 stations, and the number of its regular visitors exceeded 10000 people. The enthusiasm coming along with CFido was reflected in the spreading of CFido networks across China. In fact, many Chinese internet tycoons, including Pony Ma (founder of Tencent), Ding Lei (founder of Netease), Lei Jun (founder of Xiaomi), and Qiu Bojun (founder of Kingsoft), served as webmasters of CFido stations set up in different Chinese cities before creating their own ventures in the late 1990s.

CFido paved the way for an effervescent development of BBS in China and signalled the sweeping trend of BBS at Chinese universities in the mid-1990s. On 8 August 1995, Tsinghua University created its BBS, called Shui Mu Qing Hua (smth.edu.cn), first BBC hosted on the China Education and Research Network (CERNET). BBS entered other universities as well, and college students were the principal users. Several university-based BBS stations were highly popular: No Name Forum (*weiming luntan*) at Peking University, Luo Jia Mountain and Water

Forum (*luojia shanshui luntan*) at Wuhan University, Sunlight and Moonlight Forum (*riyue guanghua luntan*) at Fudan University, 88 Forum of Zhejiang University, Beauty Xi Zi Forum (*xizi luntan*) of Zhejiang University of Technology, Hundred Rivers Joining the Sea Forum (*haina baichuan luntan*) at Zhejiang University of Finance and Economics, to name only a few. Renmin University of China, third-best university of the country, set up the BBS much later than its peers, although its BBS quickly gained popularity shortly after its creation in 2009. In the pre-2005 period, BBS forums of Chinese universities were open to their own students and outside visitors alike. However, since 2005, these BBS stations had successively been closed to outside visitors.

Luo Jia Mountain and Water (LJMW) Forum at Wuhan University was created on 8 October 1996 by an employee called "acat" working at the Software Institute of the university. Low internet penetration level impeded the development of LJMW. It was on 14 November 1996 that LJMW welcomed its first outside visitor. In 1999, the BBS took off. According to the post of "fairy", nickname of the webmaster of LJMW, published on 20 December 1999 on the board, the first half of 1999 was the period in which "the development of the virtual world of Wuhan University embarked on a prosperous path".[5] Users could search for information and discuss social events, sports or entertainment. In 2005, outside visitors were denied access to LJMW and internal visitors must get registered with their real name. The BBS of Tsinghua University applied the same measures in 2005, causing a sharp decrease of visitors.

Despite these restrictive measures, the real name registration (RNR) was not systematically applied in cyberspace in the 2000s. It was only on 28 December 2012, with the *Decision of the Standing Committee of the National People's Congress on Strengthening Internet Information Protection* that the RNR became an obligation for ISM firms at the national level. Instead, contents continued to grow on ISM platforms. In fact, the most relevant reason behind the decline of the BBS was less the RNR than the rise of a competing platform Xiao Nei Net ("xiao nei" means "on campus") in December 2005. Created by Wang Xing, then a student at Tsinghua University, and his friends, Xiao Nei Net was open only to university students. Covering initially Peking University, Tsinghua University, and Renmin University of China in 2005, Xiao Nei Net covered

[5] "Luojia Mountain and Water BBS: Here, the Footprints of Wu Da People Are Everywhere," sohu.com, 8 October 2018, https://www.sohu.com/a/258248056_355701.

all the 100 Chinese universities belonging to the "211 project" by June 2006.[6] By July 2007, 55 universities in Hong Kong, Macao, and Taiwan were connected to Xiao Nei Net. Changing its name to Ren Ren Net in August 2009, Xiao Nei Net became the leading social network service (SNS) platform in China in the second half of the 2000s. Similar to Facebook, Xiao Nei Net encouraged students to register with their real personal information.

In the mid-1990s, since the BBS were principally used in universities, lambda netizens were not the principal producers of the UGCs. However, the slashing of internet connection fees and the business acumen of private firms helped to enlarge the size of content producers. Social media platforms sprouted up at the beginning of the 2000s and less elitist BBS created by ISM firms, represented by Tianya (created in March 1999) and Mop (created in October 1997), gained popularity. Government-patronaged forums emerged, too. In 1997, under the auspices of the *People's Daily*, Qiang Guo Forum was breathed to life. The mushrooming of less elitist platforms increased the production of online contents.

ISM firms were active in updating their services to carry UGCs. Boosted by the entrepreneurship spree in the ISM sector, the 2000s witnessed the birth of a great number of private social media platforms, in stark contrast with the current market concentration in the ISM sector. It is not my objective to give an exhaustive list of all social media platforms that appeared from 2000 to 2010. In a simpler and also clearer way, I illustrate the diversification of social media platforms by clustering them into four categories: social networking, office social networking, microblogging, and video-sharing. Besides the services provided by each social media platform, the second criterion of the clustering is the year of creation and the year of major changes of the firm, including merger, acquisition, shutdown, changes of activities. The comparison shows that in the 2000s, users enjoyed a greater number of choices than in the 2010s (Table 4.2). With an important number of platforms to choose from, users could switch from one platform to another, thus increasing regulators' costs in enforcing content regulation. Nevertheless, in the 2010s, signs of market concentration became clearer. Facing the competition of WeChat and Sina Weibo, several social media platforms were forced either to change activities or to stop providing services at all. Fetion, an

[6] "211 Project" was initiated in the 1990s. It aims to transform 100 Chinese universities into world-leading universities of the twenty-first century.

instant messaging application launched by China Mobile, was competed by WeChat at the beginning of the 2010s. In 2016, it changed its name into Hefetion (*he fei xin*) and changed its principal activities from social networking to office social networking, a sector that WeChat was not competitive. The reduced number of ISM platforms facilitated the implementation of the content regulation because regulators only needed to deal with a limited number of large firms.

The simultaneous increase in the number of netizens and ISM platforms since the beginning of the 2000s caused the explosion of online contents. BBS, Xiao Nei Net (called Ren Ren Net since 2009), and diverse forums created by portal websites helped to form the earliest online public sphere in China. Young people used ISM platforms to show their modernity (Arsène, 2011: 904–905) and personality in an increasingly individualised China. Anonymity in cyberspace facilitated the explosion of contents in the 2000s. However, this period of regulatory leniency did not last long. 2009 marked a turning point in the history of content regulation in China, because high-profile internal and external events happening around this year motivated content regulators to tighten their oversight (Negro, 2017). These events, including the riot in Xinjiang, the 20th anniversary of the crackdown of the Tiananmen pro-democracy movement, the controversy over the filtering software Green Dam Youth Escort (GDYE), and the global financial crisis, worried Beijing. Therefore, it decided to "reinforce its control over the public opinion, especially if it was hostile to the CCP" (Negro, 2017: 61). 1999 also deserves being considered as a milestone in the tightening of content regulation. The following section recounts how highly mediatised domestic and international events since the late 1990s have shaped a strict content regulation regime in China.

High-Profile Events Propelled the Political Regulation Atop the Agenda

The party-state adopts an *ex ante* approach to handling the risks threatening the regime's security. In contrast, content regulation follows an incremental and reactionary pattern, because content regulators must detect beforehand the controversial online contents or the new circumvention methods, without which they could not precisely know what needs to be regulated, which methods to use, and how effective these methods are. In other words, regulators instinctively know how to defend

Table 4.2 Ten-year comparison of the market shares of leading social media platforms

Cluster	*Social media*	*Year of creation*	*Year of major changes of the platform*
Social networking	OICQ (QQ)	1999	Still in use
	Xiao Nei Net	2005	It changed from social networking to livestreaming in 2015
	Twitter	2006	Blocked in China in July 2009
	Fan Fou	2007	Shutdown in China in July 2009
	Tao Tao (by Tencent)	2008	Merger with Qzone Shuo Shuo in 2010
	Ji Wai	2007	Shutdown in July 2009
	Di Gu	2007	Shutdown in July 2009
	WeChat	2011	Still in use
	Mo Mo	2011	Still in use
	Tan Tan	2014	Merger with Mo Mo in 2018
Office social networking	Ding Talk	2015	Still in use
	MSN Messenger	1995	Replaced by Skype in the world in 2014
Video-sharing	Youku	2006	Acquired by Alibaba in 2015
	Tudou	2005	Acquired by Youku in 2010
	Bilibili	2010	Still in use
	Douyin	2016	Still in use
	Pear Video	2016	Withdrawn from app stores in 2020
Microblogging	Fan Fou	2007	Shutdown in July 2009
	Fetion	2007	It changed name and changed activities in 2016
	Sina Weibo	2009	Still in use

Source Compiled by the author

the regime's security without necessarily undergoing a lengthy learning curve. In contrast, for content regulation, regulators mostly move later than regulatees, adjust their approaches after each battle, and test the robustness of their approaches so as to sharpen them for the next round of battle. Therefore, content regulation is mostly an *ex post* regulation. This section illustrates the incremental and reactionary content regulation. It shows how online and offline mass incidents precede and shape content regulation regimes.

Scholars have discussed the factors which influence the intensity of the content regulation in Chinese cyberspace. King and al. (2013) argue that what Chinese content regulators care the most is less the criticism against the government than the potential of the contents to cause collective mobilisations. Despite the insightfulness of this argument, it can hardly explain the tightening internet control in contemporary China, a point that has gained wide consensus among the academia (Economy, 2018; Yeo, 2020; Béjà, 2019; Bandurski, 2018). It is questionable to explain the current tight information control by arguing that over time the contents circulating in Chinese cyberspace have acquired a greater potential to cause collective mobilisations. Quite on the contrary, the Chinese government under Xi has learnt to stem the destabilising potential of online contents before that potential ushers into offline actions. The success in detecting and curbing the collective mobilization potential of online contents did not help to mitigate the intensity of information control in China today. Therefore, beyond the issue of "potential", other factors should be unearthed. This section investigates online incidents as an alternative explanation for a stricter content regulation.

Online mass incidents and major international upheavals have corresponded with the timing of adjustments to Chinese content regulation schemes. Changes of content regulation generally happen in two steps: crackdowns in the first step and rollouts of new rules in the second step. This section reviews two turning points at which the party-state's regulatory approaches undergo major shifts: 2009 and 2013. This choice is justified by the happening of politically highly relevant domestic and international events. Therefore, 2009 and 2013 are called "primary turning points" in this book. Other events also influence the history of the content regulation in China, although to a lesser extent. I take 1999 as a secondary turning point.

Negro (2017) extensively discussed the first turning point of 2009. The importance of 2009 comes from two dimensions in his eyes: citizens' rights defence endeavours and occurrence of high-profile events. For the first dimension, 2009 is the year when Chinese netizens successfully opposed the MIIT that required all PC manufacturers and vendors to pre-install the filtering software Green Dam Youth Escort (GDYE). However, 2009 was not the first time that Chinese netizens demonstrated their muscles. In 2003, without netizens' digital activism, the epidemics of SARS in China might not have been known to the world (Tai and Sun, 2007). Therefore, the importance of the first dimension deserves being revisited. In comparison, the second dimension seems to be more important because it explains the tightening internet control of 2009 with the political events representing (perceived) existential threats to the regime. 2009 is critical for the party-state "because of several important recurrences such as the 60th anniversary of the foundation of the PRC and the 20th anniversary of the Tiananmen protests. Last but not least, a series of riots against the Han people took place in Urumqi, Xinjiang province" (Negro, 2017: 61). Inspired by Negro, I investigate how high-profile political events have shaped content regulation policies in China. Different from Negro's approach, I investigate two types of turning points: primary (2009 and 2013) and secondary turning points (1999). The objective of studying high-profile cases unfolding over an extended period is to draw more systematic inferences on the contingency of Chinese content regulation on domestic and international context and its reactionary nature.

The number "9", as the letter "T", is a politically sensitive term for the CCP, because it corresponds to the happening of two important political events in China: the foundation of the PRC in 1949 and the Tiananmen massacre in June 1989. The letter "T" represents another group of taboo topics in Chinese cyberspace where the discussion of Tibet, Taiwan and Tiananmen is censored far more frequently than other words. Internet censorship is remorseless in the years ending with a "9", for instance 1999, 2009, and 2019. Therefore, compared with "normal" times which do not remind of high-profile political events, the social events and international upheavals happening in "abnormal" years usually provoke strong reactions from the leadership.

Primary Turning Point 1999: Xinjiang Riot

2009 is characterised by large-scale crackdowns in Chinese cyberspace in response to two events: the Charter 08 movement in December 2008 and the ethnic riots in Xinjiang in July 2009. The scale of the crackdown and the depth of policy changes to content regulation justify 2009 as a primary turning point.

On 5 January 2009, content regulators launched a nationwide movement to clean cyberspace. Although this movement was waged in the name of removing vulgar and pornography contents, regulators' real purpose was to eradicate the political influence of the Charter 08 movement, perceived as a dissident group by the party-state. The institutions involved in this cleansing movement included the MPS, the State Council Information Office (SCIO), the Ministry of Culture, National Radio and Television Administration (NRTA), and the General Administration of Press and Publications (GAPP). From 5 January to 27 April 2009, the Illegal and Unhealthy Information Denunciation Centre released 10 lists of punished ISM platforms. In total 125 platforms were mentioned in these lists and required to remove obscene and vulgar contents. 9 of the 10 lists were released between January and February, signalling an exceptional regulation intensity in that period. Google, Baidu, Sina, Sohu, Netease, and other smaller ICPs operating in provinces appeared on the lists. Google appeared in the first place of the first list released on 5 January 2009. It was criticised for putting quantities of pornographic links in the "web page search" and "photo search", and not taking corrective measures after being warned by regulators.

In March 2009, 8 governmental institutions launched a new nationwide cleansing movement targeting specifically obscene and vulgar contents on the mobile phones that were connected to the internet via the Wireless Access Point (WAP). Besides content regulators (SCIO, the MPS, the Ministry of Culture, NRTA, and the GAPP), economic regulators participated in the cleansing movement, too, including the MIIT, the State Administration for Industry and Commerce (SAIC), and the China Banking Regulatory Commission. Launching two rounds of national cleansing movements within three months illustrated the resolve of the party leadership to curb the influence of the Charter 08 movement.

Ethnic riots in Xinjiang also alerted content regulators and provoked nationwide internet crackdowns. On 5 July 2009, a deadly confrontation between Uighurs and Han people erupted out in Urumqi, capital city of the Xinjiang autonomous region. The incident was considered as a response to the killing of two Uighurs workers by Han people at the dormitory building of the Early Light Toy Factory in Shaoguan, Guangdong province, during the night of 25–26 June 2009. The origins of the violence were allegedly the sexual harassment of two Han women by eight Uighur male workers. 120 people, most of whom were Uighurs, were injured during the Shaoguan incident. The violent killing on the streets of Urumqi was considered as a revenge against Han people. Official data showed that 197 people, most of whom were Han, were killed, and that more than 1600 people were injured.[7] Immediately after the unrest, residents in Urumqi reported the cutoff of the internet in the city since the evening of 5 July. It was only on 14 May 2010 that the internet connection was totally re-established in Xinjiang.[8] The Urumqi unrest produced deep impact on Chinese content regulations, an impact felt both by foreign and domestic ISM firms. 2009 marked the withdrawal of most foreign ICPs from the Chinese market. Facebook was closed in mainland China after being suspected of helping Uighur protestors exchange information (Economy, 2018: 56). Twitter was blocked in the far-flung Beijing and Shanghai. The use of YouTube was restricted to only several parts of the country.

At the national level, Xinjiang riots contributed to the intensification of the regulation of domestic ICPs, and UGCs became subject to content regulators' closer oversight. Inspired by the success of Twitter, created in March 2006 by Evan Williams and Jack Dorsey, around 111 Twitter-like microblogging platforms mushroomed in the world between March 2006 and May 2007 (Cui and Zhang, 2019: 24). In China, microblogging platforms appeared forcefully from 2007, with Fan Fou being the earliest and the most widely used of all. Created in May 2007 by Wang Xing, Fan Fou allowed users to publish the posts containing as many as 140 words. Having attracted a great number of Chinese intellectuals and artists, including Chen Danqing, Ai Weiwei, Liang Wendao, and Lian Yue, Fan Fou saw its subscribers increase quickly from below 100 000 in August 2007 to more than 1 million in the mid-2009.

[7] Duncan, Max, "China Says Police Shot Dead 12 Uighurs This Month," *Reuters*, 18 July 2009.

[8] "All Internet Services Resumed in Xinjiang Today", *Chinanews*, 14 May 2010, https://www.chinanews.com/it/it-itxw/news/2010/05-14/2281723.shtml.

On 7 July 2009, two days after the eruption of the Xinjiang unrest, the updates of Fan Fou stopped. On 8 July, all servers of Fan Fou were shut down. On 22 July, other Twitter-like microblogging platforms, like Jiwai and Digu, were closed. From August to September 2009, the chatrooms of Fan Fou on Baidu were disbanded, and the entry "Fan Fou" disappeared from Baidu Baike, equivalent of Wikipedia in China. The shutdown of microblogging platforms was intended to prevent netizens from talking about ethnic riots online, thus cutting the information transmission belt. One year later, when interviewed about the forced closure of his platform in July 2009, Wang Xing admitted that Fan Fou received pressure from content regulators, and that it "has done everything that the general public could imagine" to counter the spreading of the riot. The closure of Fan Fou lasted 505 days.

Cui and Zhang (2019) consider the closure of microblogging platforms in July 2009 as a dividing line between two generations of microblogging service providers. While Fan Fou, Jiwai, and Digu belong to "independent microblogging services", the platforms rising immediately after, like Sina Weibo and Tencent Weibo, are called "portal microblogging services" (2019: 24). The difference is that the first generation, as its name indicates, is totally involved in the provision of microblogging services, while the second generation is initially portal websites. They also differ in their relations with content regulators. One month after the crushing of Fan Fou, Sina launched its microblogging service, Sina Weibo, with the help from public authorities (Negro, 2017). Sina learnt the lessons from the closure of its predecessors and tended to be more compliant. However, it is important to note that the compliance of Chinese ISM firms is multi-shaded, ranging from genuine compliance, half-hearted compliance, feigned compliance, delayed compliance, to outright disobedience. The obedience of Sina Weibo denotes, at best, the absence of overt disobedience because its content moderators do deblock the accounts of certain censored users.

The impact of the 2009 Xinjiang riot on the content regulation illustrates two patterns of the political regulation of the ISM industry. First, content regulation is conducted in an *ex post* way in normal times. In "abnormal times", including the years ending with "9", content regulators act pre-emptively to deter eventual wrongdoers. Second, high-profile domestic and international events touch off or accelerate lawmaking and policymaking processes in the realm of content regulation. In China, the codification of the real name registration (RNR) provides a good case in point. The RNR was intended to end anonymity in cyberspace. The debate around the RNR started at the beginning of the 2000s, and it had been rebuked by Chinese netizens for several times due to privacy concerns. However, the discussion was reignited after the Xinjiang riot, and content regulators vehemently promoted the RNR this time. After being codified in Beijing municipality in 2010, the RNR was generalised to Shanghai, Guangzhou, and Shenzhen, the other three first-tier Chinese cities with the largest number of internet users. The *Decision of the Standing Committee of the National People's Congress on Reinforcing Information Security Protection*, released on 28 December 2012, finally inscribed the RNR into Chinese national law.

Primary Turning Point 2013: Arab Spring Uprising

The tightening of the content regulation in 2013 also caused wide discussions (Negro, 2017; Economy, 2018; Roberts, 2018; Svensson, 2014; Strittmatter, 2020). Strittmatter argued that 2013 marked a change of epoch in China, because it was the year in which Xi Jinping showed to the whole world how one can manage to subdue the network (Strittmatter, 2020: 86). He compared the four years preceding 2013 as an era in which Chinese people considered themselves as becoming a civil society with a genuine public sphere. In that epoch, Chinese citizens "for the first time, discovered themselves, entered into contact, held public exchanges around decisive topics" (Strittmatter, 2020: 86). For him, the Xi Jinping administration brutally terminated prosperous online expressions in China. Likewise, Elisabeth Economy argued that the internet in China was no longer the vibrant virtual political space it was in the pre-Xi era. The Chinese leadership has dramatically expanded the technological and human resources to control online contents (Economy, 2018: 57).

Paul Pierson (2005) criticises two forms of ahistorical investigation in public policy research: an excessive focus on the dramatic moments of policy enactments, and the "actor-based functionalism". For him, both forms of investigations lead to highly problematical conclusions and risk reversing the causes and effects of a policy. Analysing merely the dramatic moments at which a policy is chosen rather than taking into consideration an extended period is conducive to incomplete and erroneous interpretations of the real causes of a policy choice. For "actor-based functionalism", Pierson defines it as a view according to which "policies take the form they do because powerful actors engaged in rational, strategic behaviour are seeking to produce the outcomes observed" (Pierson, 2005: 43). The danger of actor-based functionalism lies in its conflation between the effects and the causes, for it overestimates the causality between the observed outcomes and policymakers' intentions. As Pierson explains, the observed outcomes may be nothing more than the by-products of what the policymakers initially intend to reach.

Inspired by Pierson, I explain the tightening of content regulation in 2013 not only by the events having happened in 2013, but also by the events having occurred before 2013, the objective being to identify less dramatic, slow-moving, but no less decisive factors that caused policy changes in 2013. I also consider the post-2013 era to understand whether the tightening control persists or not in subsequent periods. The major upheavals in the four years separating 2009 and 2013 include the Arab Spring uprisings in 2010 and 2011 in the Middle East and North Africa (MENA). Starting in Tunisia with the toppling of Ben Ali regime within less than one month on 14 January 2011, the Arab Spring swept the MENA countries where entrenched autocrats faced unprecedented legitimacy crisis. The role of Facebook and Twitter in spreading information among protestors alerted the Chinese government of the importance of curbing information dissemination on social media platforms.

The necessity of doing so grew also due to one big challenge facing Chinese content regulators in 2013. Born on the ashes of independent microblogging services (e.g., Fan Fou), portal microblogging services (e.g., Sina Weibo) provided more performing platforms with greater capacities of information diffusion. Since Sina Weibo hosted a much larger number of intellectuals and entertainment stars than Fan Fou, each of whom had a great throng of followers, Sina Weibo built a tremendous user base within only one month after its creation. That Weibo counted 123 million monthly active users (MAU) by September 2013 accelerated the velocity of information diffusion in cyberspace. To maximise their commercial interests, microblogging service providers (MSPs) did not wholeheartedly comply with the RNR (Jiang, 2015: 210), because requiring users to provide their real personal information disincentivised them from registering on Sina Weibo. Therefore, Sina Weibo delayed implementing the RNR (Negro, 2017). In April 2012, it admitted in its annual financial report that "we have not been able to verify the identities of all of the users who post content publicly on Weibo...The microblog rules are not clear regarding the type and the extent of punishment that will be imposed on non-compliant microblogging service providers" (Negro, 2017: 77). Due to MSPs' laxist compliance with the RNR, online information diffusion was extremely quick in China. The velocity of information diffusion *per se* is not a concern. It becomes so if the spread information is against the party-state.

In parallel to MSPs' unwillingness to implement the RNR, Chinese netizens discovered the pleasure of posting their opinions on social and political events on microblogs in the 2010s. One article in *The New York Times* confirmed the rebellious zeitgeist prevailing on Sina Weibo in that epoch.[9] The author described how opinion leaders were popular in China, and how their messages were eagerly retweeted by millions of followers. What annoyed the public authorities was that many tweets criticised or ironised government officials. What could happen if opinion leaders kept spreading anti-government discourses that were retweeted among millions of followers immediately after?

To minimize the risks of regime destabilisation, a nationwide crackdown on microblogging platforms against rumour-mongering started in 2013. On 9 September 2013, the Supreme People's Court and the Supreme People's Procuratorate released their *Interpretations of Several Issues Regarding the Law Application to Criminal Cases Where Information Networks Are Used for Defamation* (thereafter "Interpretations"). The Interpretations were intended to deter opinion leaders and their followers to stir-up sensational news on microblogging platforms, in particular the news spotting the Chinese government in bad light. This document defined three notions related to rumours, including "inventing facts to slander other people" (*niezao shishi feibang taren*), "serious circumstances" (*qingjie yanzhong*) of defamation, and "serious damage to social order and national interests". Individuals who violated these rules would even face criminal charges.

The first of the four "serious circumstances" provoked wide controversy with its arbitrarily elaborated quantified indicators. It said that if one piece of defamatory information was clicked or viewed more than 5000 times, or retweeted more than 500 times, it was considered as a serious circumstance. These quantified indicators aimed to deter outspoken Big "V"s from posting controversial contents. Big "V"s were opinion leaders

[9] Buckley, Chris, "Crackdown on Bloggers Is Mounted by China," *The New York Times*, 11 September 2013.

on microblogging platforms with at least 10 000 followers. The “V” denoted that the identity of a microblogger was verified. Big “V”s functioned like radio stations and could easily reach a great number of people with their tweets. According to the Interpretations, online content constituted “serious damage to social order and national interests” if they:

1. caused collective incidents;
2. caused public disorder;
3. caused religious or ethnic conflicts;
4. produced negative social influence through slandering several people;
5. damaged the image of China and seriously endangered national interests;
6. caused negative international influence;
7. caused other circumstances which seriously damaged social order and national interests.

In fact, before the release of the Interpretations, Big “V”s had already been called upon to toe the line. On 10 August 2013, at the new CCTV headquarter, the CAC (Cyberspace Administration of China) convened a forum on the social responsibilities of Big “V”s. In the presence of more than 10 celebrities, including Pan Shiyi, Ji Lianhai, and Xue Manzi, Mr. Lu Wei, then the Director of the CAC, advanced the seven baselines that online celebrities should stick to when posting online: (1) laws and statutes, (2) Chinese socialist system, (3) national interests, (4) legal rights and interests of citizens, (5) social order, (6) morality and (7) information veracity. Ten days later, the CCTV broadcasted several news programs to criticise Big Vs for spreading rumours on microblogging platforms. It called on Big “V”s to keep their moral baseline and not to change from Big “V”s into “Big Rumours”. The arrest of Qin Huo Huo, Li Er Chai Si, and Xue Manzi on 20–23 August 2013 foretold a larger-scale crackdown on Big “V”s. This has proven to be true with the release of the Interpretations by the top judiciary instances at the beginning of the following month.

Regime changes in Tunisia and Egypt in the Arab Spring constitute the “before” of the tightening content regulation in China in 2013. The influence of the Arab Spring Uprising on the content regulation in China was profound. The coercion of opinion leaders in 2013 decreases the use of portal MSPs. The crackdown also facilitates the appearance of new ISM platforms perceived as less dangerous than microblogs. WeChat is one of these alternative platforms whose popularity rose sharply after the clampdown of Weibo in 2013. Compared with Weibo, the capacity of WeChat to spread information is limited. While Weibo allows users to consult the posts of everybody, a WeChat user only has access to the posts of the people added to her list of friends beforehand. Although users can create discussion groups on WeChat, the maximum number of members for a WeChat group is 500, in contrast with the unlimited number of persons one can follow or be followed by on Weibo. The limited capacity of WeChat to spread information satisfies content regulators' needs because it reduces the diversity of information netizens have access to.

Secondary Turning Points

Chinese content regulation rules are influenced by secondary turning points, too. Different from primary turning points, secondary ones do not give rise to new content regulation rules and do not lead to the appearance of new ISM platforms. Since it is impossible to provide an exhaustive list of all the events that have influenced the trajectory of the content regulation, this subsection discusses the 1999 crackdown on Falun Gong movement as one illustrative example.

Created by Li Hongzhi in 1992 in Jilin province in the Northeast of China, Falun Gong movement centred around the practice of Qi Gong and championed the principles of truthfulness, compassion, and forbearance. From 1992 to 1998, Chinese media reporting of the movement was largely positive, praising in particular the health benefits of Falun Gong and its practitioners' good deeds. State-backed media, including *China Youth Daily*, a mouthpiece of the Chinese Communist Youth League, praised how practitioners of Falun Gong behaved themselves during the opening ceremony of the Chinese Fitness Activities Week in the Shenyang stadium. One 44-year-old woman who suffered from the femoral head necrosis for years explained that she completely recovered from the sickness after practising Falun Gong. In 1993, Li Hongzhi received the "Advancing Marginal Sciences Award" at the 1993 Oriental Health Expos in Beijing (Greenlee, 2006: 557). In 1999, media reporting on Falun Gong changed radically, portraying the movement as an evil religious sect. In the mid-1999, Jiang Zemin, then General-Secretary of the CCP, decided to eradicate the movement in China.

The Freedom House report identified seven reasons to explain why the Chinese government fought the Falun Gong movement (Freedom House, 2017: 88–89). First, in 1999, while the movement counted 70 million members, the CCP had in total 63 million members. Moreover, the movement established 28 000 practice points across the country to allow its members to practice Qi Gong together (Leung, 2002). The sprawling network of the movement and its influence worried the party-state. Second, the worldview of the Falun Gong movement, centred around truthfulness, compassion, and forbearance, ran against the secular and the nationalist ideologies of the CCP. Third, Falun Gong not only appealed to ordinary people, but also to functionaries working in the army, state media, national security organs, or disciplinary inspection commissions. The party-state feared that letting the Falun Gong movement grow would gnaw its authority. Fourth, the party-state was suspicious of the grassroots organisations of the Falun Gong movement. Fifth, the period 1996–1999 was characterised by the intensified harassment of security institutions against Falun Gong practitioners. The 1999 crackdown was the continuity of the repression in preceding years.

Sixth, after the unexpected gathering of 10000 Falun Gong practitioners in April 1999 in Beijing, the party leadership was taken aback by the strictness of the organising disciplines of the movement and the swiftness of its information spreading. In contrast, related competent authorities of the Chinese government "did not have the slightest idea nor any awareness in advance", as Jiang Zemin admitted in his letter to the Standing Committee of the Politburo.[10] The last reason is linked to Jiang's anxiety towards the rising popularity of Li Hongzhi in China. Therefore, even though the intelligence service confirmed that Falun Gong did not seriously threaten state security, Jiang Zemin insisted to fight the movement in China.

Despite the internal division—Prime Minister Zhu Rongji disagreed with Jiang Zemin on the necessity of eradicating Falun Gong—the crackdown started in June 1999. On 10 June, the Bureau 610, in charge of preventing and fighting religious sects, was created in Beijing under Jiang Zemin's direct order. As a ministerial-level organ with around 1000 offices in provinces, municipalities, and counties, Bureau 610 was

[10] Jiang, Zemin, "Jiang Zemin tongzhi gei zhengzhiju changwei ji qita youguan lingdao de xin" [Comrade Jiang Zemin's letter to the Standing Committee Members of the Political Bureau of the CCCCP as well as to other leaders], 25 April 1999.

tasked with information collection, inter-ministerial coordination, and the elaboration of concrete measures. On 22 July 1999, the Ministry of Civil Affairs released its *Decision to Suppress Falun Dafa Study Group (minzhengbu guanyu qudi falun dafa yanjiuhui de jueding)*, marking the start of a comprehensive crackdown. The document said that after verification, Falun Dafa Study Group was not legally registered and conducted illegal activities: spreading superstitions and evil ideas, deceiving the masses, provoking incidents. Falun Dafa Study Group and Falun Gong movement under its control were defined as illegal organisations. On the same day, the Ministry of Public Security issued the "Six Injunctions" to counter the influence of Falun Gong movement. The crackdown was "the longest sustained, and since the Tiananmen Square protests in June 1989, most widely known human rights protests in the PRC" (Greenlee, 2006: 556). The websites related to the movement were all shut down. American-based Chinese language web pages, serving as the coordination point between the exiled Li Hongzhi and his China-based disciples (Harwit and Clark, 2001: 399), received several attacks. Falun Gong practitioners suspected that Chinese police was behind these harassments.

The crackdown on the Falun Gong movement influenced Chinese content regulations in two ways. First, the list of sensitive words was extended to include "Falun Gong", its founder "Li Hongzhi" and some critical comments against the CCP and its leaders, including "Jiang Zemin should be judged in front of an international court", "Heaven will crush the CCP", etc. During the crackdown and in its immediate aftermath, Chinese netizens could not find out any information on the movement in Chinese cyberspace. In September 2017, in the lead up to the 70th anniversary of the PRC, "Falun Gong" still figured on the list of the words to be censored, including on the overseas version of TikTok, supposed to be less censored than its Chinese version.

Second, that Falun Gong practitioners used ISM platforms to spread ideas and organise demonstrations goaded regulators to render internet censorship more sophisticated. Before 1999, Chinese netizens could access uncensored contents from foreign countries, as long as they could pay internet connections fees, and the MII had no real intention to counter foreign content (Harwit and Clark, 2001). In May 1999, one month after Falun Gong practitioners' "425 petition" to the National Public Complaints and Proposals Administration, the MSS reinforced internet supervision. It visited the state-owned ISPs and connected two new monitoring devices to ISPs' computers. Since emails were one of the principal means by which Falun Gong practitioners shared information (Arsène, 2012: 104), these new devices could track emails, allowing security agents to block "harmful" information and websites. The filtering technologies gained greater sophistication and the operationalisation of the GFW speeded up. However, the outcomes of the censorship of the Falun Gong movement were beyond the expectations of content regulators. Greenlee reported that "ironically, while the PRC has largely succeeded in crushing the Falun Gong movement within China, the crackdown has, if anything, strengthened Falun Gong's position abroad" (Greenlee, 2006: 566). The reason was that the crackdown within China increased international sympathy towards Falun Gong practitioners.

The above discussion shows that content regulation in China is mostly conducted *ex post*: regulators move later than regulatees, having to wait for the unfolding of "harmful" contents before taking adapted measures. Regulators have to undergo a new learning curve after each round of battle with netizens, update their know-how, and test it under real situations. Therefore, the tightening content regulation in China is closely linked to the occurrence of high-profile events, each of which strengthens the regulation intensity one extra inch. This *ex post* regulation ensures certain manoeuvring room for netizens. The dynamic is as follows: as long as the mouse runs faster than the cat and keeps updating its strategy, it has real chances to avoid being devoured.

Watershed Centralised Political Regulation

The two dimensions of the political regulation of the ISM sector—internet infrastructure and content regulation—share one point in common: the centralisation of regulatory competences. In contrast with the economic regulation, the role of local regulators in the political regulation is limited to implementing policies and measures decided by Beijing. In addition, a qualitative leap happened under Xi Jinping: party institutions have moved from backstage to frontstage on content regulation, and assume their role in an overt and assertive manner.

Centralised Regulation of Internet Infrastructure

Given the high stakes of the political regulation for the regime's security, the party-state has been centralising the regulatory competences to central regulators to ensure a streamlined control over the ISM sector.

Although centralised regulation implies heavy bureaucratic costs, in particular the costs to ensure central regulators' oversight over local ones, it establishes efficient central-local information transmission belts. This subsection examines the creation of centralised regulation in managing the internet infrastructure and online content.

The security protection of the internet infrastructure is centralised into the hands of central regulators. On 18 February 1994, the State Council issued the *Regulations of the PRC for Security Protection of Computer Information Systems (Order No.147)*. This document stipulated that the security of computers, equipment, networks, information transmitted and stocked via networks, and the environment where computer information systems operated should be protected. The MPS was tasked with this portfolio. The Ministry of State Security (MSS), the National Administration of State Secrets Protection (NASSP), and other institutions assisted the MPS in this policy domain. Any computer connected to the WWW must be registered with the MPS, rather than with provincial or municipal offices of the MPS, the MSS, or the NASSP. This portfolio passed on to the MII in 1998, and the MPS progressively turned into a content regulator. The MII Mission Statement, issued in August 1998, said that the MII "is responsible for planning, constructing and administering private networks for Chinese Communist Party and governments; manage the monitoring and controlling centres of national telecom networks and international gateways...protect national telecom security and information safety" (Foster and Goodman, 2000: 100). The MII also became the distributor and manager of "frequency spectrums, satellite slots, network IDs, domain names and IP addresses" (Foster and Goodman, 2000: 100).

The mission statement highlighted the MII's oversight on three important infrastructures: international communication gateways; backbone networks; and other fundamental resources (e.g., domain names). International communication gateways constitute a country's "virtual frontiers", because all internet communications and long-distance phone calls between China and foreign countries must pass through them. The MII enjoyed exclusive control over the international gateways of China. The *Administrative measures on the international communication gateways (guoji tongxin churukou ju guanli banfa)*, released by the MII Order 22 on 26 June 2002, stated that the MII took charge of approving the creation of gateways, supervised and administered these checkpoints (article 4). State-owned telecom operators being the only entities entitled for establishing international gateways, they could not do so without the

MII's approval (article 5). The latter was also indispensable for abolishing pre-existing gateways (article 14). The MII decided the number and location of the gateways (article 8), although it should respect objective conditions on the ground, including the intensity of economic activities, internet access conditions, strategic importance of the locations, and presence or not of submarine cables in a city. By 2020, China counts eight gateways. They are either located in economically active cities with huge need for internet connections (Beijing, Shanghai, Guangzhou), or strategically important due to their geographical positions (Kunming, Nanning, Urumqi, Fuzhou, Harbin). Local regulators have no competences on international gateways.

The MII supervised all operators of Chinese backbone networks. It supervised "the overall planning of all networks including China's public telecommunication networks... radio and TV broadcasting networks...military networks and all private networks...; responsible for overall sector administration" (Foster and Goodman, 2000: 100). It distributed the operating licences to ISPs, decided the interconnection rules between different networks and the settlement methods. Its competences passed on to its successor, the MIIT, created in 2008. The MII also decided the leadership of state-owned telecom operators. After the creation of the SASAC as the supervisor of state assets in March 2003—the SOEs being part of state assets—the MII's competence in the personnel nomination passed on to the SASAC and the Central Department of Organisation (CDO), although the MII could still express its opinions on the choice of top executives.

By January 2000, China established 10 backbone networks connecting China with the international internet. In the following two decades, no new backbone networks were built. Their geographical distribution is concentrated in Beijing, Shanghai, and Guangzhou. 6 of the 10 networks are affiliated to the MIIT. Different from the international internet data channels (IIDCs, *guoji hulianwang shuju tongdao*), that will be explained in the next paragraph, backbone networks provide services at the national scale, rather than limited to one province or municipality. In terms of users, the 6 backbone networks of the MIIT serve ordinary users and the other 4 serve a limited number of users with specific needs. For instance, the military staff in China has their own network, the CGWNET. Table 4.3 resumes the basic information of the 10 backbone networks:

The MIIT supervises the construction and the operation of the IIDCs. On 21 March 2009, one year after the global financial crisis, the MIIT

Table 4.3 Basic information of the 10 backbone networks in China

Chinese name	*English name*	*Regulator*	*Operator*	*Users*	*Year of creation*	*Nature*
中国科学技术网	CSTNET	China Academy of Sciences (CAS)	CAS	Research institutions, governmental institutions, high-tech companies	April 1994	Not-for-profit
中国公用计算机互联网	ChinaNet	MIIT	China Telecom	Public	May 1995	For-profit
中国教育与科研网	CERNET	Ministry of Education	Sai Er Network Company	Schools and research institutions	November 1995	Not-for-profit
中国金桥信息网	CHINAGBN	MIIT	China Unicom	Public	September 1996	For-profit
中国联通计算机互联网	UNINET	MIIT	China Unicom	Public	April 1994	For-profit
中国网通公用互联网	CNCNET	MIIT	China Netcom, but Netcom was merged into China Unicom	Public	July 1997	For-profit
中国移动互联网	CMNET	MIIT	China Mobile	Public	January 2001	For-profit
中国国际经济贸易互联网	CIETNET	China International Electronic Commerce Centre	Ministry of foreign trade and economic cooperation (1993–2003); Ministry of Commerce (2003)	Institutions and business dealing with foreign interlocutors	January 2001	Not-for-profit
中国长城互联网	CGWNET	Military	Military	Military	January 2001	Not-for-profit
中国卫星集团互联网	CSNET	MIIT	China Satcom (merged into China Telecom)	Public	January 2001	For-profit

issued *Guidelines on Supporting the International Telecommunications in Exemplary Service-Outsourcing Cities (guanyu zhichi fuwu waibao shifan chengshi guoji tongxin fazhan de zhidao yijian)*. To facilitate the international connection of industrial parks in 20 chosen cities specialising in service outsourcing, the MIIT allowed them to directly connect to the international gateways without passing through the three-level connection points of the backbone networks. Although both the central and local regulators supervised the IIDCs, the MIIT kept the chokehold control, because local authorities could not build their IIDCs without the MIIT's permission. Provincial Communications Administrations (PCAs), the MIIT's local outposts, must formulate formal applications to the MIIT to obtain the latter's approval. By July 2019, China counts 27 IIDCs. In February 2021, 38 cities and industrial parks are equipped with IIDCs.

Central and local regulators maintain asymmetrical relations in the regulation of the internet infrastructure. Given the latter's strategic importance to national security, central regulators (embodied, respectively, by the MPS, the MII and the MIIT in the pre-1998, the 1998-2008, and the post-2008 eras) play a much more important role than local regulators. Such an institutional layout benefits the regime's security. For instance, the central government can directly cut the connections between China and the world when it deems necessary to do so.

Centralised Content Regulation and Power Asymmetries

Content regulation in China unfolds in a centralised way, with the top propaganda organs establishing censorship guidelines (Harwit and Clark, 2001: 394). The centralised approach to content regulation remains unchanged, despite the changes of content regulators with several rounds of institutional reforms. Due to a low internet penetration rate, China had no specifically designated content regulator from the late 1980s to the 1990s. The MPS temporarily assumed the task of content regulation in 1994. In 1996, the two ministerial statutes released by the MPT and the MEI on the conditions of connection to their respective network, ChinaNet, and ChinaGBN, modified the institutional framework of content regulation. Both regulations stipulated that operators of interconnecting networks, operators of access networks, and their respective users were responsible of the content they provided online. They clarified on the types of content prohibited online. Paragraph 1, article 16, of the MEI regulation stipulated that "it is forbidden for any organism and individual to use the computer with international internet connections to undertake criminal activities, including damaging national interests and divulging state secrets. It is forbidden to use the computer with international internet connections to consult, duplicate, fabricate, and disseminate any information harmful for national security and social order or which belongs to pornography". This paragraph is identical to article 10 of the MPT regulation. These two regulations laid down a co-regulation framework in which netizens, ICPs, ISPs, and regulators assumed their respective responsibilities.

The creation of the MII in 1998 and the emergence of portal websites as the first generation of internet companies caused the Chinese content regulation rules to change again. The State Council Information Office (SCIO), created in 1991 and subject to the CDP, became the big player in the regulation of contents, mainly the news that portal websites spread online. The SCIO's leadership lasted until the creation of the State Internet Information Office (SIIO) in 2011, the predecessor of the 2014-created CAC. For reasons of clarity, I analyse the institutional framework of content regulation stabilised since 2014 to clarify on two pairs of increasingly obvious power asymmetries: between central and local regulators, on the one hand, and between party and state institutions, on the other.

Asymmetric Power of Party Institutions and State Institutions

The three principal content regulators, the CDP, the SCIO and the CAC, have substantial influence on ISM firms and the contents they carry. This institutional layout takes shape in August 2014, with the State Council nominating the CAC as the regulator-in-chief of online contents. Each regulator of the trio has a distinctive role in content regulation, the realisation of which relies on a dense network of local regulators at the provincial, municipal, and county levels. The two lowest administrative levels—boroughs and villages—have no outposts of the CDP, the SCIO, and the CAC.

The CCP's idea of creating CDP came from the USSR. In 1921, based on the propaganda organ "Агитпропа" of the USSR, the Communist International helped the CCP create the CDP after the first Party Congress in July 1921. The CDP's competences in ideological control and guiding public opinion kept reinforcing in subsequent years. In 1925, the *Decision on the Propaganda Work* of the 4th Party Congress explained the centralised power of the CDP: "to make the propaganda work perfect and systematic, the Central Committee should have a robust propaganda department to conduct diverse tasks. It should also provide guidance to local propaganda departments and build close and well-organised relations with them". The CDP directly reported to the CCCCP after the foundation of the PRC in 1949. During the 10 years of Cultural Revolution, the CDP was closed down. Since it was controlled by his rivals, Mao termed the CDP as the "palace of the king of hell" (Brady, 2018: 166). Reestablished in 1977, the CDP was sidelined again from 1980 to 1989, for the two liberal-leaning Party Secretaries, Hu Yaobang (1980–1987) and Zhao Ziyang (1987–1989) did not consider ideology as a priority (Brady, 2018: 166).

Since Deng Xiaoping interpreted the 1989 pro-democracy movement as stemming from the CCP's loosening ideological control (McGregor, 2010), the CDP was revitalised after the 1989 crackdown. Its competences extended from fixing ideological lines for state media to cover online contents. It communicates a list of to-be-censored words to influential public and private media platforms each day. Media workers should not communicate this list to other people. Otherwise, they would risk being charged with divulging state secrets. Since 2002, the CDP issued "suggestions", accessible to the general public, as "online protocol for discussing sensitive topics" (Brady, 2018). The CDP also managed the *nomenklatura* system for appointing the leaders in media, culture, and education sectors (Brady, 2018: 166). However, the CDP has no coercive power and must rely on state institutions (e.g., National Radio and Television Administration (NRTA), the Ministry of Culture, the MPS) to punish wrongdoers. In fact, before the institutional reform of March 2018, the NRTA was part of a much larger institution, the SAPPRFT (State Administration of Press, Publications, Radio, Films, and Television, 2013–2018). While the SAPPRFT was under the supervision of the State Council, the breakup in 2018 changed significantly the supervision structure of its spun-offs. While press, publications, and films were transferred to the CDP, the NRTA continued to be led by the State Council. That the CDP was given wide competences in the 2018 institutional reform demonstrates the enlarging role of party institutions over ideological control in China.

The SCIO as a Party institution is another big player in content regulation. In fact, its initial task was not to regulate cyberspace, but to present China to foreign countries. The SCIO and the CDP are highly complementary in their responsibilities. While the CDP oversees internal propaganda, the SCIO sings the praises of China to the world. For instance, on 8 June 2010, the SCIO issued the Chinese and the English versions of the widely quoted *White Paper on the Internet Development in China*. This document praised the exponential expansion of the internet in China from the late 1980s to the 2000s and explained how China regulated its ISM sector. The SCIO's role in spearheading external propaganda was reinforced after the establishment of the SIIO in 2011.

With the rise of portal websites in the late 1990s, the SCIO acquired the competences in regulating online news information. It elaborated the development plans of the internet news sector, guided and coordinated online news reporting. In the 2000s, the SCIO also distributed licences to ICPs and released several fundamental regulatory documents. On 6 November 2000, the SCIO and the MII issued the *Provisory Rules on the Management of the Internet Websites Providing News Services*. As the earliest ministerial-level content regulation, it established the SCIO as the regulator-in-chief for the publication and retweeting of news in cyberspace. It created, for the first time in China, the distinction between "news organisms" (*xinwen danwei*) and "non-news organisms" (*fei xinwen danwei*), and granted different rights to them. While news organisms could have their own teams of reporters and collect information on their own, non-news organisms, a category involving most privately-owned portal websites (e.g., sina, sohu, and netease), could only retweet the information published by news organisms. Therefore, commercial portal websites "were restricted to serving as news aggregators. They could repost and repackage, but could not themselves maintain teams of reporters" (Bandurski, 2018: 374).

The SCIO's leadership in content regulation persisted in the 2000s. On 25 September 2005, the SCIO and the MII updated content regulation rules by adopting *Administrative Provisions on Internet News Information Services*. Going one step further than the 2000 regulation, the 2005 regulation clarified the notion of "news information", stating that only news organisms could collect news information. "News information" was above all hard news on politics, economics, national defence, diplomacy and sensitive technological progress. Non-news organisms were free to collect and to report "non-news information". Moreover, the 2005 regulation extended the number of prohibited categories of contents from 9 to 11.

Despite the distinction between news and non-news organisms, commercial media operators, most of which were non-news organisms, frequently broke the rules in the 2000s by sending their own reporters to cover hot-button issues (Bandurski, 2018). Their disobedience was driven both by their personal convictions (Han, 2018) and commercial interests (Donald et al., 2002; Tai and Sun, 2007). After the earthquake on 12 May 2008 in Sichuan province, the CDP ordered mainstream media not to send reporters to the scene. Commercial media flouted the CDP's instructions and went to Sichuan immediately. Their reporting helped Chinese people follow the situation in a timely manner. It also denounced several severe social and political problems, including the collapse of shoddily constructed school buildings in Sichuan (Bandurski, 2018). Content regulators more or less tolerated the disobedience of commercial media because the latter's reporting was acclaimed by Chinese citizens (Hu, 2011).

The CAC is the youngest but the most powerful regulator of the three. Due to its pre-eminence in the Chinese content regulation landscape (for instance, the China Federation of Internet Societies and the Advisory Council on State Informatisation are all under the CAC's leadership), the CAC *de facto* spearheads an independent system (*xitong*), similar to the organisation of the propaganda system (Creemers, 2015). Created in 2014, the CAC and its 2011-created predecessor, the SIIO (State Internet Information Office), use the same staff, but wear different hats. This seemingly weird institutional layout reflects the Chinese bureaucratic culture where the state and the party are present in almost all public institutions. In China, "having one office with two nameplates is a common practice of CCP governance, enabling the Party to disguise its guiding hand within the political system" (Brady, 2018: 166). However, not all Chinese public institutions use two nameplates. When they only use the party nameplate (e.g., Central Military Commission; CDP), the overt mentioning of the Party denotes that it is a party institution, not a state one. Although the state may not be present in a party institution, the party is always present in a state institution, an institutional pattern closely related to the one-party rule in China.

The advantage of designating the CAC as the regulator-in-chief in 2014 was that the CAC addressed the situation of "nine dragons tackle the flood", a Chinese expression describing the disorder created by the involvement of badly coordinated leaders. Before 2014, content regulation was a policy domain disputed between several regulators: SCIO, SIIO, SAPPRFT, and the Ministry of Culture. For instance, both the SIIO and the SAPPRFT could oversee audio-visual contents online, and both the SIIO and the Ministry of Culture supervised online cultural products. In 2014, the State Council clarified that the CAC led the content regulation and oversaw the implementation of related laws and statutes. The SIIO was separated from the SCIO to join the CAC.

In 2015, the competences of the CAC were enlarged, for it took over certain responsibilities belonging to the SCIO previously. For instance, if news organisms created "internet news information service providers" (INISPs) to spread the information that they didn't publish offline, they needed the approval of the CAC, no longer the SCIO. Likewise, if the INISPs, created by non-news organisms, wished to retweet hard news, they must be approved by the CAC, no longer the SCIO. In addition, foreign organisms must secure the CAC's approval before providing any financial information service in China. The CAC's ascension did not stop there. In the institutional reform of March 2018, the State Council declared that the SIIO and the CAC belonged to "one institution with two nameplates", and that both were put on the list of the institutions under the direct leadership of the CCCCP. That the SIIO changed from a state institution into a party institution signaled the improvement in its rank. It also showed that party institutions were moving from backstage to frontstage in content regulation.

It is not difficult to notice the dominance of party institutions in thecontent regulation of China. The regulator-in-chief of online contents, the CAC, is a party institution. The CDP's acquisition of competences in press, publications, and films in 2018, and the shift of the SIIO to the leadership of the CCCCP in the same year show that content regulation is increasingly controlled by party institutions, to the detriment of state institutions.

Asymmetric Power of Central and Local Regulators

To build a complete picture of the Chinese particularities in content regulation, it is important to explore the second pair of asymmetric power relations, those between central and local regulators, with the former having greater power than the latter. The relationship between these two levels of regulators reminds of the regulation structure of the key internet infrastructure: "vertically structured and centralised governance" combined with horizontal fragmentation and overlapped authorities (Yeo, 2020: 14). All of the three content regulators have local outposts. However, while the CDP's local outposts range at the provincial, municipal, and county levels, the local outposts of the CAC and the SCIO do not include the counties. Local regulators are subject to dual leadership, a professional superior (*yewu shangji*) and an organisational superior (*zuzhi shangji*). The professional superior works in the same policy field as its inferiors at lower administrative levels. The organisational superior is located at the same administrative level as its inferiors, but does not necessarily work on the same issue. In most cases, the organisational superior of a local content regulator is the party secretary of the locality. The rationale behind the dual leadership is to create mechanisms of checks and balances, and to ensure rapid information transmission between the central and local governments.

For instance, the propaganda department of Guangdong province is subject to the CDP (professional superior) and the party secretary of Guangdong province (organisational superior). Likewise, the propaganda department of Shenzhen reports to the propaganda department of Guangdong (professional superior) and to the party secretary of Shenzhen (organisational superior). The CDP has no direct oversight over the propaganda department of Shenzhen. However, the authority of the two superiors varies depending on the issues at hand. On economic issues, the organisational superior prevails over the professional superior. The professional superior prevails over the organisational superior on political issues. McGregor's analysis on the changeable weight of the dual leaders is worth quoting here:

> ...Local officials are careful to stay in lock-step with Beijing on issues touching on sovereignty, like Tibet and Xinjiang. Political campaigns which come with the imprimatur of the very top, like the crackdown on the Falun Gong, the outlawed spiritual movement, are implemented with zeal. Very few officials question the fundamental structure of the system and the need for one-party rule. The everyday reality of economic administration in China, where local financial interests are involved, requires a very different calculus. Far from surging like a single river out of the capital, the transmission of economic management is more akin to a series of locks, in which each locality takes what they want out of the policy waterway. Feigning compliance with the centre...they then let the policy stream flow downwards to the next level of government. (2010: 173)

The power relations between central content regulators and their local outposts tilt in favour of the former. The power-sharing between the CDP and local propaganda departments falls within this logic:

> The Central Propaganda Department tends to look at censorship issues from a macro-level, and is concerned with overall trends in society. It has responsibility for censoring and monitoring national level media, such as CCTV and People's Daily, as well as monitoring high-profile scholarly and popular publications and cultural output. Provincial and local level propaganda departments are in charge of censoring and monitoring all non-classified sources of information in their region, including advertising. (Brady, 2018: 168)

The CDP firmly controls the sources of information with national influence, letain its local outposts deal with the contents with local influence. This institutional design where the distribution of the regulatory power matches the influence of to-be-regulated contents allows the CDP to stave off major crises at the national level. In July 2008, one month ahead of the extravaganza of the Olympics in Beijing, the CDP sent a 21-point missive to all mainstream media and provincial propaganda departments. The latter relayed the CDP's instructions to lower-level propaganda watchdogs. Point 8 of the missive stated that "all food safety issues…are off limits" in the lead up to the Games (McGregor, 2010: 186). Local-level propaganda staffers patrolled local media, such that the scandal of Sanlu tainted milk formula, which had provoked kidney stones for infants since late 2007, did not erupt out before the Games. It was only in September after the Olympics that the scandal was covered in state media, like Xinhua.

This institutional design does not always ensure smooth information exchanges between local propaganda departments and the CDP. Sanlu being the largest taxpayer of Shijiazhuang city in Hebei province, the Deputy Mayor of Shijiazhuang ordered the firm to keep the problem of tainted milk secret and, if necessary, to solve it with money (McGregor, 2010: 187). These instructions prevented the propaganda department of Shijiazhuang from informing its professional superior at the provincial level. Ultimately, the tainted milk formula of Sanlu was denounced not by propaganda departments themselves, but by the Ambassador of New Zealand in China to the Chinese Ministry of Foreign Affairs. The alacrity of the Ambassador lied in his directly contacting the central government, not the Hebei provincial government, because the latter might weigh the pros and cons of denouncing the star company of the province and might cover up the issue. By directly informing central authorities, the Ambassador of New Zealand accelerated the problem-solving process.

Likewise, from 2000 to 2014, the SCIO exercised centralised content regulation. The centralisation was reflected in the SCIO's chokehold control over which organism could collect and edit news information, irrespective of the ownership structure and the place of administrative registration of the concerned firm. The SCIO, not its local outposts, approved whether a website could retweet the news from foreign websites. The aforementioned *Provisory Regulation of the Internet Websites Providing News Services* granted the SCIO significant powers in comparison with its local outposts. Different from the three-level structure of the propaganda system, the regulation of online news information unfolded at two levels: national and provincial levels. The information offices at the municipal level and below were led by provincial information offices (PIOs). Article 4 of this regulation stipulated that while the SCIO regulated online news information at the national level, PIOs supervised online news information in their respective jurisdiction. The final decision-making power rested in the SCIO.

Central news organisms (e.g., Xinhua and *People's Daily*) and the news organisms affiliated to central governmental organs must obtain the SCIO's approval before creating their websites. At the provincial level, PIOs verified the application of provincial news organisms and decided whether to give the green light or not. However, provincial news organisms could still not create their websites if the SCIO disagreed. Therefore, provincial news organisms needed the dual permission from the PIOs and the SCIO. The SCIO and the PIOs had equal competences when it came to penalise the news organisms that created websites without autorisation, and the PIOs didn't need to inform the SCIO for inflicting sanctions. As an economic regulator, the MII was given equal competences as the PIOs for punishing the websites daring to spread undesirable news information online.

The upper hand of the SCIO over the PIOs did not change with the updating of the regulatory framework in 2005, when the SCIO introduced the licensing system for online news diffusion. The "Internet News Information Service License (INISL)" entailed different rights depending on the status of its holders. News organisms holding an INISL could collect hard news on their own, while non-news organisms holding an

INISL could only retake the elements published by news organisms. Therefore, the licensing system institutionalised the practice inscribed in the 2000 regulation which laid out the distinction between news organisms and non-news organisms. Another sign of institutionalisation was that the 2005 regulation enlarged the scope of content regulation from online news information to all platforms carrying politically sensitive information, including the BBS and the SMS (short message services). This move corresponded to the shift from Web 1.0 to Web 2.0 in China. The continuously sophisticated regulation in China demonstrated regulators' capacity to adapt their approaches to technological changes in the ISM sector. It confirmed "how masterfully dictators would learn to use [the internet] for surveillance" (Morozov, 2011: xiv; Bandurski, 2018: 372).

The 2005 regulation deepened asymmetric relations between the SCIO and the PIOs. While the competences of the PIOs remained unchanged, the SCIO granted itself larger responsibilities. The PIOs continued to oversee the holders of INISLs registered within their respective jurisdiction. In contrast, the SCIO's competences were enlarged in four respects. First, the SCIO became the only institution which could grant INISLs. News organisms and non-news organisms in provinces could only get their licence from the SCIO, rather than the PIOs, although the preliminary consent of the PIOs was required. Second, any holder of the INISL must obtain the SCIO's permission before cooperating with foreign information service providers operating in China. Third, different from the 2000 regulation which allowed both the SCIO and the MII to punish violators of content regulations, the 2005 regulation granted this responsibility to the SCIO only. Fourth, holders of INISLs should submit annual activity reports to the SCIO. Provincial holders submitted their report to the PIOs who then transmitted it to the SCIO. With the 2005 regulation, the PIOs became the intermediaries between provincial holders of INISLs and the SCIO.

The relations between the CAC and its local outposts are unequal, with the former enjoying undisputed decision-making power. The CAC's centralised control unfolds in two respects: the issuance of content regulation rules and the supervision of local ISM platforms with national influence. Local outposts of the CAC cannot issue content regulations, for the CAC has exclusive competences in this respect. From 2014 to 2022, the CAC issues in total 154 content regulations, namely 32.5% of all the content regulations from 1987 to 2022 (Ma, 2023). Local cyberspace administrations can only issue regional regulations concerning local digital economy. For instance, on 6 May 2021, Shandong province issued the *2021 Action Plan for Building Digital Shandong (Shuzi Shandong 2021 xingdong fang'an)*, the objective being to increase the number of 5G stations to 100 000 in 2021 and to improve the coverage of e-government services. This plan did not contain any content regulation rules. The centralised competences of the CAC in elaborating content regulations ensures unified implementation of content regulation across the country.

The CAC enjoys higher discretion in its relations to local ISM firms than the SCIO did in the pre-2014 period, 2014 being the year in which the CAC replaced the SCIO to issue INISLs. Although the CAC allows its local outposts to investigate wrongdoing local ISM firms, the law enforcement records of the CAC reveal that it is highly present when it comes to investigate private ISM platforms with national influence. For instance, on 15 February 2020, the CAC instructed its Beijing office to invite the head of ifeng.com for a talk, because the website retweeted news information from unauthorised sources. As the fourth largest portal website in China, ifeng.com has around 430 million users in the country. Although Beijing cyberspace administration held direct oversight over ifeng.com (the firm was administratively registered in Beijing), the CAC spearheaded the investigation. Instead, the CAC is largely absent in the investigations of local ISM platforms with local influence. Due to ambiguity of the content regulation rules in China, it is the CAC itself who decides whether an ISM platform has national influence or not.

Although the four institutional reforms in 1996, 2000, 2011, and 2014 cause the rise, demise, and coexistence of several content regulators, what remains unchanged is the centralisation of content regulation in China. In line with the party-state's need for maintaining ideological unity, central regulators have been given the chokehold control over their local outposts. Central regulators have the exclusive competence of issuing licences, elaborating content regulation rules, deciding whether ISM firms can cooperate with foreign content providers and intervening into investigations of wrongdoing firms. The central-local power asymmetry is complemented by asymmetric power relations between party and state institutions, with the former dominating the latter. The increasing power imbalance between the Party and the State in this domain makes it easy for the Party to tighten internet control within China when it deems necessary to do so.

Conclusion

The political regulation of the ISM sector puts the party-state's existential interests at stakes, namely the regime's survival, governance legitimacy, and ideological leadership. Therefore, the party-state chooses a centralised regulatory approach, in contrast with the decentralised regulation of economic issues. Chinese leaders grasped the destabilising potential of the ICTs from the beginning, the proof being the release of the earliest content regulation rules in 1996 to regulate both politically sensitive and non-sensitive contents. However, since content regulation was not the party-state's top priority in the 1990s, it was only in 2000 that dedicated content regulation rules emerged. The centralised political regulation is applied to both the internet infrastructure and the contents. While the *ex ante* approach applies to the former, the *ex post* approach applies to the latter.

This chapter studies two pairs of asymmetric power relations in the political regulation of the ISM industry: (1) power asymmetry between central and local regulators, and (2) power asymmetry between party and state institutions. These two patterns usefully distinguish the political regulation from the economic regulation. The institutional design for the political regulation grants central regulators much greater discretion than to local regulators, while the power relations between them are more balanced in the economic regulation. For instance, the MIIT has the exclusive competences over the establishment and the revoking of international gateways connecting China to foreign countries. The CAC, as the chief content regulator, can oversee all ISM firms with national influence, wherever they are administratively registered. The evolution of the regulatory framework of content regulation after 2014 distils another pair of power asymmetries: asymmetry between state and party institutions. The three major content regulators (the CDP, the SCIO, and the CAC) are all party institutions. Content regulation in China has become the *chasse gardée* of party institutions which have gradually moved from backstage to frontstage.

The academic contributions of this chapter is twofold. First, while the existing literature explains the tightening of content regulation by the regime's increasing need to ward off destabilisation, this chapter investigate micro-level factors, namely the influence of high-colour events. The rationale is that content regulation mostly follows an incremental and reactionary pattern. Therefore, the shape of the institutional framework of content regulation in China closely relates to the events that content regulators have dealt with before. The linkage between the Arab Spring uprisings and the RNR, between opinion leaders' activism and "Seven Baselines", illustrates how it is fruitful to scrutinise the impact of high-colour events on content regulation rules.

Second, this chapter broadens the scope of existing Chinese internet studies. Besides content regulation, this chapter also puts the regulation of internet infrastructure in China under the banner of the political regulation. In a similar way as contents, whether internet infrastructure is properly regulated directly influences the regime's chances of survival. As one object of the political regulation, internet infrastructure matters even more than the contents, because it influences the regime's security at a more upstream position than contents do. The vital importance of internet infrastructure explains the party-state's adoption of a pre-emptive *ex ante* approach to ensure the security of critical internet resources. Including the hitherto scantly discussed regulation of internet infrastructure into the political regulation importantly complements the current research agenda of Chinese internet studies.

References

Arsène, Séverine. 2011. "De l'autocensure aux mobilisations: prendre la parole en ligne en contexte autoritaire." *Revue Française de Science Politique* 61(5): 893–915.

Arsène, Séverine. 2012. "Protester sur le web chinois (1994-2011)." *Le temps des médias* 18(1): 99–110.

Bandurski, David. 2018. "Can the Internet and Social Media Change the Party?" In *Routledge Handbook of the Chinese Communist Party*, edited by W.W.-L. Lam, 372–390. Oxon, UK: Routledge.

Béjà, Jean-Philippe. 2019. "Xi Jinping's China: On the Road to Neo-Totalitarianism." *Social Research* 86(1): 203–230.

Brady, Anne-Marie. 2018. "'The New (Old) Normal': The Party's Propaganda System Under Jiang, Hu, and Xi." In *Routledge Handbook of the Chinese Communist Party*, edited by W.W.-L. Lam, 165–181. Oxon, UK: Routledge.

Clark, Duncan. 2016. *Alibaba: The House That Jack Ma Built*. New York: HarperCollins Publishers.

Creemers, Rogier. 2015. "The Pivot in Chinese Cyber Governance: Integrating Internet Control in Xi Jinping's China." *China Perspectives* 4: 5–13.

Cui, Ying, and Aijun Zhang. 2019. *Wei Bo Yu Lun Dao Xiang Yan Jiu* [Analysis of the Public Opinion on Weibo]. Tianjin: Tian Jin Ren Min Chu Ban She [Tianjin People's Publishing House].

Donald, Stephanie Hemelryk, Michael Keane, and Yin Hong (eds.). 2002. *Media in China: Consumption, Content and Crisis*. New York: Routledge.

Economy, Elisabeth. 2018. *The Third Revolution: Xi Jinping and the New Chinese State*. Oxford: Oxford University Press.

Foster, William, and Seymour Goodman. 2000. "The Diffusion of the Internet in China." Report of the Centre for International Security and Cooperation at Stanford University, https://cisac.fsi.stanford.edu/publications/diffusion_of_the_internet_in_china_the.

Freedom House. 2017. "Zhong Guo Ling Hun Zheng Duo Zhan: Xi Jin Ping Zhi Xia De Zong Jiao Fu Xing Ya Zhi He Di Kang" [Fight for Souls in China: Religious Revival, Repression and Resistance under Xi].

Gewirtz, Julian. 2019. "The Futurists of Beijing: Alvin Toffler, Zhao Ziyang, and China's 'New Technological Revolution'." *The Journal of Asian Studies* 78(1): 115–140.

Greenlee, Michael. 2006. "A King Who Devours His People: Jiang Zemin and the Falun Gong Crackdown: A Bibliography." *International Journal of Legal Information* 34(3): 556–584.

Han, Dong. 2016. "From Vagueness to Clarity? Articulating Legal Criteria of Digital Content Regulation in China." *Global Media and Communication* 12(3): 211–227.

Han, Rongbin. 2018. *Contesting Cyberspace in China: Online Expression and Authoritarian Resilience*. New York: Columbia University Press.

Harwit, Eric, and Duncan Clark. 2001. "Shaping the Internet in China: Evolution of Political Control over Network Infrastructure and Content." *Asian Survey* 41(3): 377–408.

Hu, Shuli. 2011. "The Rise of the Business Media in China." In *Changing Media, Changing China*, edited by S. Shirk, 77–90. New York: Oxford University Press.

Jiang, Min. 2015. "Managing the Micro-Self: The Governmentality of Real Name Registration Policy in Chinese Microblogosphere." *Information, Communication, and Society* 19(2): 203–220.

King, Gary, Jennifer Pan, and Margaret E. Roberts. 2013. "How Censorship in China Allows Government Criticism But Silences Collective Expression." *American Political Science Review* 107(2): 1–18.

Leung, Beatrice. 2002. "China and Falun Gong: Party and Society Relations in the Modern Era." *Journal of Contemporary China* 11(33): 761–784.

Ma, Aifang. 2023. "China Internet Regulations 1987–2022", https://doi.org/10.21410/7E4/R6NLOH.

McGregor, Richard. 2010. *The Party: The Secret World of China's Communist Rulers*. New York: HarperCollins Publishers.

Morozov, Evgeny. 2011. *The Net Delusion: The Dark Side of Internet Freedom*. New York: Public Affairs.

Negro, Gianluigi. 2017. *The Internet in China: From Infrastructure to a Nascent Civil Society*. Cham: Palgrave Macmillan.

Pierson, Paul. 2005. "The Study of Policy Development." *The Journal of Policy History* 17(1): 34–51.

Roberts, Margaret. 2018. *Censored: Distraction and Diversion Inside China's Great Firewall*. Princeton: Princeton University Press.

Strittmatter, Kai. 2020. *Dictature 2.0 : Quand la Chine surveille son people (et demain le monde)*. Translation by Olivier Mannoni. Paris: Tallandier.

Svensson, Marina. 2014. "Voice, Power and Connectivity in China's Microblogosphere: Digital Divides on Sina Weibo." *China Information* 28(2): 168–188.

Tai, Zixue. 2006. *The Internet in China: Cyberspace and Civil Society*. New York: Routledge.

Tai, Zixue, and Tao Sun. 2007. "Media Dependencies in a Changing Media Environment: The Case of the 2003 SARS Epidemics in China." *New Media & Society* 9(6): 987–1009.

Toffler, Alvin. 1980. *The Third Wave*. New York: William Morrow Publishing House.

Yeo, Yukyung. 2020. *Varieties of State Regulation: How China Regulates Its Socialist Market Economy*. Cambridge, MA: Harvard University Press.

CHAPTER 5

Economic Versus Political Goals

Introduction

Since the beginning of the 2000s, the combined mobilisation of the legal and extralegal means of control has enabled the party-state to weave a multi-layered monitoring network over private internet firms. With the extralegal control complementing the legal control, the party-state's tentacles have reached the decision-making core of private firms, which the weaponry of the legal control alone could not realise. However, most of the time the party-state contents itself with unleashing less than 100% of the potential of the monitoring network (Roberts, 2018). Like the People's Action Party in Singapore, which disposes undemocratic means of control but considers them only as last resort (George, 2007), it happens that the Chinese party-state leaves the control-hungry demon dormant rather than active. It avoids mobilising repressive means in an all-out manner against wrongdoing internet firms and netizens. For instance, despite the MIIT's call upon Chinese netizens to stop using VPNs by 31 March 2018, my China-based interviewees continued to use them without being harassed when interviewed in 2019 and 2020.

This phenomenon corresponds to the "porous censorship" where "many governments have the capacity to enforce censorship more forcefully, but choose not to do so. Periodic VPN crackdowns indicate that China could make the Firewall less permeable, but much of the time the government chooses not to" (Roberts, 2018: 4). This quote begs

A. Ma, *China as a Double-Bind Regulatory State*,
https://doi.org/10.1007/978-981-99-8857-0_5

the following question: why does the party-state leave interstices and cracks open in its extensive monitoring network, while it does not lack the capacity to fill them up? Based on interview data, I argue that the party-state faces built-in constraints preventing it from exploiting the full potential of the multi-layered network of formal and informal control over internet firms. The term "built-in constraints" is understood as the constraints that, designed by the party-state, are *inherent* in its governance logic. They constitute constraints for they push the party-state to self-serve and public-serve simultaneously, which limits the party-state's freedom of action. These constraints do not emerge at the same time and do not appear as constraints initially. Instead, the germs of the constraints turn up under the form of specific interests desired by the regime in specific historical periods (e.g., pursuit of modernisation of China in the post-Mao era; proactive pursuit of global leadership under Xi). When examined separately, they look banal and hardly distinguish China from any other country. When built into the governance logic of the regime as one of the numerous rings and teamed with other rings of the authoritarian machine, these germs grow into constraints and hinge on the party-state's freedom of action. In this sense, in the process where the germs grow into fully-fledged constraints, there is a part of "growing out of plan", to take Barry Naughton's expression (1995). I argue that to strike a right balance between the two sets of interests—self-serving and public-serving—both of which are indispensable for ensuring the regime's survival, the party-state's adoption of the sense of self-restraint is equally important as its endeavours to discipline private internet firms. Due to the built-in constraints, the party-state's monitoring network resembles a sleeper cell to be activated contingently rather than permanently.

There are currently two bodies of literature on the party-state's self-control, differentiating with each other on whether the party-state deliberately chooses to do so. Some scholars contend that the party-state imposes constraints to itself out of voluntary choice (Gilson and Milhaupt, 2011; Jiang, 2012). They argue that since the constraints are self-given, the party-state can adjust the constraints according to its needs. In this case, the level of autonomy of private firms depends on the party-state's goodwill. When discussing Beijing's three-stage regulatory policies of the internet industry (liberalisation, regulation, and state capitalism) from 1994 to 2012, Jiang argues that "the state has allowed for companies of various ownership types to exist, but carefully stewards them towards weaving a 'China Wide Web' to preserve the 'Chineseness' it desires"

(Jiang, 2012: 22). For her, the transition from one phase to another is voluntarily chosen by the party-state and that the latter decides the level of compromises it can accept when nurturing the domestic internet economy.

When discussing the "economically benevolent dictators", Gilson and Milhaupt (2011) emphasise the party-state's sense of self-containment. Dixit assumes that all rulers are driven by private objectives and that autocrats are less willing than democratic rulers to sacrifice rents for public goods (Dixit, 2010). Gilson and Milhaupt reject this opinion. They argue that compared with emergent democracies, dictators in autocracies with prosperous economic development are readier to contain their desire of preying on the private sector and are more capable to provide credible commitments to investors, so as to seek performance-based legitimacy (2011: 229–230). Bandurski (2018) systematically reviews the Chinese internet regulation from the 1980s to the second term of Xi Jinping. Emphasising the parallelism between the technological development of the internet and the increasing sophistication of the censorship techniques, Bandurski shows that the party-state, in particular under Xi, is perfectly able to simultaneously promote the technological development of the internet and censor inharmonious speeches. In this context, the party-state dominates the relationship with private internet firms and yields to no other constraints than the constraints it self-imposes.

This body of literature considers self-control as the party-state's deliberate choice. In contrast, the second body of literature perceives the party-state as much less prescient and agile. It contends that the party-state is obliged to do so by the self-contradiction of its sought-after objectives. Moreover, the circumstances beyond its control (e.g., international outcry against China's lack of transparency on COVID-19) pressure the party-state into controlling its desire of control. Therefore, self-containment is not deliberately chosen. It is imposed by the factors beyond the party-state's control.

This book examines the party-state's three constraints unfolding at the domestic and the international levels. Two constraints exist at the domestic level and result from two pairs of contradictions: contradiction between political and economic objectives, and contradiction stemming from fragmented authoritarianism. At the international level, the contradiction between the party-state's pursuit of global leadership and the continuation of domestic informational control becomes the third pair of contradiction. This chapter, Chapters 6, and 7 respectively examine each of the three constraints. They systematically investigate how the party-state's structural constraints open a window of opportunities allowing private ISM firms to gain margins of manoeuvring. They provide a systematic analysis of the structural opportunities which, stemming from the predicament of the party-state, augments private ISM firms' manoeuvring room. I argue that despite the Chinese authoritarian setting marked by strong state capacities, the dilemma of the party-state changes into an important source of autonomy to the benefit of private ISM firms. To my knowledge, this book is the first academic endeavour to have examined the possible transformation from the party-state's predicament into private firms' manoeuvring room.

The contradiction between the party-state's political and economic objectives in the regulation of private ISM firms creates its first constraint. The liberalisation of the Chinese internet market before the Chinese integration into the WTO (Hsueh, 2011: 71) created a new situation for the party-state. Eager to unleash the economic potential of the ICTs, the party-state voluntarily let private firms, not SOEs, operate in the market of the internet portals and contents (Jiang, 2012). In the decade following the initial introduction of the internet in China, "the Chinese internet sector has grown in a manner that is conspicuously not dominated by state-owned firms" (Jiang, 2012: 23). To champion the ICT-backed economic catch-up with advanced economies, content regulation was relegated to a secondary position. Since "VAS providers operate on network infrastructure owned and controlled by the state-owned carriers, the state extended less oversight over VAS providers, and the scope of state control did not include ownership or direct company-level intervention" (Hsueh, 2011: 71).

However, the liberalisation of the media sector caused troubles to the party-state, because it incentivised private internet firms not to comply, for the sake to survive the ferocious market competition. With liberalisation, "market forces instead of Party directives have become the primary concern for media executives, because circulation and advertising revenue constitute the lifeline for the media in this environment" (Tai and Sun, 2007: 992). To maximise profits, commercial media ran breaking news information before the rollout of censorship rules (Shirk, 2011a; Xiao, 2011), such that they could reach wide audiences. The new situation "caused the state to lose a significant degree of content control and day-to-day management of the mass media" (Tai and Sun, 2007: 992).

Lagerkvist argues that private internet firms face "two simultaneous but diverging priorities: supporting the party-state in the enforcement of social stability while providing the infrastructural base that allows China's emerging civil society to connect" (2011: 176). Inscribed in the literature on "dictators' dilemma" (Wintrobe, 1998; Francisco, 2005; Dickson, 2016), scholars highlight the party-state's dilemma in dealing with private internet firms due to the contradiction of the sought-after political and economic goals (Roberts, 2018: 23–25; Lynch, 1999: 74; Clark, 2016; Austin, 2014: 92). Roberts suggests that autocrats face three dilemmas. The first dilemma appears "when the government would like to enforce constraints on public speech but repression could backfire against the government" (2018: 23). The second is that censorship makes it difficult for the government to collect public opinions (2018: 23). Autocrats reining too far in the public space see their citizens stop telling truth (Kuran, 1997). Kuran argues that to maximise "reputational utility",[1] citizens may very well sacrifice their "expressive utility", and convey the public preferences that are supposed to bring rewards and advantages (1997). The third is that "censorship can have economic consequences that are costly for authoritarian governments that retain legitimacy from economic growth" (Roberts, 2018: 24). Austin captures the contradiction between political and economic objectives of the regime into the dilemma between "content and connectivity" (2014: 92). He highlights that while the party-state favours connectivity over content at the initial stage of the internet development in China (2014: 92), it favours content over connectivity in the 2010s.

The party-state is anything but all-powerful due to the constraints facing it. While self-imposed constraints are within its control, other-imposed constraints are much less manageable. Originating from the contradiction in the party-state's governance logic, other-imposed constraints put the party-state into dilemmas. Therefore, the party-state must muddle through different objectives the realisation of which may require contradictory courses of actions (e.g., centralised governance of political issues and decentralised governance of economic issues). Therefore, the literature on other-imposed constraints better captures Chinese reality than the literature on self-imposed constraints. The latter falters,

[1] The "reputational utility" is defined as the rewards and punishments associated with the chosen preference of one person. The "expressive utility" denotes the benefits that one individual derives from truthful expression (Kuran, 1997: 16).

because it overestimates the party-state's capacity to predict and manage the conflicts facing it.

Three weaknesses in the current literature deserve attention. First, it is important but insufficient to know what are the limits to the party-state's power. Instead, it is necessary to understand the process of emergence, demise, and re-emergence of these limits as a dynamic and forward-moving loop within which the party-state gains and loses ground without being able to triumph over private actors definitively. The advantage of adopting this holistic view is that it allows readers not to exaggerate the explanatory power of any single constraint. Second, although the existing literature recognises that the party-state faces dilemmas, it seldom investigates how these dilemmas impact private firms' pursuit of manoeuvring room in China. If the party-state is constrained by the predicament of balancing its political and economic goals, to what extent can ISM firms exploit that predicament to make themselves freer? Third, the existing literature under-explores the problems of "manufacturing" and "manufacturer" of the party-state's dilemmas. This explains insufficient attention to how the dilemma can be overcome. The importance of clarifying these dimensions is that depending on who manufactures the dilemmas and how, the party-state may face different difficulties in solving the predicament.

This chapter analyses how the party-state is engulfed into a double-bind logic in wrestling with private ISM firms. It aims to invalidate the myth of an all-powerful party-state by unveiling the structural barriers that prevent the CCP from behaving as one of the self-serving and predatory autocrats of Dixit (2010). Such structural barriers render the party-state somewhat akin to Sisyphus in the ancient Greek mythology because the party-state must repeat the process of increasing and loosening its control of private actors over and over again. Pulling the stone to the top of mountains is the punishment imposed to Sisyphus for his disobedience. For the party-state, the switching between tightening and loosening control can be compared to the punishment imposed by the "authoritarianism with Chinese characteristics". The objective to make firms simultaneously communist and commercial entails a self-contradictory logic within which firms must serve both the interests of the CCP and their own interests. While this objective can be satisfied when firms and the party-state hold convergent interests, it can hardly be achieved when interests diverge (e.g., firms make greater profits when releasing breaking news while the latter can destabilise the CCP's ruling).

To understand how the party-state jumps into and fails to jump out of this double-bind logic, This chapter, Chapters 6, and 7 analyse three source of the party-state's double-bind logic: the contradiction between the party-state's political and economic objectives, fragmented authoritarianism at the horizontal and vertical levels of Chinese governance, and the contradiction between the party-state's pursuit of global leadership and the continuity of domestic information control. The three chapters analyse how other-imposed constraints have created an endless loop entrapping the party-state, and how the latter has few chances of breaking down the loop without incurring radical political changes in China. They examine the manufacturing of dilemmas by investigating the endogeneity/exogeneity of constraints, a process which, we posit, impacts the durability of the predicament facing the party-state.

This chapter revolves around the first source of the party-state's double-bind logic: the contradiction between the party-state's political and economic objectives. The first part demonstrates the oppositionality of the two sets of objectives and the party-state's incremental adjustments to mitigate the opposition. The second part discusses the mechanism through which private firms have exploited the party-state's dilemmas to seek greater manoeuvring room. The last part concludes.

Contradiction Between Political and Economic Objectives

The contradiction between political and economic objectives in the ISM regulation is the first source of the party-state's double-bind logic. The contradiction stems from two reasons. First, the achievement of political and economic objectives implies different courses of actions. As explained in Chapters 3 and 4, the party-state respectively mobilises centralised and decentralised approaches to pursue political and economic objectives. Second, two different groups of regulators oversee the political regulation and the economic regulation of the ISM sector, and the relations between them are fraught with conflicts. Eager to defend their respective turf, political regulators (CAC, CDP, NRTA, Ministry of Culture, MPS) and economic regulators (MIIT, NDRC, Ministry of Science and Technology) compete over resources (Negro, 2017), competences (e.g., the competition between the CAC, the MIIT, and the MPS over the portfolio of cybersecurity), and institutional influence. Bureaucratic turf battles put the party-state into a delicate situation, because, ruling over the agencies, it cannot go too far in the pursuit of one set of objectives without harming the other. Therefore, it must navigate between them, stifle one set of regulators at one moment and augment their competences at another, the objective being to ensure a dynamic equilibrium between the two groups of regulators. This section explores the manufacturing of the party-state's double-bind logic by investigating the contradiction between the political regulation and the economic regulation of the ISM sector.

Contradiction Between Economic Upscaling and Information Control

The internet governance approach in China emphasises both economic growth and information control (Mueller and Tan, 1997). The "combination of capitalism and authoritarianism seeks to unleash market forces and contain their political consequences at the same time" (Jiang, 2012: 39). Struggling between economic growth and information control, the party-state has gone through three phases in its relations with private ISM firms: liberalisation, regulation, and state capitalism (Jiang, 2012: 22). Although Jiang does not give the precise dates separating one phase from another, the three-stage analysis reveals the contradiction in the party-state's regulatory logic. On the one hand, during the phase of liberalisation, internet firms were given much autonomy, and "both private and foreign companies were allowed to thrive" (Jiang, 2012: 25). Foreign investors' investment in the high-tech was welcomed. It was only on 1 December 2010 that the State Council terminated the "super-national treatments" of foreign investors. Hsueh confirmed the high autonomy of Chinese private firms in the value-added services (e.g., news portals, emails, e-commerce) before China joined the WTO. It was an epoch in which the foreign direct investment in the VAS sector was officially banned but informally encouraged (Hsueh, 2011).

Private firms' dominance over the SOEs in the first decade of Chinese internet development (mid-1990s to mid-2000s) was voluntarily designed by the party-state (Jiang, 2012). For instance, the first generation of internet firms in portals, like Sina, Sohu and Netease, and their successors in more diversified sectors (e.g., Baidu, Alibaba, and Tencent) were all created by private entrepreneurs. The State Council Information Office (SCIO), content regulator from 2000 to 2011, supported the rise of private ISM firms by granting Sina and other private news websites the qualification for spreading news information online. On 29

December 2000, Sina obtained the Telecommunication and News Information Service License (*dianxin he xinwen xinxi fuwu yewu jingying xukezheng*) from Beijing Telecommunication Administration, one of the local outposts of the Ministry of Information Industry.

It was the first time that the SCIO allowed private internet firms to enter the news sector, which had hitherto been dominated by state media and their online versions (people.cn was operationalised on 7 November 1997, one year earlier than Sina.com). In doing so, private capital entered the fiefdom of propaganda and challenged the party-state's monopoly in this area. Although article 7 of the *Provisory Regulation of the Internet Websites Providing News Services* laid out qualification-related and ownership-related restrictions of private firms' rights to collect and edit information, the enforcement of the regulation was laxist. Private ISM firms violated content regulation rules according to which private media could only retweet the hard news published by state-owned media in the first place.

The party-state was taken aback when private firms went beyond their role of economic motor and were progressively more relied upon than state-owned media as providers of hard news. Based on the media system dependency theory (Ball-Rokeach and DeFleur, 1976, 1979; DeFleur and Ball-Rokeach, 1989), Tai and Sun (2007) studied Chinese citizens' media dependency during the SARS crisis in 2003. They found that Chinese people trusted non-public sources of information much more than state media. Since netizens' dependency on private media signalled the declining credibility of state media, it arguably motivated the adoption of the 2005 content regulation, which reaffirmed that private media were prohibited from reporting self-collected hard news. Given the "credibility gap" between official and non-official discourses (Qian and Bandurski, 2011: 39), the restrictions of the 2005 regulation on private ISM firms' collection of hard news could, in principle, prevent the widening of the gap, because such restrictions reduced the sources of information available to citizens.

The 2005 regulation was not more dissuasive than the 2000 regulation. Bandurski (2018) showed how, in the 2000s, private ISM firms defied the instructions of propaganda watchdogs to scoop valuable information which embarrassed the party-state. The media coverage of the bullet train collision in Wenzhou on 23 July 2011 provided a telling case. Although the CDP ordered influential media outlets not to go to Wenzhou, private ISM platforms defied the CDP and sent reporters to the spot immediately. Media worker Beifeng (literally "North Wind") from China Sun TV in Hong Kong, released on Sina Weibo the name of the 38 victims. Although the post was deleted, the list pressured Chinese authorities to publish a more detailed list of the victims whose number was modified from 35 to 38 on July 25. Disobedience to content regulators prevailed in particular in the media outlets reporting business news. Given the party-state's recognition of the need for capital market transparency, "since the 1990s, government control of economic news in terms of licensing and supervision has been relatively loose compared with control over other types of news" (Hu, 2011: 79). Exploiting their "immunity", business media often broke the silence. Hu Shuli, co-founder of the private media Caijing and Caixin Media, reported that business media were the first to cover the SARS crisis in 2003 and the avian flu in 2005 (2011: 84), sensitive topics that propaganda watchdogs initially hid from the public. Numerous cases of disobedience from private media illustrated the limited effectiveness of the content regulation in the 2000s.

Chinese citizens' dependency on private media kept deepening in the 2010s (Tai and Sun, 2007; Stockmann, 2011; Miao, 2011). When I conducted the field research from 2019 to 2021, more than one interviewee revealed their mistrust towards state media. One informant pursued her PhD in Paris when interviewed. During her stay inside and outside China, the most important source of information for her to know what happened in China was Sina Weibo, not state media or newspapers.[2] Another interviewee also relied on private media to seek information. She explained that she had not watched Xinwen Lianbo, official news program broadcasted at 19 pm on China Central Television (CCTV), for more than 10 years.[3] Since its creation on 1 January 1978, the popularity of Xinwen Lianbo has remained relatively low due to its emphasis on propaganda rather than on informing the public. In 1999, in their book *20 Years of Xinwen Lianbo*, Yang Weiguang and Li Dongsheng explained that the principal problems of the news broadcasting of the CCTV included lack of valuable information and excessive reporting on official meetings (1999). These problems persist today (Miao, 2011). The elderly people also show disinterest into state media. The PhD student I interviewed in Paris reported that the frequency with which her parents

[2] Interview P-21/10/2019.

[3] Interview B1-19/12/2019.

watched Xinwen Lianbo has much reduced. Instead, they tended to watch local news programs.[4] The disinterest in official news reached foreign people living in China. One senior French filmmaker who had lived in China for more than 10 years, said: "in China, few people watch CCTV, few people believe in it. People simply agree to play the game."[5] In addition, among the people who do watch the news on CCTV, they seldom believe in it. Since the news is broadcast from 19:00 to 19:30, families watch Xinwen Lianbo while eating and waiting for the TV drama at 20:00 pm.[6]

Private ISM firms resemble the creature of Victor Frankenstein because they finally go beyond the initial intentions of their creator. In promoting private ISM firms as the symbols of the China Inc., the party-state has created its own problems because these economic stars progressively challenge the power of the official media to shape the public opinion. Throughout the 2000s and well beyond, private ISM firms have been winning the minds of Chinese netizens to become more relied upon than state media. When private ISM firms break the CDP rules to cover social emergencies, content watchdogs can hardly punish them without considering the public reaction. If the public is satisfied, content regulators have to turn a blind eye to the embarrassing reporting of private firms; otherwise, they would render themselves unpopular. Taking the censorship of *Freezing Point* as an example, a liberal-leaning tabloid and a supplement to the state-owned *China Youth Daily*, McGregor shows how the public feedback weighed high for content watchdogs:

> The politburo had already decided to deal with us…Previously, the propaganda department had monitored the reactions of the netizens to our articles. When there was applause around, they did not do anything. But when the reaction was bad, they felt the time had come. (2010: 251)

In the 2000s, despite the rollout of increasingly stringent content regulations, private firms continuously committed disobedience and often presented regulators with *fait accompli* whose social influence importantly shaped whether firms would be punished or not. When examined closely, the party-state's relative tolerance of firms' frequent violations could be

[4] Interview P-21/10/2019.

[5] Interview O-06/11/2020.

[6] Interview P-21/10/2019.

linked to the fact that regulators still did not know exactly how to keep private media under check cost-effectively. Nevertheless, this explanation is not entirely credible because the party-state had full capacities to coerce private entities into compliance. The fact that it did not choose this option could be more plausibly explained by the thesis of "dictators' dilemma" (Wintrobe, 1998; Roberts, 2018; Dickson, 2016). Since the 2000s was the period of economic take-off, the party-state favoured connectivity over content (Austin, 2014: 92), and it tolerated a laxist implementation of content regulations to preserve the momentum of economic development. When the task of economic catch-up became less urgent in the end of the 2000s, the party-state rebalanced economic and political objectives, which in return influenced the size of the autonomous sphere for private firms.

Rebalanced Political and Economic Functions of ISM Firms Under Xi

Under Xi, the party-state not only endeavours to reduce the role of the private media as the microphone of the civil society, but also urges them to better serve the country's economic objectives. To put it more simply, the party-state curbs private firms' political relevance while amplifying their economic relevance. One article published in *Time* in October 2021 explained that despite the tightened regulation of the digital market in the areas of fintech, online gaming, and after-school education in the post-2020 period, the administration of Xi continues to embrace the high-tech sector. The country "has identified the digital economy as a key driver for growth over the next few decades and made achieving tech self-sufficiency a top national priority. To support that growth, Beijing has doubled down on funding for strategically important industries such as semiconductors and the AI."[7] The regulation of ISM firms under Xi inverses that under Jiang and Hu, for the party-state forcefully intervenes to curb the role of the ICTs in politically empowering the civil society.

Although the economic policies under Jiang, Hu, and Xi have helped to unleash the economic potential of private ISM firms, the point of departure of these policies differs. The administration of Jiang (1989–2002) was characterised by the liberalisation of the VAS, the sector that the great majority of private ISM firms fell within. In the lead up to

[7] "Xi Jinping Eyes Innovation and Oversight to Grow China's Digital Economy," *Time*, 19 October 2021, https://time.com/6108481/china-digital-economy-technology/

2004, "the MII also did not enforce the prohibition against FDI in internet content provision. In short, the state did not scrutinise ownership and business scope until policing information flows and ordering competition became state imperatives in 2004" (Hsueh, 2011: 96). The outcome of the liberalisation was that the pursuit of economic goals came to embarrass the party-state's political interests. Therefore, economic-friendly policies under Xi, including the "Internet + ", industrial internet, and "Made in China 2025", were initiated on the basis of an imbalanced implementation of the political and the economic agendas. Xi's priority was to repair the imbalance by reinforcing information control and encouraging firms to serve the economic interests of the country.

To Curb the Political Relevance of ISM Firms

That private ISM firms have been dwarfing state media to take up an increasingly crucial role in the online ideological battlefield imbalances the party-state's equations. The way to curb private firms' influence in the ideological domain is to push state media to "grab the megaphone" (Bandurski, 2018: 378) again and to make state media present "at the places where the audience is present."[8] At the *National Conference on Propaganda and Thought Work* in August 2013 in Beijing, Xi Jinping required that state media be more combative and more active in their reporting:

> When it comes to critical issues and political principles, we must take the initiative, and fight the battle, such that cadres and masses distinguish the right from the wrong and that their vague comprehension could be clarified.[9]

To increase the competitiveness of state media in cyberspace, the party-state also pushed them to get listed in China and overseas. By turning state propaganda organs into both market regulators and market players, the party-state incentivised state media to compete with private media for audiences, or to take back the microphone. This approach is what Jiang terms "state capitalism", the quintessence of which is to.

[8] Xi, Jinping, "Speech at the National Propaganda and Thought Work Conference," CAC, 19 August 2013, http://www.cac.gov.cn/2014-08/09/c_1115324460.htm

[9] Ibid.

> ...initiate the public listing of the online propaganda apparatus. At the forefront of this process are many of China's largest propaganda organs, including the People's Daily, Xinhua News Agency and CCTV. The process is seen as "state capitalistic" because it is intended to use capitalist market mechanisms to improve efficiency and more importantly, to better promote the party's views and extend its influence in cyberspace. Unlike previous state capitalistic endeavours in China's manufacturing, industrial and media sectors, the latest attempt is ultimately about using the market to produce and sell the party's ideological goods online. (2012: 30)

To Amplify the Economic Relevance of ISM Firms

Like the "informatisation" policy under Jiang Zemin and the "informatisation-empowered industrialisation" under Hu Jintao, the "Internet +", industrial internet, and *Made in China 2025* under Xi Jinping emphasised private ISM firms' economic contributions. With the "Internet + ", the virtual economy was expected to help upscale the real economy. After two decades of expansion in the virtual economy, the moment has come for ISM firms to upgrade the traditional economic sectors in terms of quality and efficiency. The 16734-word action plan of the State Council, released on 4 July 2015, mentioned "internet firms" 25 times. Internet firms were encouraged to participate into a large range of activities: helping rural cooperatives establish efficient distribution and marketing chains; developing cloud services platforms to support the fintech; developing DNA testing in collaboration with hospitals; promoting the sharing of educational resources in collaboration with educational institutions, etc. Echoing with the State Council, ministries and provinces advanced more concrete policies. Guangdong province, known for its capacity in the manufacturing sector, released its "Internet + Advanced Manufacturing" plan on 22 March 2018. Similar policies were adopted by ministries and other local governments to concretise the policy design of the State Council: "Internet + Modern Agriculture", "Internet + Healthcare", "Internet + Regulation", "Internet + Social Services", to name only a few.

In the "Internet + Modern Agriculture", leading ISM firms embraced the call of central and local regulators with enthusiasm. In 2015, Alibaba announced its plan "One Thousand Counties and Ten Thousand Villages", saying that within three to five years, Alibaba would invest 10 billion yuan to create 1000 agricultural operating centres at the county level and 100 000 services stations in villages. The second-largest E-commerce firm in China, JD.com, started developing the "Internet + Agriculture" in December 2014 in Renshou county, Sichuan province. As of November 2015, 297 young people of the county were trained to sell home appliances to 600 villages, and to provide after-sales services. Nicknamed as "JD Gang", local marketing teams helped rural residents buy home appliances at the same price as urban residents (home appliances and smartphones sold in rural areas tend to be 10%-20% more expensive than in urban areas, due to transportation fees). Moreover, JD helped the farmers of Renshou to sell loquat fruits on its platform. In the same year, the firm declared that it would create 500 agricultural service centres in counties and enlarge the JD Gang to tens of thousands of members. Suning, another Chinese E-commerce giant, also joined the competition. On 10 May 2016, Suning Rural E-Commerce Institute (*suning nongcun dianshang xueyuan*) was created. At the inaugural ceremony, Zhang Jindong, president of Suning Holdings Group, announced the 2016 rural E-commerce strategy of the company. Suning committed to invest 5 billion yuan into the rural market, increase the number of its direct sales shops from 1011 to 2511, put 200 local-food restaurants online, and push 100 000 local talents to create business in their hometown.

The "Made in China 2025" strategy, aiming to transform China into a world-class manufacturing superpower by 2049, also features the party-state's high reliance on private firms.

Despite the achievements under Jiang and Hu, the "Made in China 2025" strategy says that the level of informatisation and the integration between informatisation and industrialisation has remained weak in China. Therefore, it emphasises again the contributions of the digital economy to improve industrialisation. The strategy opens new horizons that private ISM firms have quickly seized: cloud computing, cloud services, internet of things, big data, automation, and artificial intelligence. The deployment of the automated delivery services by ISM firms during the coronavirus in 2020 illustrates the progress firms have made in implementing the strategy. In AI, priority officially fixed in 2017, Chinese private ISM firms have turned into world-class players in less than 5 years (Ma, 2018). For the ISM firms whose activities in traditional VAS services are saturated, the opening of new horizons allows them to create new sources of revenue. The development of the new sectors with strategic importance to national economy and which have little linkage with contents matches the party-state's interest. By encouraging private ISM firms to grow in these new sectors, the party-state ensures that firms serve the economic interests of the regime and that they do not embarrass its political interests.

The Efficiency of the Rebalancing

The administration of Xi has rebalanced the economic and political functions of private ISM firms by nudging them to economically strategic but politically non-sensitive sectors. Nudging firms away from content production to technological innovation mitigates the party-state's burden with content regulation. The report of the ISC and the MIIT on the top 100 internet firms in 2019 shows that in that year more than 65% of the firms operate in online marketing and around 24% in data services.[10] Among the 17 types of services, online marketing and E-commerce have generated the greatest profits for firms, exceeding content-related social-networking, gaming, music, and video. This is in particular the case of the firms operating in both the economic and political domains, like Baidu and Tencent. Therefore, it is no surprise to spot some obvious changes in the principal sources of revenues of certain leading ISM firms. The financial reports of 2019 and 2020 of Baidu show that the revenues created by its content-related business (search and advertisements) decline from 76.46% in 2018 to 74.2% 2019. The weakening capacity of Iqiyi, a video-sharing platform that Baidu acquired in May 2013, deserves attention. The data of Forward-The Economist, a Beijing-based consulting firm, show that the contribution of Iqiyi to the revenues of Baidu has decreased from 2015 to 2019. The net profit of Iqiyi is negative in the whole period, and the loss keeps widening (Fig. 5.1). In 2020, Iqiyi's revenue reduces from 7.65 billion in the Q1 to 7.2 billion yuan in the Q3. In contrast, the contribution of AI-related activities to Baidu's revenue increases from 5.6% in Q4 of 2019 to 9.8% in Q4 of 2020.

The consequences of the rebalancing are above all the degraded quality of online information. After the Supreme People's Court and the Supreme People's Procuratorate ruled that one piece of information viewed more than 5000 times or retweeted more than 500 times might expose its creator and retweeter to criminal charges, firms and

[10] ISC & MIIT, *2019 Development Report of the Top 100 Internet Firms*, 14 August 2019, p.12, https://www.gov.cn/xinwen/2019-08/15/content_5421376.htm.

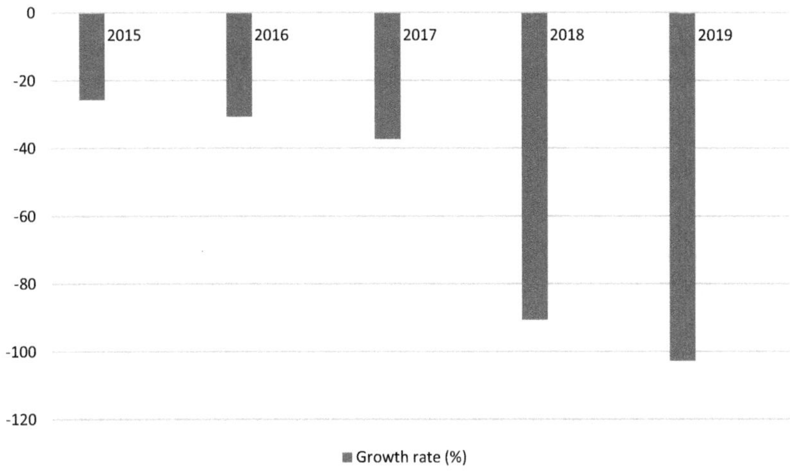

Fig. 5.1 Growth rate of net profits of Iqiyi 2015–2019 (*Source* Forward-The Economist, 2020)

netizens have been increasingly deterred from spreading breaking news information, fearing to get themselves into trouble.[11] To stay safe, one interviewee told that he tended to retweet information instead of creating new information himself.[12] His experiences show that original contents are much more likely to be censored than retweeted contents. Moreover, stricter implementation of licensing rules and the increasing presence of state media in cyberspace (Han, 2018: 104–105) make private ISM firms self-censor more. The increasingly conservative reporting on private ISM platforms reduces netizens' interest in consulting them. One interviewee explained her disinterest in using Sina Weibo when she came back from Italy to China in the Summer of 2013:

> First, I don't feel that it is a valuable source, it cannot give me what I need, not too many valuable things. Nowadays it has almost become a platform for entertainment and gossips dominated by news media and entertainment bigwigs...principally it is because I can get very few independent opinions

[11] Interview O-26/06/2020.

[12] Interview B3-21/12/2019.

> or thoughts... For Weibo, I feel like that it has become a mishmash platform where bloggers, media, or brokers behind the scenes stir up topics. I don't want to be influenced or dominated.[13]

When asked which search engine she used for obtaining information, she continued:

> ...I don't use Baidu, I climb the Wall [Great Firewall, added by the author]. The indexing of Baidu is of very low quality. I certainly climb the Wall.[14]

"Climbing the wall" or *fan qiang*, is commonly used by Chinese netizens to designate the mobilisation of VPNs to access blocked information. Since the Great Firewall (GFW) prevents Chinese netizens from accessing foreign sources of information that criticise China, netizens often use VPNs to circumvent the GFW. Some interviewees, in particular researchers and businesspeople, climb the GFW each day for consulting information and contacting business partners.[15] Young interviewees climb the GFW for consulting social media blocked in China, like Instagram. The VPNs are also used to ensure that the materials sent from abroad could be viewed normally by China-based receivers.[16]

A New Round of Rebalancing Is Underway

The "Internet + " strategy and the "Made in China 2025" strategy, inscribed into the rebalancing efforts, channel the energy of private ISM firms into economic sectors. Under Xi, the parallel promotion of information control and technological development show that the party-state has succeeded in striking what it perceives the right balance between political and economic goals. It seems that the regime finally finds the antidote to the double-bind logic. Under Xi, "the idea of innovation in communication technologies has been promoted aggressively even as controls on speech have tightened to a degree unprecedented in the reform era" (Bandurski, 2018: 383). The public declarations of Xi Jinping

[13] Interview B1-19/12/2019.

[14] Interview B1-19/12/2019.

[15] Interview B1-21/12/2019.

[16] Interview O-06/11/2020; interview B1-21/12/2019; interview B-20/12/2019.

also demonstrate that the party-state finally shakes off the double-bind logic. On 19 April 2016, presiding over the Conference on Cybersecurity and Informatisation, Xi described the relationship between the political and the economic sides of the regulation as follows: "Security is the precondition of development, and development is the guarantee of security; security and development must advance side by side. Cybersecurity and informatisation are the two wings of the same body, the two wheels of the same vehicle". This quote signals how the party-state under Xi has become quite confident in simultaneously pursuing political and economic objectives without having to suffer from the constraints generated by the double-bind logic.

However, the seemingly successful rebalancing is anything but the end of the state-firm saga, for the intensive moves of economic regulators in curbing the monopolistic practices of leading ISM firms show that the rebalancing under Xi is lurching towards the consolidation of the party-state's control over firms, namely political goals, not an equal pursuit of economic and political goals. The Deputy CEO of iDate Digi World, a French think tank specialised in digital market analysis, highlights three potentials enabling Chinese ISM firms to grow even bigger in comparison with their American counterparts: unsaturated internal market, unsaturated international market, and increasing diversification of the firm activities.[17] However, the Chinese government's intensive regulation in both the anti-trust and the data areas is likely to prevent firms from exploiting these opportunities to the fullest.

The unexpected interruption of the initial public offering of Ant Financial in Shanghai and Hong Kong on 3 November 2020 shows that Chinese regulators are trying to regulate the firms before they become too powerful to control. On 10 April 2021, the State Administration for Market Regulation (SAMR) required Alibaba, parent company of Ant Financial, to pay 18.2 billion yuan (around $28 billion), representing 4% of Alibaba's annual turnover. It is the highest administrative fine that a private firm has ever paid in China, three times higher than the 6.088 billion yuan (around $ 975 million) that the American Qualcomm paid to Chinese regulators for anti-trust charges in 2015. The rollout of anti-trust rules since November 2020 and the restructuring of Ant Financial which was to be put under the supervision of People's Bank of China

[17] Interview O-24/02/2021.

(PBOC), show that the rebalancing under Xi—the party-state's moving away from the economic side to the political side—still fit well with the double-bind analytical framework.

One article in *The Economist* in April 2021 captures the party-state's veering into greater control of tech giants: "after years of tolerating big tech's unbridled expansion, the central government is rewriting the rules, some tacit and some explicit, for how billionaires can behave, the degree of overt state control over data, and who owns the firms' other assets, including stakes in other businesses. This new master plan for Chinese big tech will transform one of the world's most innovative and valuable industries".[18] *The Economist* predicts that since tech giants have much richer pictures about their users—Alibaba holds data over more than 1 billion people—the party-state's objective to tighten the supervision over firms is "to pool data and impose more state ownership and control, which could eventually amount to a kind of nationalisation".[19] The intensification of the economic regulation of private ISM firms is unambiguous. After curbing the political power of ISM firms, the upcoming rebalancing will probably focus on curbing private firms' economic power, such that they do not become the Chinese version of GAFAM powerful enough to challenge state authority.

The purport of the upcoming rebalancing is to curb both the economic and political influence of private ISM firms, a course of action which cannot be sustainable in the long term. The reason is that through minimising the political and economic influence of internet firms, the party-state risks crowding out the non-public sector, making firms only communist, not commercial. This is the situation in which the party-state seeks one set of interests only—political interests in this case—and neglects economic interests. Reminiscent of the centrally planned economy under Mao, this situation risks stifling firms' enthusiasm in technological innovation, frightening investors away, and tarnishing the international image of China. In a nutshell, making firms communist only certainly allows the party-state to have an upper hand over the private sector. Nevertheless, the party-state risks losing almost everything else.

[18] "China's Rulers Want More Control of Big Tech," *The Economist*, 8 April 2021, https://www.businessreview.global/business/6078ff47c8ff5d651e1b01ef?dualLang=zh_CN

[19] *Ibid.*

The party-state is doomed to keep private ISM firms simultaneously communist and commercial. To achieve this objective, firms should be given greater autonomy in running business, including the autonomy in deciding where to get listed without being implicitly nudged. Granting higher autonomy to firms is likely to incentivise them to serve the party-state's economic interests and to increase the international economic clout of the country. Private ISM firms should also be given greater autonomy in running valuable news, even though such news may embarrass the party-state. Since official media have the reputation of serving propaganda (although under increasingly creative forms of "ideotainment" (Lagerkvist, 2008: 123), allowing private media certain autonomy in reporting breaking news would attenuate netizens' criticisms against information control practices in China, which indirectly improves the public perception towards the party-state and drives up its legitimacy. Adopting these options in the post-rebalancing period helps preserve the double-bind logic. Ironically, although this logic puts the party-state into dilemmas, it has trained the party-state to muddle through them, and to gain resilience.

Inverse Cooptation: Coopted Firms Coopt Regulators

The contradiction in the party-state's objectives constrains its freedom of action in regulating private ISM firms. Since economic growth underpins the performance-based legitimacy of the party-state, the latter cannot ignore firms' needs or stifle their entrepreneurship dynamic. Since the complete realisation of the party-state's economic objectives implies the incomplete realisation of the political objectives, and *vice versa*, a grey zone emerges. Aware of the party-state's trade-off between ensuring political control and driving up economic growth, firms are trying to push the limits of the grey zone by socialising with regulators and seeking their support. Entrepreneurs exploit their status of "red capitalists" (Dickson, 2003, 2007) or "coopted capitalists" to coopt policymakers, a process that I call "inverse cooptation" in this book. Private ISM firms with economic importance acquire important bargaining power vis-à-vis regulators, and are consulted for the elaboration of public policies. Centred around the "inverse cooptation", this section analyses the relations between this strategy and the party-state's double-bind logic. It demonstrates how the inverse cooptation enables private firms to reach larger margins of manoeuvring.

Maximising Corporate Autonomy in Exploiting the Party-State's Predicament

This section analyses the way in which private ISM firms in China exploit the party-state's double-bind logic to seek greater manoeuvring room. It argues that rather than being constrained by the party-state, coopted entrepreneurs actively engage themselves into "inverse cooptation", a process in which coopted entrepreneurs exploit their status of "red capitalists" to influence governmental decision-making. In other words, as a Leninist institution, cooptation is a double-way mechanism in China. It certainly helps the party-state pacify stubborn entrepreneurs, as under the Soviet era. However, it also enables private entrepreneurs to influence policymakers efficiently. That cooptation works in the inverse direction from private entrepreneurs to policymakers is more evident in contemporary China than in the defunct USSR for the straightforward reason that private economy was illegal in the Soviet era.

Private firms in China do lobby the government, regardless of their ownership structure or the economic sector they operate in (Kennedy, 2011: 113–116). The lobbying is so intensive that "the real difficulty these days is in identifying industry policies that have not been shaped by lobbying" (Kennedy, 2011: 116). Chinese firms' lobbying became prevalent after the transition from the planned economy to the socialist market economy in the early 1990s. After the 14th Party Congress in October 1992, which explicitly suggested the creation of the socialist market economy in China, the third plenum of the 14th Central Committee adopted the groundbreaking document *Decision on Several Issues in Establishing the Socialist Market Economy* on 14 November 1993. This document importantly clarified the central importance of the economic recovery rather than the class struggle, the fundamental role of the market in the allocation of resources, and bureaucratic downsizing. With the rising importance of the market, the process of decision-making was increasingly characterised by "political pluralisation", a process in which "otherwise marginalised officials, non-governmental organisations and activists of all stripes have managed to wriggle their way into the policy-making process and even help shape policy outcomes" (Mertha,

2009: 996). The thesis of political pluralisation invalidates the conventional wisdom according to which policymaking in China is dominated by governmental institutions only. Instead, it introduces nuances into the interactive modes between governmental and non-governmental actors in the Chinese political landscape. It supports Kennedy's research on Chinese private entrepreneurs' activism in pushing policy changes. In Kennedy's words, business "now typically seeks to shape the government's policy agenda as well as the content of laws and regulations before they are issued. State-owned enterprises (SOE), private companies, and foreign-invested firms are all in on the action" (2011: 116).

"Political pluralisation" in China, opportunity opened by the fragmentation of the Chinese political system, as well as the status of entrepreneurs as "co-opted entrepreneurs", facilitate business lobbying (Mertha, 2009). Comparatively speaking, business lobbying stemming from a pluralised social structure does not distinguish China from elsewhere. However, business lobbying facilitated by the status of "coopted entrepreneurs" (Dickson, 2003, 2007) constitutes a feature unseen in democracies. Inverse cooptation of officials by coopted entrepreneurs is specific to China because the possibility for achieving inverse cooptation comes from the structural predicament of the party-state, rather than the commonly observed interdependence between power and money in autocracies. One inevitable outcome of the party-state's dependence on private business is that it has to behave like a "benevolent dictator" (Gilson and Milhaupt, 2011), accepting to be "inversely coopted" in order to incentivise the private sector to serve economic goals of the regime. In this respect, that Chinese industrial associations do not enjoy full independence from the government is not a handicap for firms to defend business interests. On the contrary, since fully autonomous industrial associations in China can hardly gain the trust of and access to officials necessary to help their members (Kennedy, 2011: 117), their efficiency in defending private business is not guaranteed. Industrial associations and other cooptation venues (e.g., people's congresses and people's political consultative conferences) are semi-autonomous in China. Therefore, within these trusted venues, coopted private entrepreneurs acquire real opportunities to incorporate their preferences into new laws, statutes, or policies.

Inverse Cooptation in the Drafting of the ECL

The elaboration of the Chinese E-commerce law (ECL) shows how influential E-commerce firms can shape the content of the law by using trusted associations and their status of coopted business. From the initiation of the project in December 2013 to the final adoption on 31 August 2018, lawmakers had debated the project four times and submitted it to public consultations three times. Considering that Chinese lawmakers generally approve a legislation after three readings, the lengthy process in adopting the ECL is question-begging. The intense struggles between the stakeholders (e.g., private ISM firms, regulators, merchants) are the principal reason behind the difficult approval of the ECL. In China, ministries or ministerial-level organs usually initiate industry-specific laws. However, for the ECL, 12 ministerial-level agencies led by the Financial and Economic Commission of the NPC participated in the drafting process. That one commission of the NPC spearheaded the elaboration of an industry-specific legislation is uncommon in China. The designation of the NPC as the leader of the draft law did not prevent ministries from influencing the content of the law. From 2011 to 2013, the State Administration for Industry and Commerce (replaced by the SAMR since March 2018), the Ministry of Commerce, and the MIIT had issued several ministerial regulations in the E-commerce sector, hoping to influence the stipulations of the law.

The remarks of Yin Zhongqing, then Deputy Director of the Financial and Economic Commission of the NPC, showed how the competition of the involved stakeholders in the manufacturing of the legislation rendered the approval of the ECL a conflict-ridden process. On 31 August 2018, he explained the lengthiness of the process under such terms "[I]n comparison with other legislations, the ECL is more complicated, touching upon more stakeholders and more areas. As a new phenomenon, the E-commerce sector keeps evolving, and there exist plenty of uncertainties."[20] The conflicts of interests were so strong that more than one participant involved in the drafting of the ECL worried that the bill might be aborted half-way. Alamusi, Deputy Director of the Policy and Law Commission at the China Electronic Commerce Association (CECA),

[20] Zhang, Cong, "*People's Daily* Reviews the Legislation Process of the E-Commerce Law: Open the Door and Consult Widely the Public Opinion," The Paper, 5 September 2018, https://www.thepaper.cn/newsDetail_forward_2409960

was invited for drafting the law as the representative of professional associations. He said:

> At our first meeting, many people thought that the bill would never come into being. Ministries thought that this law was too encompassing, that it would cause conflicts of competences and interests among them. Academia thought that there was no precedent example to inspire from and that it was not necessary to legislate on e-commerce...Professional associations thought that the law would put too much burden on platforms...These concerns were corroborated by later developments.[21]

However, since the bill was highly technical and that it was not politically sensitive, it was necessary and harmless to acquire as much feedback as possible from the civil society (Deng and Liu, 2017). ISM firms and professional associations in the E-commerce industry were highly active for providing their inputs, trying to shape the contents of the ECL bill.

To understand how inverse cooptation allows private ISM firms to shape policy outcomes, I examine the lobbying by Alibaba within CECA, a government-patronaged professional association. The study on Alibaba repairs insufficient academic attention to the business lobbying by individual firms (Kennedy, 2011: 135). The reason for studying Alibaba is twofold. First, Alibaba is a coopted private firm. Since it has woven close relations with both local and national regulators over time, the case study of Alibaba can be based on available sources of information. Second, several important provisions in the adopted ECL damaged Alibaba's interests, in particular its flagship platform, Taobao, which, together with the B2C platform Tmall, created more than 85% of the annual income for Alibaba by 2018. In comparison, B2C platforms like JD.com, Yanxuan, and Tmall were less impacted by the ECL. For instance, the ECL stipulated that both platforms and individual merchants on platforms should be legally registered (article 10) and that both should pay taxations (article 11). These stipulations starkly contrasted with the old practices according to which platforms rather than merchants paid taxes. They also ended the practice in which small merchants created a shop on Taobao

[21] Zhang, Yao & Huang, Shujing, "Rollout of the E-Commerce Law: Battle among Giants, Wrestling among Institutions, Will the Current Situation of the Chinese E-Commerce Sector Change," *Caijing*, 31 August 2018.

with their ID card only, without registering the shop to market regulators. Therefore, the adoption of the ECL was expected to drain 10% of merchants away from Taobao. From the initiation of the ECL bill to its final approval, Alibaba was among the firms whose activities were most likely to be impacted. In the lead up to the adoption of the ECL, one widely talked topic was the negative impact on Alibaba. Insiders of the E-commerce sector commented that as long as the ECL was passed, it would be a setback for Ali. Therefore, compared with B2C and O2O platforms, the conflicts of interests between lawmakers and Alibaba, working on C2C, were more visible.

Xue Jun, Deputy Director of the Law School at Peking University, participated into the whole process of drafting the ECL. For him, the problems linked to Taobao were the driving concern behind the initiation of the ECL in 2013.[22] At that time, Chinese consumers complained a lot about fake products, delivery, refund policies of E-commerce platforms, complaints that finally pushed lawmakers to regulate the freewheeling E-commerce. Therefore, it was logical that Alibaba engaged itself into intensive lobbying to minimise the negative impact of the proposed bill. The aforementioned article 10 of the bill laid out the principle of "registration as the rule, non-registration as an exception" for individual merchants on C2C platforms. It seriously preoccupied Alibaba. The first version of the ECL generalised the requirement of registration to all individuals having a trading account on E-commerce platforms, implying that they had to register their business to both the market regulation administrations and the fiscal authorities in municipalities. The exception was only granted to "individuals who sell self-produced agricultural products or handicraft products and individuals who use their own skills to provide convenient services, activities the provision of which does not need any license". However, in the third reading of the ECL on 19 June 2018, the bill enlarged the exception to include "sporadic small-amount transactions" *(lingxing xiao'e jiaoyi huodong)*, thus exempting small merchants of the obligation of administrative registration when they operated on Alibaba's platforms. The enlargement of the exception was the result of Alibaba's painstaking lobbying to lawmakers.

[22] *Ibid.*

Lobbying Through Government-Patronaged Associations

Chinese private firms lobby policymakers in two ways: industrial associations and direct lobbying (Kennedy, 2011). Alibaba mobilised both to influence lawmakers. The CECA (China Electronic Commerce Association), created on 21 June 2000 in Beijing under the auspices of the MIIT, was the earliest Chinese industrial association in the E-commerce sector. Grouping almost all influential Chinese E-commerce firms, the CECA maintained special relationship with the largest of them, Alibaba. From 2004 to 2012, the CECA, Alibaba, and Hangzhou Municipal Government had co-organised 9 editions of Global Netrepreneur Conference. Besides being a forum of discussion for Chinese and foreign E-commerce companies, the conference also selected for each year 10 best E-commerce entrepreneurs of the world (Chen, 2017), considered as the highest honour of the E-commerce industry. The socialisation between firms and officials in these venues enabled regulators to take the pulse of firms' concerns. These venues also provided coopted firms with a privileged access to meet policymakers and to express their suggestions about to-be-adopted policies. For government-patronaged industrial associations, that they are trusted by the party-state changes them into reliable venues where private entrepreneurs are more likely to see their concerns addressed by government officials.

The functioning of government-patronaged associations is similar to the functioning of the internal meetings of the party committees set up in governmental institutions, universities, SOEs, or some private firms located at various administrative levels. One interviewee is a member of the CCP and worked in a local procuratorate at the time of interview.[23] She told me that although politically incorrect comments are prohibited from circulating online and offline, they can become objects of discussions at internal meetings of the CCP. To identify the factors with destabilising potential to the party-state's reign, party cadres are allowed to use frank languages in internal discussions and to express their personal thoughts, including the ideas that do not perfectly fit into the party's ideological lines. In addition, the level of tolerance towards politically incorrect remarks is positively correlated to the hierarchy of party organisations. The more an individual or an organisation is trusted by the party-state, the more they can discuss taboo issues without getting themselves into

[23] Interview B-25/12/2019.

trouble. In other words, discussions of politically sensitive issues are the privilege of trusted or coopted people.

Other factors also influence industrial associations' capacity and willingness to defend private firms. When discussing how government-patronaged business associations bridge the state and private business, Unger explains that although China fits the model of state corporatism where industrial associations are designed to champion the party-state's interests, the functioning of state corporatism can "diverge sharply from elite intentions" (1996: 817). Four reasons explain why government-patronaged associations deviate from Beijing's instructions (Unger, 1996: 813–814). First, "the constituency they have been delegated to represent is a status group that they want to identify with: to be spokespeople and facilitators for a group of people who symbolize personal success in China today" (*Ibid*). Second, the constituency of the business associations can potentially be helpful to officials' own future career interests. Third, associations under the supervision of party organs rather than administrative agencies don't take up day-to-day rule enforcement, implying that they are less likely to enter into direct conflicts with businesspeople. Fourth, business associations that are less beholden financially to higher authorities are more likely to defend their members' interests (Unger, 1996: 813–814).

The functioning of the CECA matched the characteristics advanced by Unger, except the fact that the CECA was under the supervision of an administrative agency, the MIIT, rather than a party organ. However, since the membership in the CECA was voluntary and that the CECA did not assume day-to-day rule enforcement, it was less likely to find itself at odds with its members than the industrial associations that ensured rule enforcement (e.g., the self-Employed Laborers Association) and the membership of which was mandatory. Personnel of the CECA were found to identify themselves with their members and to champion the interests of private businesses. For instance, the insertion of the "sporadic small-amount transactions" in the third and fourth versions of the ECL was partially thanks to the CECA's support. Alamusi, who represented the CECA in the drafting team, was known for his liberal stance in lowering the market entry for individual merchants on C2C platforms. He argued that the development of the E-commerce could not do away with the "quantity". Requiring all individual merchants to effectuate administrative registrations would dramatically decrease the "quantity", which would hinder the development of the E-commerce in China. He said that many offline shops didn't get registered either. Therefore, he championed the insertion of the C2C-friendly measures in the ECL, including the "sporadic small-amount transactions". In addition, he suggested that governments delegate the regulation of "trivial, basic, and simple" issues to the E-commerce platforms themselves, including procedures for administrative registration, punishment of fraudulent sales, mediation of disputes between consumers and merchants, etc. The liberal stance of the CECA contrasted with the conservative propositions from the academia—represented by Professor Xue Jun—that proposed to generalise the registration rules to the great majority of individual merchants. For Xue, whatever the amount of transactions, lawmakers and policymakers could always calculate the amount and lay out precise standards accordingly. Industrial associations ultimately obtained lawmakers' support, thus weakening the voice of the academia. During the third reading of the bill, Cong Bin, Deputy Director of the Constitutional and Legal Commission of the NPC, stated that:

> Generally speaking, concerning the business registration and the fiscal management in China, it is necessary to require all business entities to get registered to ensure equal competition between online and offline merchants. However, in practice, many individual merchants engage in low-frequency and low-amount transactions. The existing laws have already required the platforms to verify their identities. Therefore, it is not necessary to ask them to get registered. We suggest to add the following stipulation: it is not mandatory for individuals engaged in sporadic small-amount transactions to get registered.[24]

Therefore, Alibaba and industrial associations succeeded in inserting the ambiguous expression "sporadic small-amount transactions" into the ECL. However, not all claims of Alibaba were supported. Alibaba did not obtain the CECA's support concerning the generalised registration for individual merchants on Taobao. The best situation for Alibaba was to exempt all individual merchants on Taobao from being registered. However, the final outcome for Alibaba was that individual merchants on its C2C platforms, except those involved in "sporadic small-amount

[24] "The First E-commerce Law in China Is Under the Third Reading, the Threshold of Individual Online Shops for the Administrative Registration Is Likely to Be Lowered," CCTV, 19 June 2018, http://news.cctv.com/2018/06/19/ARTIZZX4Zfo222uKkWnQT3yL180619.shtml

transactions", had to register themselves with market regulators and fiscal authorities. Two reasons explained the disagreement between the CECA and Alibaba: (1) the opposition from top-level leaders to unregulated sales by individual merchants on C2C platforms, and (2) widely reported violations of consumer rights. On Sina Weibo, Li Keqiang, then Prime minister, directly addressed Jack Ma, saying "Comrade Jack Ma, frankly speaking, none of your companies [on Taobao] is lawful, if they are to be measured by the laws. How come that one can become a company simply after creating a trading account on your website?...".[25] The opposition from top-level leaders was a signal for the CECA. Interviewed by *Legal Evening News*, a state-owned news outlet, Meng Xiangbin, then Deputy Secretary of the CECA, parroted Li Keqiang. He explained that the registration of online shops, although a difficult choice for Taobao, served the orderly development of the E-commerce in the long term (*Ibid*).

On the violations of consumer rights, the non-registration of individual shops on Taobao complicated law enforcement. Without knowing the actual address of the shops, regulators could hardly locate and penalise wrongdoing merchants, thus leaving damaged consumers helpless. Therefore, the registration of individual merchants was proactively pursued by industry and commerce administrations during the deliberation process of the bill. The difficulties of the CECA demonstrate the pitfall of the business lobbying through semi-official industrial associations, because the latter can difficultly defend firms if the top-level leadership expresses unambiguous policy preferences or if the framing of the issue (e.g., protection of consumer rights) achieves an insurmountable height.

Direct Lobbying

Direct lobbying is the primary form of business lobbying in China (Kennedy, 2011: 118). Public appearance of charismatic firm leaders at public consultations is mobilised to increase the relevance of firms' voice. In the deliberations on the ECL, while Tencent, JD.com, Netease, and Meituan generally sent deputy CEOs and researchers to lobby lawmakers, Jack Ma personally attended public consultations more than once to make

[25] Yang, Shifan, Zhang Xin, and Wang Xiaofei, "Deputy Secretary of the China Electronic Commerce Association: the Registration of the Shops on Taobao to the Industry and Commerce Administration Is an Inevitable Trend," Economic Information, 12 November 2013, http://jjckb.xinhuanet.com/2013-11/12/content_475939.htm

Alibaba better heard. During the last consultation on 16 August 2018, although Jack Ma had to fly to Shanghai in the afternoon to sign a strategic cooperation agreement with the Party Secretary of Shanghai, he attended the NPC's meeting in Beijing on the same day. Alibaba was actually the only firm represented by its highest leader in person. Realising that Alibaba could not seek better deals, he still regretted the "immaturity" of the proposed bill: "the ECL should be international and forward-looking, it should have included the stipulations favourable to the development of E-commerce. The moment for legislating on E-commerce is not mature".[26]

Lack of coordination among E-commerce platforms weakened the efficiency of the business lobbying and reduced firms' capacity to seek favourable policies from regulators. E-commerce firms less impacted by the ECL sided with lawmakers and asked for harsher rules to disadvantage Alibaba. As the biggest competitor of Alibaba, JD.com lobbied in the opposite direction. Operating in the B2C sector, JD.com was much less impacted by the ECL, because merchants on JD were registered business entities rather than unregistered individuals. Therefore, JD.com's lobbying for reinforcing the regulation of the C2C was synonym of amplifying the negative impact of the ECL on Alibaba and other C2C firms. Interviewed by *Caijing*, Beijing-based news magazine known for its independent reporting, Cai Lei, then Deputy CEO of JD.com, explained that the ECL was still not deterrent enough to make individual merchants pay taxes. For him, "Today, the internet economy is strong, the real economy is weak. The ECL should be rolled out and enforced as soon as possible. Rules on market access should be implemented, interim regulation be reinforced, such that the tax loss...could be solved."[27] JD.com was opposed to the exceptions laid down in the ECL according to which individual merchants involved in "sporadic small-amount transactions" were exempted from administrative registrations. Since the exception would allow around 90% of the individual merchants on Taobao to continue their business without formal registration, JD.com criticised that the term "sporadic small-amount transactions" was too ambiguous, urging Chinese lawmakers to fix explicit criteria. It lobbied for closing the

[26] *Ibid.*

[27] Zhang, Xinyu, "Focus Analysis: the E-Commerce Law that Jack Ma Has to Worry About, Didi and Pinduoduo Should Also Pay Attention to," 36kr, 5 September 2018, https://www.36kr.com/p/1722806026241

loopholes for Alibaba which might benefit from the ambiguous wording of the law.

Comparing Business Lobbying in China and France

While Chinese private ISM firms lobby policymakers either through trusted industrial associations or through direct lobbying, ISM firms in France or in the EU, more generally, lobby public authorities in a more institutionalised manner either by the mediation of independent professional guilds or by public hearings with parliamentarians and policymakers. Informality importantly distinguishes Chinese business lobbying from its Western counterparts. In China, firms' reliance on informal lobbying is partly explained by the window-dressing nature of institutionalised channels designed for formal lobbying. Most business associations, "particularly those at the national level, operate in name only, and public hearings are held primarily for show. As a result of these constraints, attempts at influence peddling have been driven to take other forms. Lobbying is often highly fragmented, and companies rely on various informal tactics, from privately meeting with policymakers to influencing the media" (Kennedy, 2009: 213). On the contrary, lobbying in advanced economies unfolds under more formal forms. To compare Chinese ISM firms' lobbying with the lobbying in France, I interviewed two people working in the Orange Group, one of the largest telecom operators in France. In recent years, Orange has enlarged its activities from telecom services to the provision of contents, like video. By the time of the interview, while one interviewee headed the department of governmental relations, the other person worked at the legal department of the Group. They explained how their company sought favourable policies in formal ways:

> "In fact, the deputies to the National Assembly often invite us for hearings, to which we reply, either through the Fédération Française des Télécoms (FFT, French Federation of Telecoms) which represents telecom firms like us, either directly in order to bring our contributions, our expertise, such that deputies have an idea about what they can or cannot propose."[28]
>
> Researcher: "Does the voice of your firm is well respected in the law-making?"

[28] Interview P1-21/11/2019.

> "It depends on the subjects. Some subjects are very complicated, somewhat technical [and we are heard, added by the author]. It is true that it is difficult to draw the attention of the general public to such issues...The society is probably waiting for some apparently very simple things, but they are very complicated in practice.[29]
>
> On technical and juridical issues that many people cannot understand at first sight..."[30]

According to them, the voice of Orange Group became particularly relevant when deputies legislated on highly technical issues. A Beijing-based French interviewee, who worked in the consulting area, emphasised the role of professional associations through which private ISM firms in France could lobby the government:

> The methods include alliances, professional associations according to their sector...In the digital area, there are associations in video games, one association for telecom companies, etc. Associations work on these subjects, they provide consulting services, they send memos, they organise meetings with deputies and officials, such that their message is transmitted to policymakers. This is the official channel [of business lobbying]: on the regulation, on the management of human resources, on taxations, on industrial standards, etc.[31]

French ISM firms also exploit informal channels to make their voice heard. The way in which they defend business interests by informal networking is similar to their Chinese counterparts. Grossman and Saurugger (2012) mention three types of resources that businesses in democracies mobilise to influence policymakers: financial resources, social resources, and societal resources. Social resources are particularly mobilised for supporting informal lobbying and have diverse characteristics. They are characterised by "the degree and the mode of group organisation, the nature of the group's elites, the group's degree of institutionalisation inside the politico-administrative apparatus, more or less close contacts between groups and actors, the mobilizable know-how and expertise, and finally, the group's capacity to define its own interests in a more or less autonomous way" (Grossman and Saurugger, 2012: 13).

[29] Interview P1-21/11/2019.

[30] Interview P2-21/11/2019.

[31] Interview O-24/02/2021.

Among these multiple characteristics, the relationship between interest groups and governmental agencies is an important asset for shaping the outcomes of business lobbying. In a similar way to Chinese private firms, business groups in liberal democracies mobilise relational resources to obtain their "*entrée au ministère*" (*entry to the ministry*) (Grossman and Saurugger, 2012: 14). Business groups also try to reinforce their influence by increasing their presence in media. This view is corroborated by my field research. For instance, the aforementioned consultant did not see obvious differences between Chinese and French private firms on informal lobbying. He explained how French ISM firms conducted informal lobbying:

> Well, it is somewhat similar...Each firm can do the lobbying in its own way. If the firm is well equipped, it can use consulting firms, or its own team, to discuss with members of the government, or a deputy; they can organise conferences or other things allowing the companies to write reports and to transmit the reports to deputies who prepare the laws. When French deputies prepare the laws, they receive a great quantity of notes written by the firms trying to push their interests through. Sometimes, it works. Sometimes, it does not work. This is the classical lobbying work in France and elsewhere in the world. It is less well framed in France than in Anglo-Saxon countries ...[32]

As explained by this interviewee, informal business lobbying takes on similar forms in China and France. What distinguishes business lobbying in China is inverse cooptation analysed in this part. Echoing the argument of Kennedy (2011), business lobbying by coopted firms within trusted venues presents a significant advantage over business lobbying by non-coopted firms through independent associations. Since Chinese policymakers have greater trust in coopted entrepreneurs and government-patronaged industrial associations, coopted firms are more likely to obtain business-friendly policies than non-coopted firms. The successful lobbying by Alibaba through the government-patronaged CECA in convincing lawmakers to insert "sporadic small-amount transactions" into the ECL illustrates the efficiency of "inverse cooptation". This example shows that it is not because the party-state succeeds in coopting private firms that it can easily dictate them. On the contrary, since the cooptation allows

[32] Interview O-24/02/2021.

private firms to be closer to policymakers, firms can capitalise on this proximity to shape policies and regulations. In this sense, the efficiency of the cooptation in augmenting the party-state's mightiness vis-à-vis private business is to some extent counterbalanced by firms' "inverse cooptation".

Conclusion

The conflict between the party-state's economic and political objectives creates a grey zone where private ISM firms can exploit their status of "red capitalists" (Dickson, 2003, 2007) or "coopted capitalists" to inversely coopt regulators to seek favourable regulations. To ensure the loyalty of influential entrepreneurs, the party-state has been coopting them by increasing the coverage of grassroots party cells in private firms and by incorporating young entrepreneurs into government-patronaged industrial associations and advisory bodies. However, the analysis on whether the top 100 Chinese ISM firms in 2020 mention "party" or "party -building" on their official websites shows that more than 80% of them avoid mentioning these terms (Ma and Hu, 2024). Moreover, we find that the more an ISM firm is influential, the more it is unlikely to mention its linkage with the CCP on its official website, implying that in their mutual relations, private ISM firms are as utilitarian as the party-state. Therefore, the cooptation strategy, instead of ensuring private entrepreneurs' loyalty, is exploited by entrepreneurs to seek favourable policies, a process called "inverse cooptation" in this chapter. The example of Alibaba's lobbying by mobilising government-patronaged industrial associations to insert Alibaba-friendly stipulations into the ECL illustrates private firms' exploitation of the party-state's predicament. Coopted firms exploit their being coopted to gain easier access to policymakers. Therefore, coopted firms are abler to defend their interests than non-coopted ones.

Despite a carefully woven multi-layered network of control and influence, the party-state faces structural constraints under which the monitoring network cannot produce its full effect. I have argued that the party-state faces three constraints when regulating private ISM firms: the contradiction between political and economic objectives, fragmented authoritarianism, and the contradiction between the pursuit of global leadership and the continuity of domestic information control. This chapter has analysed the first constraint. It also analyses the implications of the party-state's predicament on the pursuit of manoeuvring room by private ISM firms in China. Having to pursue both political and economic objectives throws the party-state into a dilemma, for it goads the party-state to turn to one set of objectives at one moment and to the other set at another moment. The periodicity in the party-state's shift between the two sets corroborates Jiang (2012) and Hsueh (2011) who highlight the party-state's differentiated priorities in different epochs. This pattern is particularly instantiated by Hsueh (2011). For her, the party-state has been shifting between liberalisation and reregulation in an iterative manner. The shift from one phase to another is explained by the changes of challenges facing the regime in different periods.

The introduction has discussed three weaknesses of the existing literature on the regulation of private ISM firms in China: (1) lack of a holistic view of the constraints facing the party-state; (2) insufficient discussions of the impact of the dilemmas on private firms' pursuit of autonomy, and (3) insufficient attention to the "manufacturers" and the "manufacturing" of dilemmas. This chapter addresses the first and the third points, showing that the party-state is anything but all-powerful in regulating private firms, for its freedom of action is constrained by the factors and actors located beyond its control. Caught between the objectives the realisation of which implies contradictory courses of action, the party-state cannot focus on one set of objectives only and must switch side alternatively. Compared with the single-factor analysis, the holistic and dialectic perspective adopted in this book helps zoom large state-firm interactions. It disaggregates the state-firm battle into a wrestling of several rounds in which the success or failure at the end of one round is not definitive. As long as the regulation of private ISM firms continues in China, the battle will persist, implying that both the party-state and firms may win and lose and that nobody wins definitively. The holistic perspective thus invalidates the simplified thesis of al all-powerful party-state.

This chapter sheds light on the "manufacturer" and the "manufacturing" of the double-bind logic. In a similar way to Frankenstein, a tragic hero in Mary Shelley's classic novel, the party-state manufactures the double-bind logic and gets in trouble due to it. The trajectory trekked by post-Mao China is fraught with predicaments. It is designed by the party-state itself: controlling the autonomy of private firms while trying to make them innovative; consolidating the power reach over localities while unleashing their dynamism in driving up economic development; pursuing global acceptance while reinforcing internet censorship inside

China. Although the three pairs of contradictions unfold in different forms (e.g., economic, political) and at different levels (e.g., subnational, national, and international), they share the commonality of being built in the governance logic of the party-state in the post-Mao era. The endogeneity of the contradictions implies that the double-bind logic cannot be shaken off easily without provoking internal disequilibrium of the regime. In this sense, the party-state is doomed to live with the double-bind logic, which, if managed skilfully, brings not only authoritarian power, but also what autocrats normally dare not aspire for simultaneously: economic clout, international prestige, and governance legitimacy.

The contradiction between political and economic objectives in the regulation of private ISM firms constitutes one of the built-in constraints for the party-state. It explains why the latter must leave its control-hungry machine asleep rather than active all the time, thus transforming the party-state's unfreedom into a window of opportunity enabling private firms to be more autonomous. Private ISM firms make use of their status of coopted entrepreneurs to inversely coopt government officials. The example of Alibaba's lobbying during the drafting process of the ECL has illustrated that private ISM firms are well aware of their strategic economic value to the party-state, and that they don't hesitate to capitalise on it to facilitate their pursuit of commercial profits.

References

Austin, Greg. 2014. *Cyber Policy in China*. Cambridge, UK: Polity.

Ball-Rokeach, Sandra J., and Melvin L. Defleur. 1976. "A Dependency Model of Mass Media Effects." *Communication Research* 3(1): 3–31.

Ball-Rokeach, Sandra J., and Melvin L. Defleur. 1979. "A Dependency Model of Mass Media Effects." In *Inter/Media*, edited by G. Gumpert, 81–96. New York: Free Press.

Bandurski, David. 2018. "Can the Internet and Social Media Change the Party?" In *Routledge Handbook of the Chinese Communist Party*, edited by W.W.-L. Lam, 372–390. Oxon, UK: Routledge.

Chen, Deren. 2017. "Cong Shi Jie Wang Shang Da Hui de Ba Da Bian Hua Tan Jiu Xin Shi Ti Jing Ji" (Explore New Real Economy Based on the Eight Changes of Ten Editions of Global Netrepreneur Conference). Ali Research Institute, https://read01.com/zh-my/oLLo6NJ.html#.Yga_B9_MLIU

Clark, Duncan. 2016. *Alibaba: The House That Jack Ma Built*. New York: HarperCollins Publishers.

Defleur, Melvin L., and Sandra J. Ball-Rokeach. 1989. *Theories of Mass Communication*. Fifth Edition. New York: Longman.

Deng, Jinting, and Pinxin Liu. 2017. "Consultative Authoritarianism: The Drafting of China's Internet Security Law and E-Commerce Law." *Journal of Contemporary China* 26(107): 679–695.

Dickson, Bruce. 2003. *Red Capitalists in China: The Party, Private Entrepreneurs, and Prospects for Political Change*. Cambridge, UK: Cambridge University Press.

Dickson, Bruce. 2007. "Integrating Wealth and Power in China: The Communist Party's Embrace of the Private Sector." *The China Quarterly* 192: 827–854.

Dickson, Bruce. 2016. *The Dictator's Dilemma: The Chinese Communist Party's Strategy for Survival*. Oxford: Oxford University Press.

Dixit, Avinash. 2010. "Democracy, Autocracy and Development." *Journal of Globalisation and Development* 1(1): 1–45.

Francisco, Ronald A. 2005. "The Dictator's Dilemma." In *Repression and Mobilisation*, edited by C. Davenport, H. Johnson, and C. Mueller, 58–81. Minneapolis, MN: University of Minnesota Press.

George, Cherian. 2007. "Consolidating Authoritarian Rule: Calibrated Coercion in Singapore." *The Pacific Review* 20(2): 127–145.

Grossman, Emiliano, and Sabine Saurugger. 2012. *Les Groupes d'intérêt: action collective et stratégies de représentation*. Paris: Armand Colin.

Han, Rongbin. 2018. *Contesting Cyberspace in China: Online Expression and Authoritarian Resilience*. New York: Columbia University Press.

Hsueh, Roselyn. 2011. *China's Regulatory State: A New Strategy for Globalisation*. Ithaca, NY: Cornell University.

Hu, Shuli. 2011. "The Rise of the Business Media in China." In *Changing Media, Changing China*, edited by S. Shirk, 77–90. New York: Oxford University Press.

Jiang, Min. 2012. "Internet Companies in China: Dancing between the Party Line and the Bottom Line." *Asie Vision* 47: 1-50.

Kennedy, Scott. 2009. *The Business of Lobbying in China*. Cambridge, MA: Harvard University Press.

Kennedy, Scott. 2011. "Fragmented Influence: Business Lobbying in China in Comparative Perspectives." In *Beyond the Middle Kingdom: Comparative Perspectives on China's Capitalist Transformation*, edited by S. Kennedy, 113–135. Stanford, CA: Stanford University Press.

Kuran, Timur. 1997. *Private Truths, Public Lies: The Social Consequences of Preference Falsification*. Cambridge, MA: Harvard University Press.

Lagerkvist, Johan. 2008. "Internet Ideotainment in the PRC: National Responses to Cultural Globalization." *Journal of Contemporary China* 17(54): 121–140.

Lagerkvist, Johan. 2011. "New Media Entrepreneurs in China: Allies of the Party-State or the Civil Society?" *Journal of International Affairs*. Fall/ Winter 2011: 169–182.

Lynch, Daniel. 1999. "Dilemmas of 'Thought Work' in Fin-De-Siècle China." *The China Quarterly* 157: 173–201.

Ma, Aifang, and Yong Hu. 2024. "Zhong Guo Dui Min Ying Hu Lian Wang Qi Ye De Fei Zheng Shi Jian Guan: Yi Min Ying Hu Lian Wang Qi Ye Dang Jian Wei Li" [Informal Regulation of Private Internet Firms in China: A Case Study of Party-Building]. *Twenty-First Century*. Forthcoming.

Ma, Aifang. 2018. *L'intelligence artificielle en Chine: Un état des lieux*. Paris: Fondation pour l'Innovation Politique.

McGregor, Richard. 2010. *The Party: The Secret World of China's Communist Rulers*. New York: HarperCollins Publishers.

Mertha, Andrew. 2009. "'Fragmented Authoritarianism 2.0': Political Pluralization in the Chinese Policy Process." *The China Quarterly* 200: 995–1012.

Miao, Di. 2011. "Between Propaganda and Commercials: Chinese Television Today." In *Changing Media, Changing China*, edited by S. Shirk, 91–114. New York: Oxford University Press.

Mueller, Milton, and Tan Zixiang. 1997. *China in the Information Age: Telecommunications and the Dilemmas of Reform*.Westport: Praeger.

Naughton, Barry. 1995. *Growing Out of the Plan: Chinese Economic Reform 1978–1993*. Cambridge: Cambridge University Press.

Negro, Gianluigi. 2017. *The Internet in China: From Infrastructure to a Nascent Civil Society*. Cham: Palgrave Macmillan.

Qian, Gang, and David Bandurski. 2011. "China's Emerging Public Sphere: The Impact of Media Commercialization, Professionalism, and the Internet in an Era of Transition." In *Changing Media, Changing China*, edited by S. Shirk, 38–76. Oxford: Oxford University Press.

Roberts, Margaret. 2018. *Censored: Distraction and Diversion Inside China's Great Firewall*. Princeton: Princeton University Press.

Shirk, Susan. 2011a. "Chapter 1: Changing Media, Changing China." In *Changing Media, Changing China*, edited by S. Shirk, 1–37. Oxford: Oxford University Press.

Stockmann, Daniela. 2011. "What Kind of Information Does the Public Demand? Getting the News During the 2005 Anti-Japanese Protests". In *Changing Media, Changing China*, edited by S. Shirk, 175–201. Oxford: Oxford University Press.

Unger, Jonathan. 1996. "'Bridges': Private Business, the Chinese Government and the Rise of New Associations." *The China Quarterly* 147: 795–819.

Wintrobe, Ronald. 1998. *The Political Economy of Dictatorship*. Cambridge: Cambridge University Press.

Xiao, Qiang. 2011. "The Rise of Online Public Opinion and Its Political Impact." In *Changing Media, Changing China*, edited by S. Shirk, 202–224. Oxford: Oxford University Press.

CHAPTER 6

The Dynamics of Fragmented Authoritarianism

Introduction

Although the private ownership of Chinese internet firms does not shield them from governmental predation, they can exploit the predicament of the party-state to minimise the negative influence of the party-state's multi-layered network of supervision. The previous chapter investigates the first source of the party-state's double-bind logic: the contradiction between the party-state's political and economic objectives. This chapter examines the second source: fragmented authoritarianism. Although fragmented authoritarianism is a decades-old research topic, the innovation of this chapter is that it goes beyond the predicament created by the fragmented Chinese governance system. Instead, it investigates the implications of the predicament for the margins of manoeuvring of private internet firms in China.

Much ink has been spilled on the motivations of private ISM firms behind their noncompliance. Scholars have also extensively discussed firms' strategies for committing noncompliance without being punished (Miller, 2018; Cairns, 2016; Negro, 2017). The existing literature analyses corporate noncompliance principally from three perspectives: the pursuit of commercial interests, the defence of the corporate reputation, and fragmented authoritarianism. The assumptions behind the first body of literature correspond to the neoclassical economic theory which portrays private firms as wealth-maximising actors (Jevons, 1879;

A. Ma, *China as a Double-Bind Regulatory State*,
https://doi.org/10.1007/978-981-99-8857-0_6

Stigler, 1971). In contrast, the second body of literature denies the simple expected utility model as the driving explanatory framework (Negro, 2017; Han, 2018), for "under certain conditions, traits like honesty, integrity and living up to a reputation pay off in strictly wealth-maximising terms" (North, 1990: 42). Private ISM firms may well commit noncompliance due to the liberal-leaning inclinations of their founders or managers (Han, 2018).

"Fragmented authoritarianism", coined by Oskenberg and Lieberthal (1988) and developed by Andrew Mertha (2009), complements the first two bodies of literature. Instead of putting firms at the conceptual centre of the analysis, fragmented authoritarianism analyses the noncompliance of ISM firms with the particularity of the Chinese political system (Miller, 2018: 6–7). "Fragmented authoritarianism" denotes the "*de facto* veto power of local governments when implementing central policies and the tangled lines of authority in China's vast bureaucracy" (Miller, 2018: 7). According to Lieberthal and Oskenberg, "what on paper appears to be a unified, hierarchical chain of command turns out in reality to be divided, segmented, and stratified. Indeed, the fragmentation of authority is a core dimension of the Chinese system" (1988: 137). Within fragmented authoritarianism, policy outcomes are shaped "by the incorporation of interests of the implementation agencies into the policy itself. Fragmented authoritarianism thus explains the policy area as being governed by incremental change via bureaucratic bargaining" (Mertha, 2009: 996).

Authoritarian fragmentation in China unfolds at both the vertical and the horizontal levels of governance. Vertical fragmentation is concerned about the territorial organisation of the Chinese political system, and denotes the conflictual relations between local governments and the central government. McGregor highlights that in order to preserve the interests of local firms, local governments could well deviate from the instructions of Beijing (2010). Horizontal fragmentation is concerned about the bureaucratic fragmentation/overlapping in the division of competences among regulatory agencies. For instance, it is said that in China around 50 agencies can send instructions to ISM firms to require the suppression of harmful contents. Lack of coordination among the agencies enables content providers to implement the least constraining regulation without being harassed (Miller, 2018; Cairns, 2016).

The literature on fragmented authoritarianism explains corporate noncompliance at a more structural level. It argues that firms could exploit the fragmentation in the Chinese political system to circumvent undesirable regulations. It significantly contributes to turn down the image of the party-state as an all-powerful decision-maker and to establish more balanced power relations between private firms and the party-state. However, when it comes to systematise the opportunity structure of private ISM firms in China, the existing literature need to be fleshed out more explicitly. The literature on "fragmented authoritarianism" touches upon the defining characteristics of the Chinese political and economic governance and, indeed, provides a relatively structural explanation of corporate noncompliance. However, the empirical evidence used in this body of literature focuses on individual firms (Miller, 2018; Crandell et al., 2013; Crete-Nishihata et al., 2016; Knockel et al., 2017). To systematically evidence the interplay between "fragmented authoritarianism" and corporate noncompliance, it is necessary to replace the incident-centric approach with a more generalisable analysis: instead of investigating the interactions between one firm and one agency, it can be more valuable to examine the interactions between multiple firms and multiple regulators.

Based on my empirical data (interviews and analyses of policy releases), this chapter explores the second source of the party-state's double-bind logic: fragmented authoritarianism. Going one step further than the existing literature, this chapter not only analyses the extent to which the vertical and horizontal fragmentation creates governance dilemmas for the party-state. It also unveils the implications of fragmented authoritarianism for the margins of manoeuvring of private ISM firms in China. Like the previous chapter, this chapter argues that the margins of manoeuvring of private ISM firms in China are shaped by more structural factors related to the governance logic of China. More concretely, the contradiction in the party-state's sought-after objectives in the ISM regulation makes the tolerance of corporate noncompliance an *inevitable* choice.

Vertical Fragmentation: Bifurcated Agenda

The Chinese political system is one of the most decentralised in the world (Fu, 2018: 51; Birney, 2013), and the interjurisdictional competition is one of the driving factors of the Chinese economic growth (McGregor, 2010: 175). The economic decentralisation and the fiscal reforms in the 1990s have been reported to be the origins of the decentralised governance in China (Oi, 1992; Milhaupt and Zheng, 2015: 695; Lin et al.,

2013). For instance, the fiscal reforms in 1994 were intended to reinforce the authority of Beijing over localities. However, since the new fiscal system required the latter to grant a greater share of their fiscal income to the former, it incentivised local officials to create new sources of revenues to finance local development projects (Milhaupt and Zheng, 2015).

The role of local governments in regulating the ISM industry tends to be neglected. When asked who regulated the Chinese ISM industry, both Chinese and non-Chinese interviewees spontaneously replied that it was the central government, the CCP, or the CAC. Almost none of them mentioned local regulators. The dialogue with one interviewee showed the centrality of Beijing in the collective representations of the Chinese citizens:

> Researcher: Do you have any general idea about who regulates Chinese cyberspace?
>
> Interviewee: Who regulates? Of course it is done by the people with the highest power. However, what I mean by "regulation" does not mean that these guys directly conduct day-to-day work, they can use the hands of other people, like private firms, Alibaba, Tencent...Undoubtedly, China is building a highly centralised, controllable, traceable, localisable, coercible [regulation system], although the rules are not completely clear. This reality is taking shape.[1]

The intersubjectively shared perceptions inside and outside China on the relations between Beijing and localities is that central regulators systematically dwarf local regulators, whatever activity is concerned. The portrayal of unequal power relations between the centre and the periphery is a tacit consensus, although, as it will be demonstrated shortly, local governments are as important as Beijing in the economic regulation of the ISM industry. The reason for which the role of local regulators has been less visible than Beijing is threefold. First, this situation partly relates to the top-down organisation of the propaganda system. Since propaganda is directed from the centre to the localities, the activities of central leaders are more likely to be mediatised than those of local leaders. Differentiated intensity of media coverage creates the impression that the Chinese political system is highly centralised, while in the day-to-day politics, China applies "*de facto* federalism" (Zheng, 2006). The second reason relates

[1] Interview P-01/12/2019.

to the political ethos centred around national unity. In this framework, the consensus among elites and citizens is that "so long as the government is consciously making efforts to foster national identity, localism or regionalism will not become excessive and impose a major threat to China as a nation-state" (Zheng, 2007: 6). The third reason interestingly relates to the semantics of the word "regulation". In the Chinese language, "regulation" is composed of two words "*jian*" and "*guan*", denoting respectively "supervision" and "control". Since "regulation" entails imbalanced relationship between the regulator and the regulatee, with the latter being acted upon by the former, "regulation" becomes an embodiment of power relations: whoever regulates has more power than whoever is regulated. Regulation being an incarnation of unequal power, it is generally considered to be spearheaded by the instance with the highest capacity to act upon other actors, namely the central government.

The vertical fragmentation of the Chinese political system unfolds under the form of conflicts between local governments and the central government. Given that local governments wield power principally on economic issues, not on political ones (McGregor, 2010), the opportunities opened by the vertical fragmentation for firms are principally related to the economic regulation. This part analyses how against the backdrop of the decentralised economic regulation, the local ownership of private firms and the "competitive liberalisation" between jurisdictions (Yang, 1997) create the second predicament for the party-state. One additional note relates to the relations between the two pairs of contradictions: political versus economic goals, central versus local regulators. The two dimensions are not independent from each other and the conflict between central and local regulators is subordinate to the conflict between political and economic imperatives. The reason is that it is on the economic axis that central regulators face the challenges from local regulators. On the political axis, local regulators can hardly challenge Beijing.

Fiscal Reform and Decentralised Economic Regulation

The local ownership of private ISM firms in China results from the decentralisation. Although the latter started well before the Open Doors policy in 1978 (Lin and Liu, 1998), it was the fiscal reforms in the 1980s which provided a decisive impetus to the deepening of the decentralisation. In the 1980s, the Chinese fiscal system importantly changed. It changed from a unitary system, in which Beijing had absolute control over revenue collection and budget appropriation, to a relatively decentralised arrangement, in which revenues were shared by the central and local governments (Lin and Liu, 1998: 2). Beijing simultaneously initiated "the transfer of spending responsibilities for public works, education and welfare subsidiaries to local governments" (Yang, 1997: 43–44). Fiscal decentralisation transformed local governments into the principals in their respective jurisdiction, and incentivised and empowered them to pursue local objectives (Yang, 1997: 44; Eggertsson, 1990: 40–45). Fiscal decentralisation halted for a while after the Tiananmen Square massacre, and fiscal austerity prevailed from 1989 to 1991.

The fiscal reform in 1994 in particular required local governments to transfer an important share of their fiscal income to Beijing (Milhaupt and Zheng, 2015). With the 1994 reform, county governments could only take respectively 15 and 25 million yuan for each increase of 100 million yuan in their value-added taxation and income taxation (Xiao, 2008). Simultaneously, greater spending responsibilities in education, healthcare, family planning, and salaries of civil servants were put on the shoulders of local governments (Xiao, 2008). However, initially intended to reinforce the authority of Beijing over localities, the 1994 reform counterintuitively goaded local authorities to seek new bases of taxation, deepening rather than contracting fiscal decentralisation (Milhaupt and Zheng, 2015). Mertha suggested that the centralisation efforts in the 1990s produced at best a "soft centralisation" under which the bargaining power shifted from lower levels to the provincial level, without increasing the authority of the centre (Mertha, 2005a, 2005b). In the 1980s and the early 1990s, half of the county governments in China suffered from budgetary deficits, pushing local governments to find out new sources of revenue and to increase their income through development or development-cum-predation (1990).

The fiscal decentralisation led to the decentralisation of the policy-making on economic and social matters, the general objective being to reduce the budgetary burden of Beijing. In the ISM industry, although the telecom infrastructure and the basic telecom services had been directly managed by the MPT and the MEI since the 1980s, the ISM sector was delegated to local authorities in the 1990s and was initially regulated as an ordinary economic sector. According to the regulatory parlance in China, the ISM firms also operated in the telecom services. However, while state-owned carriers provided "basic telecom services", private ISM firms were providers of "value-added services" (VAS). The *Telecommunications Regulations*, adopted on 25 September 2000 by the State Council Order 291, defined the VAS as "the telecom services and the information services the provision of which relies on the public network infrastructure" (article 8). On the division of competences between Beijing and local governments, paragraph 2, article 9, stipulated that:

> For value-added telecom services, if the coverage of the provided services **involves no less than two** provinces, provincial-level municipalities, or autonomous regions, it is mandatory to obtain the approval from the competent authorities of the State Council in charge of the information industry. It is mandatory to obtain the Inter-Regional VAS Operating License (*Kuadiqu zengzhi dianxin yewu jingying xukezheng*). If the provision of services is **limited within one** province, one provincial-level municipality, or one autonomous region, it is mandatory to obtain the approval of the provincial-level telecommunication administrations, and to obtain the VAS Operating License (*Zengzhi dianxin yewu jingying xukezheng*) (emphasis added by the author).

The regulatory regime of the VAS sector, to which the great majority of private ISM firms belonged, gave the provincial communications administrations (PCAs) discretionary licensing power. During the reregulation phase from 2004 to 2009, after the discretionary power of local regulators shrunk with the centralisation of the licence approvals to the MII, local communications administrations exploited other channels to serve local goals. Although the MII took charge of granting and revoking the licence of the VAS providers operating in no less than two provinces, the PCAs scrutinised the applications in a preliminary way, enabling them to indirectly serve local goals. In their business plan (part of the application documents to submit), firms must put each service they intended to provide and submit a justification for each service. Empowered to scrutinise the services and the justification of the services, the PCAs could "exercise this power and other discretionary powers to achieve personal or local goals" (Hsueh, 2011: 100). PCAs could also contribute to realise local priorities by exercising their scrutiny over local public procurement contracts.

Local Ownership of Private ISM Firms

With the decentralisation of the VAS sector, the administrative registration of private firms is conducted at the local level. Before 2018, private firms registered themselves to the Municipal Administrations for Industry and Commerce (MAIC), local outposts of the State Administration for Industry and Commerce (SAIC). Since the institutional reforms in March 2018 replaced the SAIC with the SAMR (State Administration for Market Regulation), private ISM firms changed to register with the Municipal Administration for Market Regulation (MAMR, local outposts of the SAMR). In the policy document released on 30 April 2001, the State Council required that provinces vertically administer the industry and commerce administrations at the municipal levels. Although provincial administrations for industry and commerce (PAIC) were the big winner of the 2001 institutional reform, this reform did not change private firms' local ownership. The pre-eminence of the PAIC started declining with the second institutional reform in 2011. On 10 October 2011, the State Council abolished the vertical leadership of the PAIC over the MAIC, leaving the MAIC self-administer the private firms registered within their jurisdiction. By 30 November 2017, the SAIC had completed the delegation of all its functions on firm registrations to municipalities. It declared that from 1 December on, it would no longer register any new firm. During the institutional reform in March 2018, the newly created economic regulator SAMR replaced the SAIC. To reduce bureaucratic overlapping, the SAMR absorbed the SAIC, the GAQSIQ (General Administration of Quality Supervision, Inspection and Quarantine), and the CFDA (China Food and Drug Administration). In addition, the SAMR took over the anti-monopoly bodies of three institutions—the NDRC, the State Council Anti-Monopoly Commission, and the Ministry of Commerce. Therefore, since March 2018, the municipal outposts of the SAMR have started registering private firms, overseeing the pricing policy, and combating monopolistic practices.

The ownership level of private ISM firms contrasts with that of telecom operators, all owned and overseen by central regulators. In contrast, private ISM firms are not governed as state assets, which explains that Beijing has no formal rights to intervene in their decision-making. As legal entities, private ISM firms nominate and change firm leaders, manage their assets and debts on their own.

Conditional Symbiotic Relations between Local Regulators and ISM Firms

Beijing can certainly influence the strategic orientations of private ISM firms by indicating in the FYPs the priority sectors it aims to develop. The sectors that Beijing supports in priority are supported by local regulators, too. However, the priorities fixed by local governments have more tangible influence on the day-to-day operations of private firms. The reason is that local governments can elaborate locality-specific development plans and make sectorial regulations based on local realities, thus producing immediate impact on firms. For instance, the *Regulation of Shenzhen Special Economic Zone on the Improvement of the Business Environment*, taking effect on 1 January 2021, presents local firms with the opportunities or constraints that do not exist elsewhere. Certain provisions in this regulation do not exist in national regulations either, including the "delisting system" (*Chuming zhidu*) and "*ex officio* cancellation system" (*Yizhiquan zhuxiao zhidu*). Firms in Shenzhen must above all comply with local regulations written in a more precise and workable way than national ones. In addition, local regulators decide the sectors to develop in priority, the supportive policies to take, and the form of governmental support.

Private ISM firms are registered with municipal market regulators (for instance Baidu, Meituan, and Bytedance are registered in Beijing, Tencent in Shenzhen, and Alibaba in Hangzhou). Although local authorities must respect national policies, they enjoy important discretion on policy implementation. For instance, in 2019, Shanghai dedicated more than 11 billion yuan to support strategic emergent industries and integrated circuits, including among others 5G, artificial intelligence, and industrial internet. In contrast, in the same year, Gansu province spent 20.76 billion yuan on poverty eradication and 11.45 billion yuan on ecological preservation, two big concerns of the province. The spending on these two items was second only to welfare spending. Since the development of emergent industries with strategic importance was not the priority of Gansu, spending on this policy field was even not mentioned in the 2019 report on the execution of the budget. The contrast between Shanghai and Gansu shows that depending on whether the ISM industry is perceived as a priority by local governments, the latter use different approaches to regulate the ISM firms registered in their jurisdiction.

The differentiated local treatment of the ISM industry has caused a highly uneven geographical distribution of the leading ISM firms in China. The provinces where the ISM industry drives economic growth are readier to support ISM firms than the provinces relying on other industries. Generous governmental support in return nurtures the growth of new ISM firms. Since the provinces in the east of China are abler to provide favourable conditions to ISM firms and that they have developed higher reliance on the high-tech to drive local development, they perform quite well in boosting the ISM sector. According to the report of *Forward* in 2019, a Shenzhen-based consulting firm, 82 of the top 100 Chinese ISM firms are located to the east of China. Four first-tier cities, Beijing, Shanghai, Guangzhou, and Shenzhen, concentrate 69 of them. Beijing alone is home to 38 firms.[2] Given the decentralised economic regulation of the ISM sector in China, the contributions of local governments to the formation of national digital giants are as important as, if not more than, the central government.

The uneven distribution of Chinese ISM firms is plotted on Figs. 6.1 and 6.2. The two pies respectively capture the distribution of ISM firms according to regions and city-tiers.

The extent to which local governments rely on the internet sector to realise local objectives influences the margins of manoeuvring of local ISM firms. If the local government considers the internet sector as a driver of the economic growth in its jurisdiction, its regulation of ISM firms will be intensive. Simultaneously, the reliance on ISM firms pushes local governments to create a favourable policy environment and to make concessions to keep promising firms within their jurisdiction. Instead, localities relying on other sectors than the ISM tend to regulate less ISM firms and to grant less policy support accordingly. Therefore, the paradox with the local regulation of the ISM industry is that the strategic importance of the internet sector to local authorities augurs both intensive regulation and intensive policy support. The uneven regional distribution of the top 100 Chinese ISM firms in 2019 illustrates the paradox: although the regulation in the eastern provinces is much stricter than in the western and

[2] Forward (2021). "Analytical Report on the Development Prospects of the Chinese Internet Industry and Its Leading Firms (2020–2025)". One resume of the report is available at https://dy.163.com/article/FRSD1F0G0519811T.html.

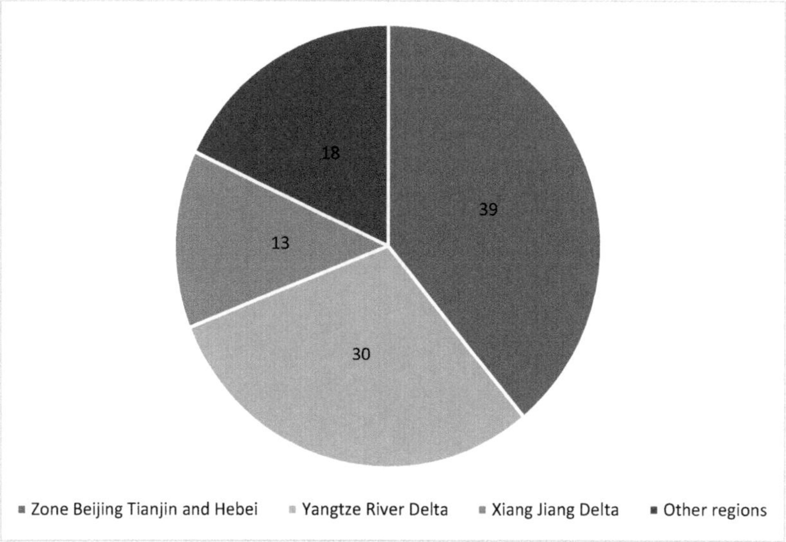

Fig. 6.1 Distribution of the top 100 internet firms in 2019 based on regions (%) (*Source* Compiled by the author based on the data of Forward consulting firm in 2019)

central provinces, firms are clearly concentrated to the east. The strategic importance of the ISM sector positively correlates with the regulation intensity and the policy support at the local level.

"Competitive Liberalisation" Weakens the Control of Beijing

The local ownership of private ISM firms is a logical outcome of the economic decentralisation, for local governments rely on private businesses to achieve local objectives. The second outcome of the economic decentralisation is "competitive liberalisation", coined by Yang Dali (1997) in his seminal work on the dynamics and the consequences of the decentralised governance in the post-Mao China. The importance of discussing competitive liberalization is not only because it crucially influences the Chinese economic catch-up in the 1980s and the rise of the ISM industry in the late 1990s. It is also because this concept unveils the

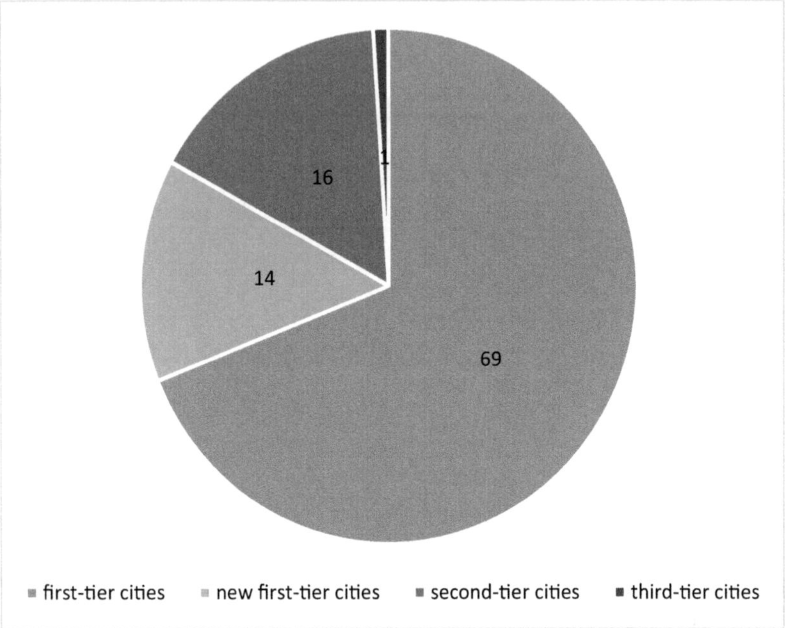

Fig. 6.2 Distribution of the top 100 internet firms in 2019 based on city-tiers (%) (*Source* The same as in Fig. 6.1. In this pie, first-tier cities include Beijing, Shanghai, Shenzhen, and Guangzhou; new first-tier cities include Chongqing, Chengdu, Hangzhou, and Wuhan; second-tier cities include Fuzhou, Xiamen, Hefei, and Guiyang; the third-tier city is Wuhu, prefecture-level city of Anhui province)

factors that shorten the power reach of Beijing in the economic regulation of the ISM industry.

"Competitive liberalisation" is a process in which the liberalisation of one locality and the profits thus generated incentivise other localities to liberalise in their turn. As Yang explains:

> If one area (such as a special economic zone) in a country adopts preferential policies towards investors ahead of the other areas of the country, then this one area will probably get a disproportionately large share of the total investment and remain ahead for a long period to come. Investors are simply unlikely to pack up their factories and move when another area

> catches up in liberalisation. Because of the dynamics described here, localities are expected to emulate and compete against each other to liberalise and to offer more preferential policies as long as they are politically and ideologically allowed to do so. The competition to offer ever more attractive terms will tend to make the entire country a giant special economic zone, making no area more attractive than the others in policy terms... (Yang, 1997: 46)

The quote shows that peer pressure between localities stands at the heart of competitive liberalisation in which localities compete to make their business environment more liberal than their neighbours, such that they are abler to attract investors and businesses. Beyond the pricing and the investment studied by Dali Yang (1997), competitive liberalisation also matters in the rise of the ISM industry, for localities have competed to roll out preferential policies for attracting talents, investors, businesses, and technologies. To realise high-speed economic growth, local governments put the pursuit of capital above all (Yin and Feng, 2014). Since the fiscal reform of 1994, local governments have to transfer more than 50% of their fiscal income to Beijing (Xiao, 2008; Yin and Feng, 2014), which reduces their sources of income. To generate extra income for local use, local governments have been ready to offer preferential policies for wooing promising private business. McGregor describes how local governments are capable to do so:

> The officials brought a lot to the table in their dealings with business. They could open and close their factories at will at a time when the businesses were highly profitable. They could calibrate the level of pollution allowed as well, and how and when it could be discharged. They controlled the police, who could keep any protesters at bay, or not. They were in charge of the judges in the local courts. They might even have been able to help arrange loans through the local bank... (McGregor, 2010: 93–94).

Although decentralisation implies more delegation of competences from Beijing to localities, it has been producing considerable benefits for Beijing, too. For instance, if competitive liberalisation increases the competitiveness of local economies, it will drive the overall competitiveness of the Chinese economy, thus consolidating the performance-based legitimacy of the party-state. Therefore, Beijing has been encouraging decentralisation rather than hindering it. In China, economic policies made at the central level are mostly formulated in vague terms. That

Beijing refrains from instructing in detail how local authorities proceed allows them quite large margins of manoeuvring. Local governments work out local versions of policies after studying national policies.

Economic decentralisation and competitive liberalisation in the ISM sector can also embarrass Beijing in two ways. First, local governments are jealous of their neighbours if the latter obtain better deals from Beijing. Yang reported that after Beijing turned its attention from interior provinces to coastal provinces in the 1980s, "leaders of interior provinces have engaged in policy *panbi*, lobbying central leaders for the extension of preferential policies to the interior" (1997: 43). Second, the pressure of outperforming their neighbours incentivises local governments to resist unfavourable policies from Beijing, leading to noncompliance from local regulators (McGregor, 2010: 155–167). Using the example of the regulation of the telecommunication sector, Hsueh argued that "local and regional powerbrokers are thus strengthened by formal deregulation and increasingly operate without much reference to formal and constitutional rules" (Hsueh, 2011: 15). Moreover, local leaders defending local interests have greater chances to obtain a promotion than those who sacrifice local interests to please Beijing (McGregor, 2010). Therefore, it is no surprise that local officials, in a "race-to-the-bottom" manner (Hsueh, 2011: 15) fiercely compete among them, even though such competition sometimes pushes them to operate beyond the parameters fixed by Beijing.

Interjurisdictional competition makes it important for local governments to engineer creative methods to mitigate the negative effects of the unfavourable regulations from Beijing. Local governments either delay implementing Beijing's policies or squarely disobey (McGregor, 2010), the objective being to ensure that their jurisdiction is not more disadvantaged than other jurisdictions. Refusing to be outdone by their neighbours, local governments tailor Beijing's policies by inserting local concerns and skipping irrelevant content. The outcome is that when the policy of Beijing is finally implemented at the local level, it often becomes something different from its initial version. The boldness of local authorities originates from decentralisation. To reap the benefits of competitive liberalisation, Beijing must make reliable commitments by showing that local leaders will not be admonished for being bold, creative, or deviant in their day-to-day management.

Yin and Feng (2014) analyses how the "race-to-the-bottom" competition at the local level has weakened the control of Beijing:

> To attract capital, in particular foreign capital, local governments have elaborated on their own different types of preferential fiscal policies, including the reduction and exemption of local income taxes, partial return of the value-added taxes to firms. They even resort to financial subsidies to firms in order to avoid falling into troubles during the inspections by central government officials...To reduce the cost of the investors, many local governments decide on their own to reduce the fees of the land use...-Some localities even loosen the standards of environmental protection and labour protection. (Yin and Feng, 2014)

Competitive Liberalisation Causes Disobedience of Local Governments

The pressure for surviving competitive liberalisation has generated local regulators' noncompliance or "feigned compliance". Based on Yin and Feng, I study the competition of 4 neighbouring new first-tier cities (Xi'an, Wuhan, Chengdu, and Chongqing) for attracting ISM firms, and how their competition violates the instructions of Beijing. The rationale of the case selection is twofold. First, geographically speaking, the 4 cities are located to the west and the centre of China. The provinces where they are located (Shanxi, Hubei, Sichuan, and Chongqing) are neighbours. Therefore, it is more direct to observe the effects of competitive liberalisation on local governments' compliance and noncompliance. Second, in recent years, leading ISM firms have been creating their branches and regional headquarters in these four cities, for they provide high-quality human resources and cheaper land.

To have the same criteria of comparison, I focus on the fiscal policies that the four cities have been applying to local ISM firms since 2012. ISM firms generally pay more than 10 types of taxations, including value-added tax, income tax, city construction and maintenance, education, consumption, natural resources, real estate, land use, stamp duty, etc. While Beijing and local governments share the taxation on the value-added tax (VAT) and the income tax, the remaining items count as local fiscal income. For instance, article 4 of the *Companies Income Tax Law (Qiye suodeshui fa)* stipulates that the rate of the income tax is 25%. However, since the great majority of the ISM firms are granted the status of High-Tech Enterprises (*Gaoxin jishu qiye*), the income tax rate for these firms is actually 15% (article 28). The taxation rate of the VAT is 6%. For the VAT, Beijing and local governments respectively take 75% and 25%. For the income taxation, while the 1994 fiscal reform stipulates that the income taxations of local firms belong to local governments, the situation changed in 2002 and 2003. To support economically backward western regions, the new rules in 2002 required that the central and local governments respectively take 50% of the income taxation. The 2003 rules increase the percentage of the central government to 60%. Both the *Income Tax Law* (article 20) and the *Temporary Provisions on the VAT* (article 2) stipulate that it is the State Council that can modify the tax rate and the taxable items. The *Income Tax Law* also clarifies that the State Council, rather than local governments, decides preferential fiscal policies (article 35).

However, local government do not always respect these rules. Since preferential fiscal policies produce immediate effects, local governments often make promises of fiscal reductions or exemptions to local businesses, even though they have no competence to do so (Ye, 2017). The Chinese National Audit, a ministerial-level institution, publicised an audit report in 2012, showing that from 2008 to 2011, 53 of the 54 surveyed counties had released 221 policy documents which contained preferential fiscal policies that violated the instructions of Beijing. It also showed that 53 of the 54 counties voluntarily refrained from collecting 7043 billion yuan of taxations. To curb this trend, the State Council, the Ministry of Finance, and the State Taxation Administration, had urged localities to toe the line several times in 1994, 1998, 2000, 2014, and 2015, without achieving tangible effects (Ye, 2017). Although the attitude of Beijing has always been clear, the enthusiasm of localities in mobilising preferential fiscal policies has never receded. Moreover, since local governments have discretion over local economic policies, it is *de facto* impossible to completely prevent them from abusing their power. In December 2014, the State Council stipulated that except in ethnic autonomous regions, "local governments are forbidden to elaborate preferential fiscal policies; without the approval from the State Council, any law, any statute, any regulation, any development plan or any regional policy should not include detailed preferential fiscal policies". Nevertheless, local governments have developed innovative ways to disobey, such that they continue to attract profitable businesses by promising preferential fiscal policies.

Local governments compete in mobilising both fiscal and non-fiscal policies to reduce the financial burden of local ISM firms. Table 6.1 resumes the noncompliance of the four municipal governments with the central government in their wooing of local ISM firms. Interestingly enough, the way in which they support their ISM firms show certain similarities, implying policy imitation and "competitive liberalisation" at the local level. For instance, all the four cities grant significant subsidies to reduce firms' expenditure in land use.

Table 6.1 shows that Chengdu, Chongqing, Wuhan, and Xi'an have been engaged into fiscal liberalisation since 2012 against the instructions of Beijing. Although the fiscal regulations explicitly mandate the State Council and the central-level fiscal authorities to adjust the tax rate and preferential fiscal policies, municipalities have not hesitated to violate the rules or work out innovative circumvention strategies (e.g., cheap rent of public land, subsidies to R&D) that produce the same effects as preferential fiscal policies. Interestingly, very few local governments have been summoned thus far, for the central government takes a significant part of the revenues created at the local level. The comparison of the preferential fiscal policies in the four cities shows the party-state's dilemma. Greater intervention from Beijing in the local regulation could certainly increase its supervision over localities. However, when Beijing reins in too far, it risks reducing the enthusiasm of local governments in policy innovation. In contrast, when Beijing grants certain autonomy to localities, the latter unleash higher dynamism in elaborating locality-adapted supportive policies, which in return increases the economic and technological clout of the whole country. Paradoxically, Beijing is better-off when it controls its desire of control.

"Fragmented Authoritarianism" Opens Opportunities

"Fragmented authoritarianism" (Oskenberg and Lieberthal, 1988; Mertha, 2009) in China invalidates the image of the central government as a unified decision-maker. Fragmentation is reflected in two pairs of conflicts: (1) conflicts between central regulators and local regulators, and (2) bureaucratic fragmentation. Although the two pairs of conflicts concern different actors, both crucially influence the configuration of opportunities for ISM firms to enlarge their margins of manoeuvring. The conflicts between central and local regulators enable firms to ally themselves with local regulators to seek favourable policies from Beijing. Bureaucratic fragmentation within which regulatory agencies share overlapped competences (Miller, 2018; Pearson, 2005) enables firms to navigate between the rules of different agencies and to implement

Table 6.1 Noncompliance of local governments

City	*Noncompliance*	*Justification*
Chengdu	• Subsidies to the rent of local ISM firms for 3 years: 40 yuan per square metre for the first year and 20 yuan per square metre for the second and the third year. This policy is applied to the ISM firms that rent the land managed by the local government; • Subsidies to well-known entrepreneurs who create new ISM firms in the high-tech zone of Chengdu: free use of the public land for 5 years	• Rent public land at a cheap price; • Rent public land at a cheap price;
Chongqing	• Return 50% of the income tax paid by the ISM firms for three years; • Subsidies of 5 million yuan to leading Chinese or foreign ISM firms that choose Chongqing as their regional headquarter; • Subsidies to local firms for 2–5 years: reduction or exoneration of the rent for ISM firms that use public land	• Tax return is forbidden; • Subsidies with the same effect as preferential fiscal policies; • Rent public land at a cheap price;
Wuhan	• Subsidies to the R&D activities of ISM firms: subsidies reach 6% of firms' expenditure in R&D; • Allow firms to rent public land in the first place and to use it for free in the second place (*xian zu hou rang*); • Wuhan returns all income taxation that Big Data firms have paid for three years; for Big Data firms having operated in Wuhan between 3 and 5 years, Wuhan returns 50% of the income taxations; • For influential firms, Wuhan adopts "one issue, one discussion" (*yi shi yi yi*) and "one company, one policy" (*yi qi yi ce*)	• Subsidies with the same effects as preferential fiscal policies; • Rent public land at a cheap price; • Local governments cannot decide tax return policies; • Local governments cannot adopt "one issue, one discussion" and "one company, one policy"

City	*Noncompliance*	*Justification*
Xi'an	• Xi'an government applies "one issue, one discussion" to global top 500 firms, well-known multinationals, and big Chinese firms that create R&D centres in Xi'an; • Exemption of the sales tax for the revenues coming from technological consultations, technology transfer, R&D • Influential firms having been in Xi'an for more than 10 years and having paid more than 10 million yuan of income tax, VAT, and business tax, Xi'an returns 30% of the paid taxations to firms;	• Local governments cannot adopt "one issue, one discussion"; • Local governments cannot decide the exemption of taxations; • Local government cannot decide tax return policies;

the least constraining directives. Similar to the party-state's predicament between the political interest of preserving the regime's survival and the economic interest of promoting economic upscaling, fragmented authoritarianism is part of the party-state's structural constraints. In return, it turns out to be a structural opportunity for private ISM firms to augment their level of autonomy against the Chinese authoritarian backdrop.

While the dynamic of bureaucratic/horizontal fragmentation for accruing the autonomy of Chinese private ISM firms will be analysed in the next section, this section explores how geographical/vertical fragmentation of the Chinese political system has enabled private ISM firms to seek greater margins of manoeuvring. The autonomy-enabling opportunity of vertical authoritarian fragmentation is made possible by the alliance between private ISM firms and local regulators, alliance contingent on the economic contributions that firms are able to bring. The opportunity for private ISM firms to seek room to manoeuvre—contingent alliance between firms and local regulators—is instantiated in this section by two contrasting cases: generous policy support to ISM firms in the eastern provinces of China and poor governmental backing of ISM firms in the west of the country.

Contingent Alliance between Private ISM Firms and Local Governments

The party-state's use of opposite regulation approaches to the ISM industry—centralised political regulation and decentralised economic regulation—illustrates the dynamics of the double-bind logic in shaping the configuration of the Chinese political economy. The ISM industry is one of the very few sectors that represent both crucial economic and political interests to the party-state. The policy implication of the particularity of the ISM industry is that the party-state must constantly distinguish what belongs to the political regulation from what belongs to the economic regulation and regulate them differently. For the economic regulation, local regulators are the principal interlocutors of private firms. The proximity between firms and local regulators, and, in particular, the reliance of the former on the latter, in return influences the bargaining power of ISM firms vis-à-vis local regulators.

Since China created the system of regular evaluation of local cadres in 1979 (Chen and Zhu, 2019: 3), quantifiable economic indicators became the crucial basis for deciding whether local officials could obtain professional promotions (Zhan, 2015). The choice of quantifiable economic indicators was due to two factors: (1) the imperative of realising economic catch-up with advanced economies, and (2) lack of sophisticated evaluation methods (Zhan, 2015). In addition, the similarity in local economic structures created a comparable basis to evaluate local officials (Qian and Xu, 1993; Maskin et al., 2000). Although Beijing had attempted several times to replace the GDP-centred performance evaluation with human-centred evaluation criteria (Zhang et al., 2020), the regions with better economic performance continued to see their cadres be more likely to obtain a promotion than the regions failing to do so (Chen and Zhu, 2019; Qiao, 2013; Li and Zhou, 2005). At the county level, each time the economic growth rate advanced by one rank, the county leader was 1.3% more likely to be promoted (Chen and Zhu, 2019). At the provincial level, each 6% increase in the economic growth rate allowed the provincial governor to be 33% more likely to obtain a higher post (Li and Zhou, 2005).

Economic liberalisation has turned private firms into the leitmotiv of local economic growth, and accordingly, one game changer in the professional career of local officials. Therefore, it is not surprising that local officials have been more than cooperative with influential firms in their jurisdiction. On the contrary, their attitude towards the firms with weak contributions to local development is visibly less cooperative. One entrepreneur I interviewed qualifies local officials as the "waiter" of influential firms,[3] highlighting that they are ready to use different sets of standards to deal with big and small firms. Government officials in competing jurisdictions try hard to convince promising private firms to move to their own fief. Interjurisdictional competition between local governments increases the bargaining power of entrepreneurs. One passage in *Chinese Supermen* (*Zhongguo niuren*) (Liu, 2014), recounted how regional leaders in Shanghai and Guangdong demonstrated unbounded friendliness towards Jack Ma and lobbied him to operate in their respective jurisdictions:

> From 2002 to 2007, during Xi's reign in Zhejiang, Alibaba grew up into a powerful firm in Hangzhou... In 2007, Xi Jinping was muted to serve as the Party Secretary of Shanghai. During his visit to Alibaba in Hangzhou,

[3] Interview B-18/12/2019.

> he asked Ma: 'Is it possible for you to come to Shanghai'? At the beginning of 2008, Yu Zhengsheng was the newly appointed Party Secretary of Shanghai. He met Ma and appreciated him much. Subsequently, Wang Yang, Party Secretary of Guangdong province, came to visit Alibaba and invited Ma to develop his company in Guangdong.

Economic contributions of Alibaba to the local development explained the warm invitations from the party chiefs of the richest municipalities and provinces in China. In April 2016, Li Qiang, then Governor of Zhejiang province, highly praised the contributions of Alibaba to the fiscal income of the province:

> In 2014, Alibaba paid in total 10.9 billion yuan of taxes. On average it paid more than 30 million yuan on each day. This is the only internet company in China whose payment of taxes exceeds 10 billion yuan: its contributions are immense.[4]

Under the current fiscal system, Zhejiang province could not take every penny of the 10.9 billion yuan. Fiscal offices of the Yuhang district of Hangzhou, headquarter of Alibaba, explained that it could only take 4.38 of the 10.9 billion yuan, with the remaining part going to Beijing. Given the contributions of private ISM firms to drive local economic growth, the failure to retain successful firms could cause disputes among local officials of the same jurisdiction. Alibaba was initially created in Shanghai. In 1999, Jack Ma moved Alibaba to Hangzhou. In one speech, he explained why he made this choice: "In the past, I planned to put the headquarter of Alibaba in Shanghai. I rented a very big office in Huaihai Road. The result was that one year later, I felt very tired psychologically, and I had big difficulties in recruiting people".[5] In 2008, when Yu Zhengsheng replaced Xi Jinping as the new Party Secretary of Shanghai, he required his men to reflect on why Shanghai failed to keep Alibaba. Disappointed that none of the three Chinese tech giants, Tencent, Baidu, and Alibaba, was headquartered in Shanghai, Yu Zhengsheng called on Shanghai's officials to find out the reasons.

[4] "Ali Imitated the Mother of Menfucius Who Changed Address Three Times, Why It Does Not Choose Beijing or Shanghai as Its Headquarter," sohu.com, 22 September 2016, https://www.sohu.com/a/114845861_157451.

[5] Ibid.

Aware of its strategic economic value to Zhejiang province, Alibaba acquired strong bargaining power vis-à-vis local regulators. For instance, it was able to convince the Governor of Zhejiang to lobby the central government on its behalf.

Before the adoption of the E-Commerce Law (ECL), individual merchants on C2C platforms did not have to register themselves with municipal administrations for industry and commerce (MAIC). In 2010, the SAIC, predecessor of the SAMR, released the *Provisory Measures for Regulating Online Transactions of Goods and Services (wangluo shangpin jiaoyi ji youguan fuwu xingwei guanli zanxing banfa)*. Since this regulation stipulated that individuals selling goods or services on E-commerce platforms should get registered only if they satisfied certain conditions, it *de facto* exempted individual merchants from the obligation of administrative registration. With this regulation, to open up a shop on Taobao, it sufficed for individual merchants to provide their personal information to platforms. This victory was achieved thanks to Jack Ma's lobbying of Lü Zushan, Governor of Zhejiang from January 2003 to August 2011. Lü recalled how Jack Ma asked him for help:

> In the growth of Alibaba, I clearly remember two incidents. One incident is that one time Jack Ma came to me, grumbling that the local department of industry and commerce required individual merchants on Taobao to get an administrative registration... It was said that the SAIC planned to make online shops get registered, for this, it asked provincial administrations for industry and commerce (PAIC) to prepare a document...After knowing this, Jack Ma became anxious. He came to me.
>
> When I went to work the next day...I asked the PAIC to maintain the existing measures and to transmit our opinions to the SAIC. Later, after listening to our opinions, the SAIC promulgated the *Provisory Measures for Regulating Online Transactions of Goods and Services*, stipulating that it sufficed that shops ran by individuals (on C2C platforms) provided real identity information... (Chai, 2018)

Governor Lü offered another big service to Alibaba, for his lobbying of the central government helped to exempt individual merchants on Taobao from the obligation of paying taxes. The exemption helped retain individual merchants on Taobao and drive up Alibaba's business. In June 2011, Taobao hosted already 40 000 merchants, becoming the largest B2C and the C2C platform in China. Governor Lü recalled how he had helped Alibaba:

> The second incident is that I accompanied Prime Minister Wen Jiabao and several ministers to visit Alibaba. After the visit, some leaders said that it was so cool, other leaders said that individual merchants did not pay taxes, they hoped to make them pay taxes. Immediately, I 'disputed' with them. I said that a great number of online shops were opened by people in the lowest social stratus...In addition, many laid-off workers became re-employed through opening an online shop. Many farmers found it difficult to sell their produces in the past, now they could sell agricultural products very easily. Alibaba solved so many employment problems that the government should have been in charge, it reduced so much burden for the government, why didn't encourage it? If we harmed online shops (by asking them to pay taxations), who could solve the employment problem of so many people? (Chai, 2018)

The two incidents that Lü mentioned reveal that influential private ISM firms can ally themselves with local officials to push back Beijing's unfavourable regulations. The contributions of private firms to local development provide local regulators with an important incentive to satisfy firms' needs. When the satisfaction of local firms' needs implies to wrestle with Beijing, local regulators like Lü Zushan don't hesitate to lobby the central government on behalf of local firms. As illustrated above, Chinese fragmented authoritarianism on the geographical dimension opens a window of opportunities for private firms. The symbiosis between local regulators and private firms in which the former provides business-friendly policies to the latter and the latter drives up the socio-economic performance of the former incentivises both actors to resist unfavourable policies of the central government. Since private firms and local regulators depend on each other, the former can well play out the leverage they hold upon the latter to augment their margins of manoeuvring. The window of opportunities opened by vertical fragmentation makes state regulation in China somewhat deviate from existing regulation theories. While the latter generally pit firms against regulators, vertical fragmentation in China shows that conflicts between regulators at different administrative levels can change regulator-firm power relations, pushing one group of regulators to side with firms to resist another group of regulators.

Like Alibaba, other ISM giants also benefit from the protection of local regulators. In other provinces than Zhejiang, in other municipalities than Hangzhou, governors and mayors negotiate with higher-level regulators on behalf of their ISM firms. Even at the lowest administrative level—villages and districts—the protectionism of local governments is unambiguous. It is well illustrated by the extremely high rate of success that ISM firms have obtained for the lawsuits judged by the courts in their home jurisdiction. Chen Hangping, Associate Professor at the Law School of Tsinghua University, comments that "the chances of tech giants of winning lawsuits in their home courts tend to be very high. Besides the 'Must-Be-Victorious Quest of Nanshan' (namely Tencent), there are also 'The Invincible of Longgang' (namely Huawei), Futian Hooligan Head (namely Pingan). Alibaba, JD.com, ByteDance, and many other firms are no exceptions..." (Chen, 2021).

Tencent is a particularly telling case. Registered in the Nanshan district of Shenzhen, Tencent is nicknamed "Must-Be-Victorious Quest of Shanshan" (*Nanshan bisheng ke*) by Chinese netizens. The nickname stems from the fact that each time a lawsuit pits Tencent against the internet firms in other localities, Tencent overwhelmingly wins as long as the court of Nanshan handles the case. For this reason, when involved into a lawsuit, Tencent often requires that the case be transferred to the Nanshan district. In September 2019, ByteDance took Tencent to the Intermediary Court of Fuzhou, capital city of Fujian province, accusing Tencent of unfair competition and requiring a compensation of 90 million yuan. Tencent contested that Fuzhou intermediary court had no jurisdiction over the case and required that the case be transferred to Nanshan of Shenzhen. After the intermediary court of Fuzhou agreed, Bytedance dropped the lawsuit in March 2021, knowing that it could hardly win over Tencent after the transfer.

On the contrary, when the lawsuits are judged outside Shenzhen, Tencent is much less likely to defeat the other party. In June 2020, when Tencent lost the lawsuit against Qingdao Ruanmei, a software and IT service company registered in Qingdao, Shandong province, Chinese netizens made fairly satiric remarks. Since the case was judged in Shandong, the home province of Qingdao Ruanmei, one article published in Sohu was titled "Tencent, the Must-Be-Victorious Quest of Nanshan, fails again: it is so hard to win a lawsuit outside the Nanshan District".[6]

[6] "Tencent, Which the Folks Call the Must-Be-Victorious Quest of Nanshan, Fails Again: It Is So Hard to Win a Lawsuit Outside the Nanshan District," sohu.com, 10 June 2020,
https://m.k.sohu.com/d/457628622?channelId=4&page=3.

The capacity of ISM firms to seek compromises from local regulators stems from structural reasons specific to the Chinese political economy. With fragmented authoritarianism, localities have discretionary power on economic policies. Local regulators mollycoddle influential ISM firms in their jurisdiction, for they depend on firms to outperform their neighbours. Being relied upon by local governments constitutes one of the sources of firms' bargaining power. The cases of Alibaba and Tencent show the potential of firm-regulator alliances to expand the manoeuvring room of private firms. However, the support of local regulators is not unconditional. It depends on how much the ISM sector helps to realise local economic objectives. In the localities where the ISM industry drives the local economy, local regulators are ready to support firms because such support can handsomely pay off. In contrast, in the localities where the ISM industry plays a minor role in driving local development, local regulators deploy significantly less endeavours to defend ISM firms. I turn to this point immediately.

Contrasting Case of Lanzhou, Gansu Province

The previous section corroborates Yeo (2020). Her research on the regulation of the telecommunication and the automobile industries in China finds that firms' level of ownership closely relates to the way in which they are regulated. When firms are registered at the central level—in the case of central SOEs (*yang qi*)—the central government exerts formal oversight over them. In contrast, when firms are registered at the local level—in the case of local SOEs and private firms—local regulators enjoy formal oversights over them. Since all private ISM firms in China are locally registered, local regulators decide the intensity and the forms of the regulation. However, one point that is missing in Yeo (2020) is the implications of the strategic importance of one industry for local regulators. Although Hsueh (2011) has defined the "strategic value" and analysed the impact of one industry's strategic value on the choice of regulatory approaches, her research mainly focuses on how the strategic value is perceived by central regulators, not local regulators. Therefore, based on Yeo (2020) and Hsueh (2011), I argue that different perceptions of local regulators towards the strategic value of the same industry explain their adoption of different regulatory approaches. Extending the analysis of the previous section, this section investigates the ISM regulation in Gansu province where the ISM industry is not the mainstay of the local economy.

Since the competences of local regulators are concentrated in the economic domain, it is conceptually more accurate to talk about local regulators' perception of the "strategic economic value" of the ISM industry, rather than the "strategic value" as a whole. I argue that the more the ISM industry contributes to local economic upscaling, the more the local regulation of the ISM sector takes on a supportive flavour. It can be inferred that the ISM firms registered in the localities, whose economic upscaling depends on other industries than the ISM sector, have less bargaining power. As a result, it is difficult for them to ally with local regulators to lobby Beijing, because such alliances produce less benefits to local regulators than alliances with the firms that significantly contribute to local economic growth. Therefore, local governments facing diverse priorities differ from each other in their regulation of private ISM firms. Their perceptions of the strategic economic value of the ISM industry in return influence the capacity of private ISM firms to seek room to manoeuvre.

Having shown how the economic importance of Alibaba and Tencent has enabled them to seek protection from Zhejiang and Guangdong regulators, I now turn to the contrasting case of Gansu province, located to the North-West of China. While the ISM industry is the pillar for Zhejiang and Guangdong, it is not a priority for Gansu. Since the first Five-Year Plan in 1953, heavy industry has dominated the local economy of Gansu. Economic liberalisation in the 1980s and subsequent industrial adjustments, enshrined into the Great Western Development strategy in 1999, did not change the reliance of Gansu on the heavy industry. Until today, equipment manufacturing and colour metal are still the two most important economic sectors in Gansu. From 2010 to 2014, the heavy industry has created between 83.65% and 87.43% of the local economy, largely dwarfing technological innovations (below 50%). Although the contribution rate achieves 52.8% in 2018, it remains below the national average of 58.5%, announced by Wang Zhigang, Minister of Science and Technology in 2019. After the global financial crisis in 2008, the State Council released its decision on developing the emergent industries with strategic importance (EISI) on 10 October 2010 and officially fixed seven EISI. Based on local realities, provinces could decide which sectors they wished to develop in priority. The ISM industry, enshrined into the EISI, did not figure on the list of the priority sectors of Gansu. In its 13th FYP on the development of the EISI, issued in August 2016, Gansu government ranked the seven EISI according to their respective importance to local economy. In the section titled "Development Objectives", the provincial government said:

> Strengthen our competitive industries. In the strategic emerging industries, we will build 50 industrial chains... By 2020, new energy, new materials, advanced equipment, and intelligent manufacturing will become the **pillar industries of the province**; biomedicine and information technology industries will become the province's **leading industries**; energy conservation and environmental protection, new coal chemical industry, and modern service industries will become the province's **new economic growth point** (emphasis added by the author).

The ISM industry, included into the category of information technology, was intended to be nurtured as one of the "leading industries" of Gansu, a status much lower than that of "pillar industry". During the 13th FYP period from 2016 to 2020, the priority sectors in Gansu included new energy, new materials, advanced equipment, and intelligent manufacturing. The choice of these sectors corresponded to the traditional emphasis of the province on the heavy industry it had been good at. The reliance on the heavy industry was due to the under-development of the ISM firms in Gansu. Since the Internet Society of China (ISC) started publishing the list of the top 100 internet firms of China in 2013, none of the listed firms has come from Gansu (Fig. 6.3). During the ten years from 2013 to 2022, while certain provinces in the West of China have been able to nurture influential ISM firms, provinces like Gansu, Inner Mongolia, Qinghai, Tibet, Xinjiang, and Shanxi were unsuccessful in creating their ISM champions. That the ISM industry is not the mainstay of the local economy did not incentivise local officials to support this sector. The vicious circle of "low strategic economic value, low local support" widens the gap between the provinces relying on the ISM and the provinces not relying on it.

I collected the ten lists of the top 100 internet firms compiled by the ISC and the MIIT from 2013 to 2022, and built a database. Some firms have been listed several times in this period. For instance, Baidu, Tencent, Alibaba, Sohu, Sina, and many other firms have been listed 10 times among the top 100 internet firms of China. To ensure the exhaustivity of the data, I eliminated redundant information. After this, the database contained 350 internet firms which have been listed among the top 100

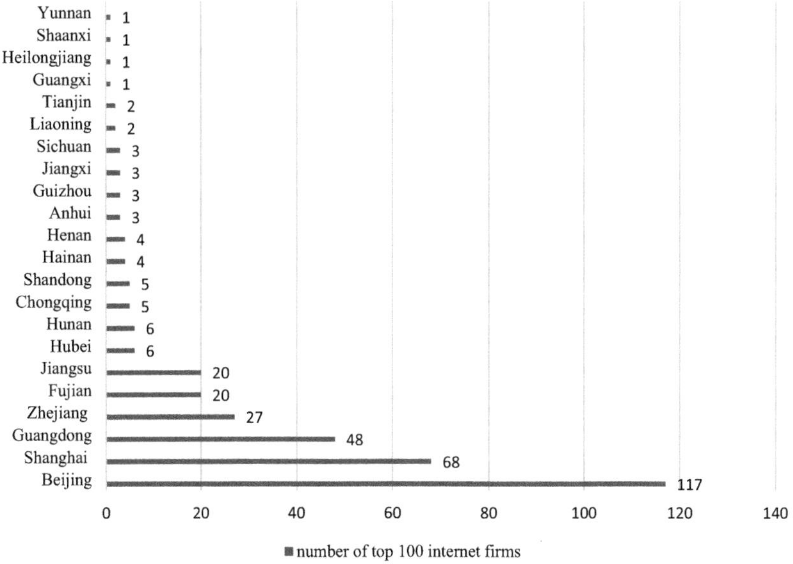

Fig. 6.3 Geographical distribution of the top 100 ISM firms 2013–2022 2023 (*Source* Compiled by the author based on the annual reports of the Internet Society of China and the Information Centre of the MIIT on the top 100 internet firms in China from 2013 to 2022. See Ma (2023); ***Unit: province and provincial-level municipalities)

internet firms at least once from 2013 to 2022. I classified these firms into different clusters based on the provinces and provincial-level municipalities where they were registered. The distribution of the firms is plotted in Fig. 6.3. It shows that six provinces and provincial-level municipalities, including Beijing, Shanghai, Guangdong, Zhejiang, Fujian, and Jiangsu gather 300 of the 350 (85.7%) strongest internet firms in China. All of them are geographically located to the East of China. Provinces like Gansu have had no internet firm strong enough to be listed among the top 100 in the past decade.

In contrast with the Governor of Zhejiang, Mr Lü Zushan, who headlong lobbied the Prime Minister and central bankers on behalf of Alibaba, successive Governors in Gansu have rarely made the headlines for speaking up for their ISM firms. Since the ISM sector is not the pillar of the Gansu economy, it is less profitable for local regulators to lobby for the interests of private ISM firms than for the firms in the sectors of energy, colour metal, or equipment manufacturing. In addition, as shown in Fig. 6.3, by 2020, there are no influential ISM firms in Gansu that local regulators can speak for. The star companies of Gansu operate in other sectors than in the ISM.

That the ISM industry is not perceived as having "strategic economic value" in Gansu lowers their bargaining power in dealing with local regulators. Similarly, the absence of the strategic economic value *de facto* shrinks the manoeuvring room of private firms for defending corporate interests, let alone for shaping policy outcomes. Chinese netizens complain that there is a lack of internet mindset in Gansu, including in its capital city Lanzhou where numerous administrative red tapes prevent the progress of the ISM industry. They compare the operating conditions of the ISM firms in Gansu as "creating soil for a region having no soil", meaning that the internet mindset has to be nurtured by firms little by little, rather than by local governments. One post on the forum Zhihu, viewed by 39063 people by 17 February 2022, points out the backwardness of Gansu in comparison with its neighbours. It regrets that provincial officials of Gansu do not pay sufficient attention to the ISM industry:

> ...In the past, Gansu, Guangxi and Guizhou were the three most backward provinces. Recently, while the other two provinces have performed well, Gansu has become the worst. If the backwardness of Gansu is due to its mountainous geographical conditions, the Yunnan-Guizhou plateau in the South-West is even more rugged than Qinling [of Gansu]. However, Big Data and Cloud Computing have been so well developed there. In recent years, different from the traditional manufacturing industry which depends on geographical and transport conditions, the internet sector has clear potential to be developed. Unluckily, for unknown reasons, local officials have not paid attention to the internet sector...[7]

Gansu officials are more willing to protect local firms operating in the profitable mining and energy sectors, even at the price of violating the

[7] "Based on the Reality of the Province, Which Special Industries Should Gansu Develop to Achieve a Better Development," Zhihu.com, 16 October 2019, https://www.zhihu.com/question/275247595.

instructions of Beijing. On 20 July 2016, the General Office of the CCCCP and the State Council released a notice on the degrading ecological system of Qilian Mountain, located to the West of Gansu. Several provincial officials, including three Vice Governors, were named and shamed. The wording of the notice was extremely harsh. The notice explicitly attributed the degrading of the natural environment of Qilian Mountain, classified as a National Nature Reserve in 1988, to the inaction, power abuse, noncompliance, and deliberate cheating of local officials. However, put aside the anger of the State Council and the CCCCP, it was the desire of pulling up local economic development that Gansu officials tolerated wrongdoing firms and turned a blind eye to their illegal activities. Several facts mentioned in the Notice are worth quoting:

> In September 2018, regarding environmental protection in the National Nature Reserve, the Ministry of Environmental Protection and the National Forestry Administration invited the Department of Forestry of Gansu and the Municipal Government of Zhangye for a talk. Gansu Province did not pay sufficient attention to this talk. The proposed reparation plan **voluntarily omitted 31 mining exploitation projects**...By the end of 2016, the production equipment in 72 places was still not evacuated as required.
>
> The concerned departments of Gansu and certain municipalities selectively and neglectfully implemented the decisions of the CCCCP...After the provincial government put the Xiaquangou Mine of Mayinggou Coal Mine, located in the protected zone, on the list of the to-be-closed mines, the Safety Production Inspection Administration of Gansu **still granted the previous production capacity, and agreed to resume the production (emphasis added by the author).**

It was for economic concerns that the Gansu government tolerated the illegal activities of local mining companies in the National Nature Reserve. Between environmental protection and economic gains, Gansu officials chose the latter. They went as far as to cheat Beijing to defend local firms. In Gansu, the treatment of the mining firms contrasts with the treatment of ISM firms, the proof being that the Gansu government has not been reported violating the laws or statutes for defending their ISM firms. In a nutshell, whether firms are perceived as having strategic economic value for local development essentially explains the way they are treated.

"Bureaucratic Fragmentation" Augments Firms' Room to Manoeuvre

Horizontal authoritarian fragmentation is exploitable by private ISM firms for achieving greater room to manoeuvre. The opportunities opened within horizontal fragmentation work in similar ways to vertical fragmentation: firms exploit the division of regulators, ally with one group of them, and resist the other groups. As will be illustrated below, the differences between the two levels are twofold. First, the group of regulators with which firms create alliance within the two structures is different. With vertical fragmentation, private ISM firms augment their room to manoeuvre by allying with local regulators to resist central regulators. With horizontal fragmentation, regulators that firms ally with or oppose to are often located at the same administrative level and share overlapping competences. Second, the regulatory domains related to the two levels of authoritarian fragmentation are different. As shown above, vertical fragmentation mainly opens autonomy-enabling opportunities for private firms in the economic regulation. On the contrary, mainly concerned with the political regulation, horizontal fragmentation allows private ISM firms to carry boundary-spanning expressions with lower costs.

Horizontal Fragmentation

Horizontal authoritarian fragmentation denotes the phenomenon in which jurisdictions at the same administrative level and which share overlapping competences manage content regulation on their own and lack coordination among themselves. Regulators exercise direct oversight only over the firms registered in their respective jurisdiction. In this context, "because there is no leadership relationship between mass media in one province and the political administrative power in another, there are no qualms on the part of media professionals from one province to pursue sensitive issues in another province" (Tai, 2006: 113). Likewise, media outlets located at higher administrative ranks could report the misdeeds of lower-level officials without getting themselves in trouble. This system is called "(public opinion) supervision by different jurisdictions" (*yidi jiandu*) (Zhao, 2008: 44). Before the State Council cancelled this system in 2005 upon the request of angered local governments, the denunciation of the misbehaviour of the officials in one locality by the media in another locality or a higher-level jurisdiction had "been widely recognised as a positive step toward a more critical media in China" (Zhao, 2008: 44).

It is important to recall that in the late 1990s and the early 2000s, the party-state tried to centralise the political regulation of the media sector by creating large conglomerates in the television, broadcasting, radio, and press sectors. The objective was to "enable party organs to consolidate a powerful economic base through the market mechanism and ensure the better fulfilment of the party's propaganda objectives". However, the efforts to centralise media control largely failed. In the broadcasting sector, for instance, the communication scholar Yuezhi Zhao highlighted:

> In fact, one of the unintended consequences of the recentralisation campaign was further fragmentation...The media conglomerates have been fraught with profound conceptual, structural, and operational challenges. The first issue concerns these conglomerates' ambiguous, hybrid, and indeterminate corporate identity. The national-level China Radio, Television, and Film Group, for example, exists only on paper. Unlike provincial broadcast groups such as those in Guangdong and Zhejiang, or Shanghai's SMC, there has not been any integration of the group's affiliated broadcast operations. (2008: 101)

Several factors explained the resistance to the endeavours of Beijing at recentralisation: inefficient allocation of resources, losses of organisational dynamism, and more fundamentally, lack of sufficient senior positions for high-level managers (Zhao, 2008: 107). For this reason, the regulation of the press sector in China still features fragmentation. Local governments at the provincial, municipal, and county levels all set up their own TV channels and use them to propagate local news. Except the obligation of retransmitting the news program of the CCTV at 19:00 pm, localities are sovereign in deciding which news to report in priority. Their competences are parallel to each other, for one locality cannot instruct the other concerning the making of TV programs.

Fragmentation crucially characterises content regulation in China, for the regime of content regulation of the country has long been dubbed as "nine dragons tackle the flood". This is meant to say that a multitude of agencies take charge of this portfolio. Although the State Council tasks the restructured CAC with the regulation of the internet information content at the national level, as well as the supervision and the law enforcement since August 2014, other agencies, regardless of whether they are located at the same administrative level or not, can also send instructions to ICPs. Even within the same jurisdiction, the competences of content regulation are dispersed among several agencies eager to enlarge their respective turfs. For instance, registered in Beijing, Sina Weibo must observe the instructions of the Beijing Cyberspace Administration and Beijing Public Security Bureau. National regulators (e.g., the CAC, the CDP, and the MPS) and local regulators in other localities than Beijing can also ask Sina Weibo to remove specific contents.

The interview of one former content moderator of Sina Weibo shows how bureaucratic fragmentation has messed up the implementation of content regulation of the firm:

> In China, more than 50 governmental organs oversee cyberspace. The MPS, the National Administration of Radio and Television, the Ministry of Education, the CAC, the MIIT, etc., all these organs can regulate. It is a typically three-dimensional power system. It is not as simple as a point-to-surface system ranging from the network management office or the cyberspace administration's office to online platforms…I give you one example. For instance, there is a film, if a school appears [in the film], you should submit the film to the Ministry of Education for verification; if two policemen appear, you should transfer the film to the MPS; if a mining worker appears, you should transfer the film to the Ministry of National Resources. Overall, the appearance of related contents implies the submission of the film to different organs for checking…[8]

[8] "Special Interview with the Former Content Moderator at Sina Weibo: How China Is Building an Online 'Ministry of Truth'," VOA, 12 August 2020, https://www.voachinese.com/a/internet-censorship-20200812/5540475.html.

Corporate Autonomy for ISM Firms

With bureaucratic fragmentation, the opportunity structure for private firms to seek margins of manoeuvring can be explained by the principal-agent framework. Inside this framework, corporate noncompliance can be taken as a kind of agent shirking in the principal-agent relations. After being delegated the routine management of the internet, private ISM firms deviate from the goals of their principal to pursue their business interests. The shirking and drifting of ISM firms are facilitated by two factors. In the first place, the pressure to survive the market competition pushes ISM firms to implement content regulation with some reserves, for, as explained in the previous chapter, firms that fully satisfy regulators' requirements risk losing popularity among users and being considered as abetting internet censorship in China (Han, 2018). ISM firms perceived as such will be disadvantaged on the market. The root reason of firms' defection lies in the misaligned preferences between regulators (or the "principal") and firms (or the "agent"), misalignment leading to agency loss and exacerbated by the competition between ISM firms (Miller, 2018: 6).

In the course of implementing content regulations, ICPs ensure that they are disadvantaged less than their competitors. In the case of Sina Weibo, since it did not wish to censor more strictly than its competitor Tencent, it disobeyed to 16% of government directives (Miller, 2018: 2). The employees of Sina Weibo were instructed to "not be stricter than Tencent" and to "hold off on implementing government directives until urged to block content a second time" (Miller 2018: 4). Internet businesses push back government controls in order to appear "freer" than their competitors, leading to expanded space for contention (Miller, 2018: 2). The reputational concern has motivated firms not to take the instructions of content regulators at their face value, let alone to implement them in the way desired by regulators. Instead, ICPs have shown to be apt to identify the loopholes in content regulation rules, to implement unpopular rules later than their competitors, and to implement them less strictly. The shirking and drifting of ISM firms are made possible by their technical superiority over regulators. For instance, Sina Weibo invoked technical reasons to justify its laxist implementation, while it actually had no technical difficulties in meeting the requirements of regulatory agencies.

In the second place, bureaucratic fragmentation enables private firms to be freer, for it creates the problem of "multiple principals" (Gailmard, 2009). This institutional design weakens the oversight of principals (Gailmard, 2009; Ting, 2003) and produces bureaucratic drift (McCubins et al., 1987; Hammond and Knott, 1996). Lack of coordination among multiple principals allows firms to enforce the instructions of competing agencies selectively. The general rule for firms is to enforce the directives representing the lowest enforcement costs (Miller, 2018). Given that regulatory agencies hold different objectives (Calvert et al., 1989), Miller highlights that inconsistencies in the implementation are not uncommon in China (2018). Inconsistent implementation exists in the firms operating in search engines (Villeneuve, 2008), blogging services (MacKinnon, 2009), chat apps (Crandell et al., 2013), live-streaming (Crete-Nishihata et al., 2016), and mobile games (Knockel et al., 2017), to name only a few.

The proximity between private ISM firms and local regulators influences firms' decision on which agency to follow. For Miller,

> ...The power to enforce directives is locally biased, i.e., power is concentrated in the locality of an ICP's headquarters. Local agencies where media companies are headquartered have more control over what gets censored as they have access to the most proximate and effective levels of power to ensure compliance with directives. (2018: 7)

Miller illustrates selective implementation with the case of Sina Weibo. Although all local governments are in principle allowed to send instructions to the Beijing-headquartered Sina Weibo, the firm tends to comply with the instructions of Beijing regulators more than with those from other jurisdictions, like Shanghai or Guangzhou. Since only the Beijing municipal government has direct leadership over Sina Weibo, it is not uncommon to see the firm leave negative reporting on other provinces and municipalities uncensored (Miller, 2018).

Potential Shrinking of the Opportunity Structure

It is noteworthy to highlight that the political regulation has progressively been centralised in the CAC since 2014. Large ICPs with national influence (e.g., Sina Weibo, Tencent, Bytedance, ifeng) are actually under the direct whim of central regulators, even though they are locally registered.

That is why in the law enforcement records of the CAC, the invited talks with influential wrongdoing ICPs are often presented as being conducted by local cyberspace administrations under the guidance of the CAC. For instance, during the coronavirus period, several private platforms, including Baidu and ifeng, were reprimanded by content regulators for having carried self-collected news. On 25 February 2020, instructed by the CAC, Beijing Cyberspace Administration "invited" the managers of ifeng for a talk. Likewise, on 7 April 2020, Beijing Cyberspace Administration talked with Baidu, in accordance with the instructions of the CAC.

Centralisation of the content regulation risks narrowing down the window of opportunity opened by bureaucratic fragmentation in the future. By reducing bureaucratic fragmentation, the centralisation move currently at play in China mitigates the problem of "multiple principals" with overlapping competences. Instead, it introduces clear hierarchy among agencies, making it difficult for ISM firms to play the division among regulators.

The most recent centralisation move occurred in the institutional reform of March 2018. The SAPPRFT was broken into three agencies: the NRTA, the National Press and Publication Administration (NPPA), and the China Film Administration (CFA). While the NRTA was put under the responsibility of the State Council, the CDP acquired the oversight over the NPPA and the CFA. The institutional reform of 2018 enlarged the oversight of the CDP to publication and films. That the same agency takes on transversal responsibilities and leads several other agencies reduces the number of "veto players" that firms deal with (Tsebelis, 1995). With centralisation, the policy process becomes more streamlined and top-down, and the number of actors with final decision-making power shrinks. The limits to bureaucratic fragmentation in the content regulation is expected to narrow down the opportunity structure where private ISM firms seek manoeuvring room.

Conclusion

This chapter analyses the second source of the party-state's double-bind logic: fragmented authoritarianism. The latter puts limits to the party-state's freedom of action and invalidates the image of an all-powerful party-state in China. This chapter has shown how fragmented authoritarianism has opened a window of opportunities for firms. In a similar way to Chapter 5, this chapter presents another dialectic cycle where the double-bind logic transforms the party-state's constraints into margins of manoeuvring of private ISM firms. Within the double-bind regulatory state, private firms exploit the party-state's predicament to reclaim room to manoeuvre and defend corporate interests. The double-bind regulation thus helps firms repair, to some extent, unbalanced state-firm relations in China and acquire greater bargaining power vis-à-vis authoritarian regulators.

"Fragmented authoritarianism" unfolds along two dimensions in China: territorial/vertical fragmentation and bureaucratic/horizontal fragmentation. Territorial fragmentation is captured into "*de facto* federalism" of China (Zheng, 2006) where local regulators enjoy significant discretion in deciding which economic sectors to support and what forms of support to provide. Three interrelated dynamics explain the predicament of the central government: competitive liberalisation, local ownership of private ISM firms, and the weight of the local economic performance for the professional career of local officials. Under the pressure of competitive liberalisation, localities compete for offering preferential policies to private firms. The "race-to-the-bottom" interjurisdictional competition ushers in their deviations from the parameters of the central government. Local governments nurture ISM firms in the ways they see fit, thus weakening Beijing's influence over firms. Considering ISM firms as the drivers of local economic development, local governments tend to create alliances with them. They provide favourable policies, try to attract promising firms registered in competing jurisdictions, and recommend local firms to participate into national projects.

The close alliance between local governments and private ISM firms significantly shortens the power reach of Beijing. Since the economic performance of a locality plays a crucial role in deciding the professional career of local officials in the post-1979 China (Chen and Zhu, 2019), local regulators support the industries with the greatest potential to pull up local development. Local governments' dependence on private ISM firms enables the latter to lobby local regulators, and to lobby the central government through allying with local regulators. The comparison between Zhejiang and Gansu provinces, both vehemently defending local champions, shows the significant margins of manoeuvring of influential private firms in wrestling with local regulators.

Bureaucratic fragmentation of Chinese authoritarianism also opens a window of opportunity for private firms for their pursuit of autonomy.

That around 50 agencies can send instructions to ISM firms creates a problem of "multiple principals" and common agency, enabling firms to selectively implement content regulations. As a result, firms tend to implement the instructions representing the least costs for them, weakening the power reach of some agencies and augmenting the power reach of others. Bureaucratic fragmentation and territorial fragmentation often work in concert. Since private ISM firms are locally registered and that no other administration than the local government has direct oversight over them, private firms tend to implement the instructions of the local government and to neglect the instructions from competing agencies, including central-level regulators. However, that content regulation has been progressively centralised into the CAC since 2014 is likely to reduce the number of principals, thus narrowing down firms' manoeuvring room.

This chapter has analysed fragmented authoritarianism as the second source of the party-state's predicament. It has investigated both the manufacturing of the predicament and its implications on the margins of manoeuvring of private ISM firms in China. Fragmented authoritarianism results from the structural dilemmas of the party-state. For this reason, the autonomy-enabling opportunities it creates will persist. The durability of the room to manoeuvre for private ISM firms in China relates to the durability of the governance dilemmas of the party-state. As long as the territorial and bureaucratic organisation of the regime remain unchanged, private ISM firms in China are expected to continuously obtain autonomy by exploiting fragmented authoritarianism.

References

Birney, Mayling. 2013. "Decentralisation and Veiled Corruption Under China's 'Rule of Mandates'." *World Development* 53: 55–67.

Cairns, Christopher. 2016. "Fragmented Authoritarianism? Reforms to China's Internet Censorship System Under Xi Jinping." Presented at the American Political Science Association Annual Meeting in Philadelphia, PA on 3 September 2016.

Calvert, Randall L., Mathew D. McCubbins, and Barry R. Weingast. 1989. "A Theory of Political Control and Agency Discretion." *American Journal of Political Science* 33(3): 588–611.

Chai, Yanfei. 2018. *Zhejiang Gaige Kaifang 40 Nian Koushu Lishi (1978–2018)* (Oral History of 40 Years of Reform and Opening-Up in Zhejiang). Hangzhou: Zhejiang Science and Technology Press.

Chen, Hangping. 2021. "Qi Di 2020 Nian Teng Xun 82 Fen Pan Jue Shu: Nan Shan Bi Sheng Ke Shi Zhen De Ma?" [82 Court Rulings Since 2020: Is the Phenomenon of 'Nanshan Must-Be-Victorious Quest' True?]. 2 March. WeChat public account "Zhong Guo Fa Lü Ping Lun" (China Legal Review).

Chen, Shuo, and Zhitao Zhu. 2019. "Xiang Dui Ji Xiao Kao He Yu Zi You Cai Liang Quan: Zhong Guo Xian Ji Guan Yuan Zhi Li Kao Cha" [Relative Performance Evaluation and Discretionary Power: Investigations into Chinese County-Level Cadres' Governance]. Fudan University Working Paper, 1–24.

Crandell, Jedidiah R., Masashi Crete-Nishihata, Jeffrey Knockel, Sarah McKune, Adam Senft, Diana Tseng, and Greg Wiseman. 2013. "Chat Program Censorship and Surveillance in China: Tracking Tom-Skype and Sina UC." *First Monday* 18(7).

Crete-Nishihata, Masashi, Andrew Hilts, Jeffrey Knockel, Jason Q. Ng, Lotus Ruan, and Greg Wiseman. 2016. "Harmonized Histories? A Year of Fragmented Censorship Across Chinese Live Streaming Applications." Citizen Lab, 1–13, https://citizenlab.ca/2016/11/harmonized-histories-year-fragmented-censorship-across-chinese-live-streaming-applications/.

Eggertsson, Thrainn. 1990. *Economic Behaviour and Institutions*. Cambridge: Cambridge University Press.

Fu, Diana. 2018. *Mobilizing without the Masses: Control and Contention in China*. Cambridge: Cambridge University Press.

Gailmard, Sean. 2009. "Multiple Principals and Oversight of Bureaucratic Policy-Making." *Journal of Theoretical Politics* 21(2): 161–186.

Hammond, Thomas, and Jack Knott. 1996. "Who Controls the Bureaucracy? Presidential Power, Congressional Dominance, Legal Constraints, and Bureaucratic Autonomy in a Multi-Institutional Model of Policy Making." *Journal of Law, Economics, and Organization* 12: 119–166.

Han, Rongbin. 2018. *Contesting Cyberspace in China: Online Expression and Authoritarian Resilience*. New York: Columbia University Press.

Hsueh, Roselyn. 2011. *China's Regulatory State: A New Strategy for Globalisation*. Ithaca, NY: Cornell University.

Jevons, W. Stanley. 1879. *The Theory of Political Economy*. Second Edition. London: Macmillan and Co.

Knockel, Jeffrey, Lotus Ruan, and Mashashi Crete-Nishihata. 2017. "Measuring Decentralization of Chinese Keyword Censorship via Mobile Games." Presented at the 7th USENIX Workshop on Free and Open Communications on the Internet (FOCI 17). Vancouver, BC: USENIX Association.

Li, Hongbin and Li-An Zhou. 2005. "Political Turnover and Economic Performance: The Incentive Role of Personnel Control in China." *Journal of Public Economics* 89(9): 1743–1762.

Lieberthal, Kenneth, and Michel Oksenberg. 1988. *Policy Making in China: Leaders, Structures and Processes*. Princeton, NJ: Princeton University Press.

Lin, Justin Yifu, and Zhiqiang Liu. 1998. "Zhong Guo De Cai Zheng Fen Quan Yu Jing Ji Zeng Zhang" [Shared Budgetary System and Economic Growth in China]. Presented at the Third Pan-Pacific Joint Conference of Economic Organizations, 13–18 January, Bangkok, Thailand, http://vod6.pkudl.cn/xzgl/czss/contents/thesis/the_010_01.pdf

Lin, Li-wen, and Curtis Milhaupt. 2013. "We Are the (National) Champions: Understanding the Mechanisms of State Capitalism in China". *Stanford Law Review* 65(4): 697–759.

Liu, Mianmian. 2014. *Zhong Guo Niu Ren* [Chinese Superman]. New York: Mirror Books.

Ma, Aifang. 2023. "Geographical Distribution of the Top 100 Internet Firms in China from 2013 to 2022", https://doi.org/10.21410/7E4/PWQ6WF.

MacKinnon, Rebecca. 2009. "China's Censorship 2.0: How Companies Censor Bloggers." *First Monday* 14(2), https://firstmonday.org/ojs/index.php/fm/article/download/2378/2089.

Maskin, Eric, Yingyi Qian, and Chenggang Xu. 2000. "Incentives, Scale Economies and Organization Forms." *Review of Economic Studies* 67: 359–378.

McCubins, Mathew D., Roger G. Noll, and Barry R. Weingast. 1987. "Administrative Procedures as Instruments of Political Control." *Journal of Law, Economics, and Organization* 3: 243–277.

McGregor, Richard. 2010. *The Party: The Secret World of China's Communist Rulers*. New York: HarperCollins Publishers.

Mertha, Andrew. 2005a. "China's Soft Centralization: Shifting Tiao/Kuai Authority Relations." *The China Quarterly* 184: 792–810.

Mertha, Andrew. 2005b. *The Politics of Piracy: Intellectual Property in Contemporary China*. Ithaca, NY: Cornell University Press.

Mertha, Andrew. 2009. "'Fragmented Authoritarianism 2.0': Political Pluralization in the Chinese Policy Process." *The China Quarterly* 200: 995–1012.

Milhaupt, Curtis, and Wentong Zheng. 2015. "Beyond Ownership: State Capitalism and the Chinese Firm." *The Georgetown Law Journal* 103: 665–722.

Miller, Blake. 2018. "The Limits of Commercialized Censorship in China." Working Paper, http://eprints.lse.ac.uk/101417/1/miller_limits_of_commercialized_censorship_in_china.pdf.

Negro, Gianluigi. 2017. *The Internet in China: From Infrastructure to a Nascent Civil Society*. Cham: Palgrave Macmillan.

North, Douglass. 1990. *Institutions, Institutional Change and Economic Performance*. New York: Cambridge University Press.

Oi, Jean C. 1992. "Fiscal Reform and the Economic Foundations of Local State Corporatism in China." *World Politics* 45: 99–126.

Pearson, Margaret M. 2005. "The Business of Governing Business in China: Institutions and Norms for the Emerging Regulatory State." *World Politics* 57: 296–322.

Qian, Yingyi, and Chenggang Xu. 1993. "Why China's Economic Reforms Differ: The M-Form Hierarchy and Entry/Expansion of the Non-State Sector." *Economics of Transition* 1(2): 135–170.

Qiao, Kunyuan. 2013. "Wo Guo Guan Yuan Jin Sheng Jin Biao Sai Ji Zhi: Li Lun Yu Zheng Ju" [Tournament Mechanism in the Promotion of Chinese Government Officials: Theories and Proves]. *Jing Ji Ke Xue* [Economic Science] 1: 88–98.

Stigler, George J. 1971. "The Theory of Economic Regulation." *Bell Journal of Economics and Management Science* 2: 3–21.

Tai, Zixue. 2006. *The Internet in China: Cyberspace and Civil Society*. New York: Routledge.

Ting, Michael M. 2003. "A Strategic Theory of Bureaucratic Redundancy." *American Journal of Political Science* 47(2): 274–292.

Tsebelis, George. 1995. "Decision Making in Political Systems: Veto Players in Presidentialism, Parliamentarism, Multicameralism and Multipartyism." *Bristish Journal of Political Science* 25(3): 289–325.

Villeneuve, Nart. 2008. "Sear Monitor: Toward a Measure of Transparency." Citizen Lab Occasional Paper#1: 1–29.

Xiao, Lihui. 2008. "Xian Wei Shu Ji Yan Zhong De Zhong Yang Yu Di Fang Guan Xi" [Central-Local Relationship in the Eyes of the Party-Secretaries of the Counties]. *Jing Ji She Hui Ti Zhi Bi Jiao* [Comparison of Economic and Social Systems] 4: 145–150.

Yang, Dali. 1997. *Beyond Beijing: Liberalization and the Regions in China*. New York: Routledge.

Ye, Shan. 2017. "Di Fang Zheng Fu Shui Shou Jian Mian Cheng Nuo De Xiao Li Xia Ci" [Flaws in Local Governments' Promises to Reduce and to Exempt Taxations]. *Dang Dai Fa Xue* [Contemporary Law Studies] 6: 116–125.

Yeo, Yukyung. 2020. *Varieties of State Regulation: How China Regulates Its Socialist Market Economy*. Cambridge, MA: Harvard University Press.

Yin, Laisheng, and Bangyan Feng. 2014. "Cong Cheng Shi Jing Zheng Dao Qu Yu He Zuo" [From Competition between Cities to Cooperation between Regions]. Jing Ji Ti Zhi Gai Ge [Reform of Economic Systems] 5.

Zhan, Xuying. 2015. "Di Fang Zheng Fu Ji Xiao Ping Gu De Bei Lun Jie Xi: Re Dian Ju Jiao" [Analysis of the Paradoxes in Evaluating Local Governments'

Performance: Focus on Hot Points]. *Zhong Guo Xing Zheng Guan Li* [Public Administration in China] 11: 45–49.

Zhao, Yuezhi. 2008. *Communication in China: Political Economy, Power, and Conflict. Lanham*. MD: Rowman & Littlefield Publishers.

Zhang, Jun, Haichao Fan, Zhiwei Xu, and Longfei Zhou. 2020. "GDP Zeng Su De Jie Gou Xing Xia Tiao: Ji Yu Guan Yuan Kao He Ji Zhi De Shi Jiao" [Structural Downward Adjustment of the GDP Growth: From the Perspective of the Evaluation System of Officials]. *Jing Ji Yan Jiu* [Economic Studies] 55(5): 31–48.

Zheng, Yongnian. 2006. *De facto Federalism in China: Reforms and Dynamics of Central-Local Relations*. Singapore: World Scientific Publishing.

Zheng, Yongnian. 2007. *Technological Empowerment: The Internet, State and Society in China*. Stanford, CA: Stanford University Press.

CHAPTER 7

Global Leadership Comes at a Price

Introduction

The analysis so far has been concerned with the domestic sources of the party-state's predicament. It has illustrated, how, in the increasingly authoritarian context of China, private ISM firms have been able to seek certain manoeuvring room by exploiting the party-state's predicament. The dilemmas of the Chinese government unfold at the international level, too. This chapter extends the analytical scope from the domestic to the international level, exploring the autonomy-enabling opportunities open for private ISM firms in the international sphere.

The international context also produces constraints to the party-state's exercise of power inside China. The existing literature has extensively discussed the role of the international context on domestic politics in non-democratic regimes. That impact comes from the inter-penetration between the national and the international environments. As Légaré argues, the national political system can work both as the receiver and the sender of norms; it can be both the origin and the result of a modification of the international system (2007: 504). The existing literature recognises that the international context can push democratisation in non-democratic regimes (Légaré, 2007; Huntington, 1991; Oberschall, 1996). Huntington's conceptualisation of "waves of democratisation" is one of the earliest academic works on this matter (1991). The international context can cause political changes in semi-authoritarian regimes in

A. Ma, *China as a Double-Bind Regulatory State*,
https://doi.org/10.1007/978-981-99-8857-0_7

four ways: (1) strategic ideological influence of international actors on the national context; (2) normative exchanges between domestic and international civil societies; (3) transfer of norms from Western organisations to non-democratic environments; and (4) strategic interaction between national and international spheres (Légaré, 2007: 505).

The existing literature has discussed the impact of several high-profile international events on internet censorship in China: the Chinese membership into the WTO (Harwit and Clark, 2001; Hsueh, 2011), the international questioning of the Chinese handling of the SARS epidemics in 2003 (Tai, 2006; Tai and Sun, 2007), the pro-democracy movements in the Middle East and North Africa (Negro, 2017; Lagerkvist, 2012; Economy, 2018: 69), and the Orange revolution in Ukraine (Strittmatter, 2020). Although not all international events can pressure the party-state to loosen internet censorship, some of them have produced such effects. Tai and Sun (2007) find that the international pressure pressured the Chinese government into disclosing information on the SARS in 2003 and into stopping censoring SARS-related information online. The 2008 Olympics Games in Beijing provides another case in point. After discussing with the International Olympic Committee in the end of July 2008, certain websites, including those of Amnesty International, Human Rights Watch, and the Chinese version of BBC, became available in some parts of Beijing.

However, not all international events can mitigate internet censorship in China. The pro-democracy movements, initiated in Tunisia, Egypt, and Ukraine respectively via Twitter, Facebook, and YouTube, did not disturb the censorship endeavours of the Chinese government (Strittmatter, 2020: 84–85). Instead, the internet-empowered Arab Spring in 2010 and 2011 were among the most relevant factors to explain the heavy-handed Chinese internet crackdown in 2013 (Negro, 2017). Economy considered the harsh internet control in 2010 and 2011, epitomised by the all-out censorship of the word "Jasmine" in Chinese cyberspace, as a form of overreaction which denoted "both the strength of the country's security apparatus and the leadership's acute insecurities" (2018: 69). She mentioned that the Chinese government deployed a much greater number of soldiers than protestors to suppress the Jasmine Revolution on 20 February 2011 in a dozen of Chinese cities. Examples like the Arab Spring illustrate the mixed effects of the international events on Chinese internet censorship.

Despite precious insights it brings, the existing literature has two weaknesses. First, it construes the influence of the international context mostly from the perspective of regime changes. Without investigating how international factors impact the day-to-day management in autocracies, some scholars directly move to discuss how international actors can change autocrats into democrats. The tropism of regime changes originates from the perception of authoritarianism as a residual or transitional situation. However, what happens in well-established autocracies like China and Russia shows that regime changes are anything but a fatality, and that authoritarianism deserves to be analysed as a political regime in its own rights. Instead of reflecting on how international factors nurture regime changes, it makes greater sense to examine why regimes changes have not happened despite the presence of international actors. Second, when it comes to China, although the existing literature recognises that international factors can work in opposite directions (reinforce or mitigate internet censorship), few of them have systematically investigated why the party-state reacts differently to international upheavals.

This chapter enlarges the analytical scope from the domestic to the international level. It investigates the extent to which the interplay between national and international politics influences state-firm power relations in China. It argues that, when combined with national factors, international factors can amplify the party-state's double-bind logic. In a similar way to the domestic sources of the party-state's predicament, the party-state's predicament at the international level also generates autonomy-enabling opportunities for private ISM firms. The source of the party-state's predicament scrutinised in this chapter is the contradiction between the party-state's pursuit of global leadership and the continuity of domestic information control.

This chapter addresses the two weaknesses mentioned above. On the one hand, it rejects the tropism of regime changes. Instead of taking high-profile international events and international audiences as boosters of democratisation of China, I take them as factors and actors that Chinese regulators or regulatees of the ISM industry make use of to improve their position in the state-firm wrestling inside China. On the other hand, this chapter does away with the incident-centric approach, for the latter more often than not leads to incomplete findings. This chapter does not content itself with describing how one international event narrows down or enlarges the room to manoeuvre of private firms. It goes one step further by unveiling how one international event generates one specific configuration of state-firm wrestling, rather than another one.

Reputational Concern and Corporate Noncompliance

With the entering of international factors and actors into the Chinese state-firm wrestling, the reason private ISM firms in China can hope to seek certain room to manoeuvre resides into the importance they attach to their corporate reputation. That Chinese private ISM firms care about their brand name makes their coalition with netizens (irrespective of national or international) possible and stable. Taking netizens on board, ISM firms acquire significant bargaining power vis-à-vis the party-state. The bargaining power accrues in parallel to the increase of the size of the netizens siding with firms. To put it more simply, the firm-netizen coalition is such a crucial source of power that it incentivises firms not to obey.

It has been argued previously that firms may disobey to regulation rules for economic reasons. While this is true, focusing exclusively on instrumental intentions is overly reductionist. Non-material motivations count, too. As Colin Hay argues, "…if we are genuinely to understand the conduct of political actors, it is important that we do not presume to reduce all motives for political action to the level of pure instrumentality…To see self-interest as the primary motivation for behaviour is, of course, a simplification and a distorting simplification at that" (Hay, 2009: 272–273). Portraying actors as interest-maximisers is problematic not only because "interests are social constructions and cannot serve as proxies for material factors" (Hay, 2011: 68), but also because "actors cannot simply be assumed to have fixed (and immutable) preference set, to be blessed with extensive (often perfect) information and foresight and to be self-interested and self-serving utility maximisers" (Hay & Wincott, 1998: 954). Reducing the motivations of firms to pure self-interest does not suffice to build a full picture for explaining why they deviate from regulators' design. It is important to dig into the immaterial considerations behind corporate disobedience (Negro, 2017; Jiang, 2015; Tai, 2006).

North (1990) figures among the earliest scholars who studied the influence of non-material motivations in influencing the courses of actions of actors. He argues that "under certain conditions, traits like honesty, integrity and living up to a reputation pay off in strictly wealth-maximising terms" (1990: 42). Non-material reasons, rather than material ones, constitute the fundamental motivation behind human actions. The imperative of being respected within a society or a tribe "must exert a continuous pressure on the individual to eliminate economic self-interest from his consciousness to the point of making him unable, in many cases...even to comprehend the implications of his own actions in terms of such an interest" (Polanyi, 2001: 40). As economic organisations, private ISM firms differ from individuals described by Polanyi, for it is logical that firms put economic interests at a higher place than altruistic concerns. However, what individuals and economic organisations share in common is that neither can fully flourish without public recognition. In China, a private ISM firm that citizens perceive as independent from governmental influence wins a greater user base and higher respect. This is the root reason behind the popularity of private ISM firms in contrast with dull state-owned media. The reputational concern of private ISM firms explains why they resist to the regulations the compliance with which damages their public image. The concern of private firms with their public image contradicts the conventional wisdom according to which Chinese private ISM firms are reduced to be the economic instruments of the party-state, seeking to please the latter even at the cost of sacrificing users' interests.

That ISM firms care about their reputation illustrates a simple truth that, unfortunately, tends to be overlooked by some observers of Chinese politics. It is the fact that even in an authoritarian regulatory setting like China, users matter as much as (even more than) the party-state for the business of private firms. Despite its considerable state capacities, the party-state has not been successful in crowding out netizens' influence on private ISM firms. The latter care about how they are perceived by netizens and try their best to improve such perceptions. By introducing netizens as a co-arbitrator of the destiny of private firms, the literature on the corporate reputational concern weakens the particularity of Chinese private ISM firms. In a similar way as their counterparts in democracies, Chinese ISM firms must ensure that they are "liked" by citizens, without which they could hardly flourish in the long run.

The pre-installation of the filtering software Green Dam Youth Escort (GDYE) to all the personal computers sold or manufactured in China, a measure required by the Ministry of the Industry and Information Technology (MIIT) on 19 May 2009, is one of the most widely discussed cases

(Negro, 2017; Yang, 2009). It illustrates how reputational concerns have motivated private firms not to comply with the regulations that are likely to damage their corporate image. The GDYE was designed to monitor the contents on the computer and to block politically sensitive contents (Wolchok et al. 2009). Negro argued that public contestations against this measure led to the *de facto* refusal of PC makers to pre-install the GDYE (Negro, 2017: 55–60). One online survey by Sohu on more than 12 500 users showed that 83.95% of the respondents considered that the GDYE would violate users' privacy. 74.42% thought that the GDYE could not achieve its objective to cleaning cyberspace for adolescents. In addition, 93.93% explained that after one year of trial, they would not pay the software for continuing to use it.[1]

Public contestations against the GDYE explained PC makers' reluctance to comply with the MIIT. On 30 June 2009, one day before the deadline fixed by the MIIT, none of the computers sold in Beijing, regardless of domestic or foreign brands, had pre-installed the GDYE. For Lenovo, although it pre-installed the GDYE to its computers, it let its consumers decide whether to activate the software or not. Several foreign firms, including Hewlett-Packard and Dell, did not install the software into any computer they sold in China. In the end, Li Yizhong, then minister of the MIIT, changed the pre-installation of the GDYE from a mandatory requirement into an optional choice, exempting PC makers, internet cafés and school devices from the obligation of pre-installing the filtering software (Negro, 2017: 60). China being the second largest PC market in the world in 2009, the compliance with the MIIT regulation logically would bring tremendous economic benefits to firms. However, that influential domestic and foreign PC makers did not comply with the regulation showed that they were ready to exchange short-term economic benefits for maintaining consumers' trust (or, in other words, for securing longer-term economic benefits).

The analysis of ISM firms' reputational concerns almost always involves netizens, because reputational concerns are more likely to encourage corporate noncompliance if the latter better serves netizens' interests. In this situation, netizens and firms form a win-win coalition and corporate noncompliance makes both netizens and firms better off. To incentivise

[1] Liu, Shanyun & Li, Bin, "The Filtering Software on the Computer Can Be Disinstalled, More Than 80% of the Netizens Consider the Software as Violating Privacy," Sohu.com, 10 June 2009, http://news.sohu.com/20090610/n264437676.shtml.

firms not to comply, the preferences of netizens should be explicitly expressed. The expression of such preferences needs to unfold at a sufficiently large scale, so as to reduce firms' costs in committing noncompliance. In the GDYE case, that netizens intensively expressed their dissatisfaction towards the new regulation constituted a clear signal which encouraged firms not to comply, for netizens' wide support would lower firms' cost of noncompliance.

The contributions of the international audiences are that they scale up and amplify the expression of preferences by domestic audiences, lowering still further the costs of private ISM firms for committing noncompliance. The prerequisite for such an incentive to be effective is that the preferences of the domestic and the international audiences must converge, for the convergence of preferences creates a relatively comfortable situation for firms. It suffices that they join netizens to enlarge even further the size of the coalition, thus marginalising regulators (content regulators above all). Fearing to render themselves unpopular, content regulators are expected to make concessions and to mitigate the sanctions inflicted to wrongdoing content providers. That private ISM firms position themselves with netizens illustrates the centrality of the corporate reputational concerns for firms. As profit-seeking entities, firms tend to live up to the expectations of those who matter the most for the increase of their commercial interests. Since it is citizens that drive up the online traffic of ISM firms, it is logical that firms seek to please netizens more than they please the party-state. In this sense, the corporate reputational concerns and the pursuit of commercial interests mutually serve, which renders the firm-netizen alliance much more stable than the state-firm alliance.

Regime Survival Versus Norms-Based Legitimacy

This section turns to the international factors which prevent the party-state from exercising effective information control. It reveals the third source of the party-state's double-bind logic in the regulation of ISM firms: the necessity to shape a respectable image of China on the global stage obliges the party-state to conduct "double justifications" to the domestic and international audiences. The process of "double justifications" creates extra constraints to the freedom of action of the party-state.

Although the latter has attempted to get rid of international pressure by mobilising a jingoistic rhetoric, the limits of this method are being felt.

To Win International Support: Priority of External Propaganda

Two institutions with complementary responsibilities oversee the propaganda work in China. While the CDP has been charged with propaganda within China since 1921 (*nei xuan*, see Brady, 2018), the State Council Information Office (SCIO), created in January 1991, is tasked with the propaganda work outside China (*wai xuan*). In fact, the *wai xuan*, or external propaganda, starts well before the creation of the SCIO. At the Central Meeting of Cadres on 16 January 1980, Deng Xiaoping delivered a speech entitled *Current Situation and Missions (muqian de xingshi yu renwu)*. The speech emphasized three principal tasks facing China in the 1980s: maintaining global peace by fighting hegemony in international affairs; realising national reunification by retaking Taiwan; and promoting economic development. Since the first two missions linked China to the international environment, the Secretariat of the CCCCP started discussing the idea of creating a propaganda organ to coordinate the external propaganda work. On 8 April 1980, the CCCCP decided to create the Central Foreign Communication Small Group (CFC-SG) which would act as the coordination and the planning agency for propaganda directed to foreign countries and non-mainland Chinese territories, including Hong Kong, Macao, and Taiwan. In February 1982, ministries and provincial governments institutionalised the spokesperson system. In terms of hierarchy, both the CFC-SG and its successor, the SCIO, were subject to the CDP.

Propaganda represents vital importance to the party-state. Liu Zhongde, Deputy Director of the CDP from June 1990 to February 1998, once said that "[I]n China, the head of the Central Department of Propaganda is like the Secretary of Defence in the United States and the Minister of Agriculture in the former Soviet Union. The manner by which he brings leadership will affect whether the nation can maintain stability" (McGregor, 2010: 229). The importance of external propaganda co-variated with the Chinese integration into the international system and the rise of the economic strength of the country. In parallel with the economic take-off since the 1980s, the importance of external propaganda accrued. At the National Conference on External Propaganda Work on 26 February 1999, Jiang Zemin pointed out that external propaganda should match the international influence and status of China, urging the propaganda personnel to report the positive results of the Open Doors policy to the world.[2] Hu Jintao reiterated the importance of achieving the match between the external propaganda and the Chinese international influence at the National Conference on Propaganda Work in December 2003 in Beijing. Under Hu (2002–2012), China continued to make worldwide dissemination of the achievements it had made in economy, social affairs, and politics. The staff of external propaganda also endeavoured to explain to the world China's foreign policies. At the National Conference on External Propaganda Work from 4 to 5 January 2012, Wang Chen, then Director of the SCIO, declared that external propaganda should focus on spreading the ideas of "path of peaceful development" and "building a harmonious world".

The administration of Hu went one step further than his predecessor in institutionalising external propaganda, a process closely linked to the preparation of the Beijing Olympics Games in August 2008. During his first presidential term (2002–2007), a three-level spokesperson system was established, including the central level, the ministerial level, and the provincial level. The spokesperson system was created within 13 organs of the CCCCP, more than 80 governmental institutions, and 31 provincial-level entities. At the provincial level, since 2006, each province has created its External Propaganda Leading Small Group (EP-LSG) to plan and coordinate external propaganda. Provinces created several work mechanisms in different areas: news reporting, exchanges with foreign media, diffusion of the news information on the internet, reporting on the Olympics Games, etc. At the same time, state media reinforced their capacity to reach the international audience under Hu Jintao. Jiang and Zhang (2019) reported how China built a modern system for external propaganda under Hu. In 2004, two years after Hu came to power, the Chinese government advanced the strategy according to which Chinese

[2] "At the National External Propaganda Work Conference, Jiang Zemin emphasized that we should stand on a higher starting point and improve the external propaganda work; we must form a strong propaganda force which is commensurate with the international status and prestige of our country and which better serves the Open Doors policy and the modernization; Hu Jintao, Li Lanqing and others attended the conference," *People's Daily*, 27 February 1999, p. 1.

media should "go abroad". From 2008 onwards, China has implemented two FYPs for the regulation of external propaganda (2008–2013 and 2014–2019). In June 2009, the government issued the *Overall Planning on Building the International Communication Capacity of the Core Media*, saying that:

> China will incorporate the capacity-building of the core media into the national socio-economic development plan: we should take the core media as the backbone of external propaganda, and the new media as the entry point...promote simultaneously the infrastructure-building and the construction of information contents. We should comprehensively improve the capacity to collect information, spread information, and commercialise products. We need to accelerate the construction of a modern system of external propaganda characterised by large coverage, rich information, and edgy technologies. The capacity-building in external propaganda should correspond to the socio-economic development level and the international status of our country. (Cheng, 2017, quoted by Jiang and Zhang, 2019)

Under Xi Jinping, the international political influence of China keeps growing in parallel to the expansion of the economic size of the country. On 21 October 2014, the online version of *People's Daily* released the comments of Gong Fangbin, Professor at the PLA National Defence University and an expert on the thoughts of Xi Jinping (the latter are captured into the lengthy expression of "Xi Jinping thought on socialism with Chinese characteristics under the new era"). Gong declared that "Mao Zedong enabled Chinese people to stand up; Deng Xiaoping made Chinese people rich; Xi Jinping will make Chinese people strong".[3] Under Xi, China has initiated several projects with international reach: Belt and Road initiative in 2013, Asia Infrastructure Investment Bank in 2014, and World Internet Conference in 2014. That China brought the covid-19 under control earlier than the West and dispatched vaccines to under-developed countries (e.g., Serbia, Columbia, Belarus, Zimbabwe, Cambodia, Laos) further increased China's international influence. By February 2021, China had provided vaccines to 53 countries, in stark contrast with the hoarding of the expensive vaccines by certain developed countries. One article in *The Economist* commented that China and Russia

[3] "Gong Fangbin: Mao Zedong Enabled Chinese People to Stand Up, Deng Xiaoping Made Chinese People Rich, Xi Jinping Will Make Chinese People Strong," People.com.cn, 21 October 2014, http://theory.people.com.cn/n/2014/1021/c49150-25876113.html.

filled the "vaccine vacuum" left by Europe and the USA, and that they used the "vaccine diplomacy" to secure new alliances in Africa and Latin America.[4]

Under Xi, the increase of the Chinese international influence is accompanied with higher emphasis on external propaganda. As an illustration, Xi Jinping personally presides over the national conferences on external propaganda, in contrast with the past practices where the chief of the external propaganda agency presided over such meetings (Zhao Qizheng under Jiang; Wang Chen under Hu). In addition, although the fundamental objective of external propaganda has remained unchanged—to shape a positive reporting of China in the world—the administration of Xi makes tremendous efforts to improve the efficiency of external propaganda. At the CCP's Conference on Press and Public Opinion Work on 19 February 2016, Xi emphasized the importance of the discourses in conducting external propaganda and the importance of using China's own discourses to tell Chinese stories. State-owned media compared the relationship between Chinese theories and Chinese practices as the relationship between "shoes" and "feet". Using foreign theories to explain Chinese practices is like forcing someone to put on inappropriate shoes. At the same meeting, Xi recommended the storytelling as a new approach to conducting external propaganda. Based on foreign audiences' fondness for details and demonstrations, Chinese media workers were encouraged to do away with their habit for generalisation and abstraction and to adapt Chinese propaganda to foreign audiences.

On 31 May 2021, at the 30th collective study session of the Politburo on reinforcing Chinese capacity to conduct external propaganda, Xi Jinping explained what China should tell to foreign audiences and how to do the storytelling:

> We should accelerate the construction of a Chinese discursive and narrative system, use Chinese theories to explain Chinese practices, use Chinese practices to improve Chinese theories. We should create new concepts, new fields, and new expressions that speak both to the domestic and foreign audiences to show the Chinese stories and the spiritual strength behind them. We should...help foreign public realise that the CCP

[4] "Vaccine Diplomacy Boosts Russia's and China's Global Standing," *The Economist*, 29 April 2021, https://www.zawya.com/en/economy/vaccine-diplomacy-sees-russia-boost-its-geopolitical-leverage-sfp2mwic.

genuinely strives for the well-being of the Chinese people. We should make foreign public understand why the CCP is competent, why the Marxism works, why the socialism with Chinese characteristics is remarkable. We should deepen the research on Chinese spirit, Chinese values, and Chinese strength from several perspectives…to provide theoretical support for conducting international communication… We should present to the world an excellent culture representing Chinese particularities, Chinese spirit, and Chinese wisdom. We should adopt a right attitude – open and confident on the one hand, humble and gentle on the other hand – and deploy our efforts to create a reliable, friendly, and respectable image of China."[5]

Ensuring international support stands at the heart of the Chinese leadership eager to take up global leadership. The more China wishes to become the global leader, the more it needs to ensure that it is not disliked internationally. In November 2014, at the Central Foreign Affairs Meeting, Xi explained that one important task for the current period and the near future would be to seek the understanding and the support of foreign countries for the ideal of "Chinese Dream".[6] However, the pursuit of the international fame comes at a price. If China can, to some extent, afford to "buy" the diplomatic support of under-developed and developing countries by creating economic and trade dependence (e.g., the support of Cambodia and Laos to China on the South China Sea issue), it can hardly win the support of developed countries by playing the same song. Ideological and political differences put aside, China indeed needs to "build a reliable, friendly, and respectable image" on the global stage, as Xi put it at the external propaganda meeting on 31 May 2021.

[5] "At the 30th Collective Study Session of the Politburo of the CCCCP, Xi Jinping Emphasizes the Reinforcement and Improvement of International Communication to Show a True, Tridimensional and Complete China," Xinhuanet, 1 June 2021, http://www.xinhuanet.com/politics/leaders/2021-06/01/c_1127517461.htm.

[6] "Taking the Right of International Discourses and Effectively Disseminating the Voice of China—an Analysis of Xi Jinping's Ideas and Concepts on External Propaganda," Xinhuanet, 6 April 2016, http://www.xinhuanet.com/politics/2016-04/06/c_1118542256.ht.

Supporting "Virtues" with "Double Justifications"

The double-bind logic on the political axis of the ISM regulation can be de-constructed in two steps. First, the increase in the economic and political importance of China goads the party-state to seek what it deems overdue: global leadership. Second, the pursuit of global leadership pushes the party-state to behave in a more accountable way, for Chinese leaders face greater pressure for improving governmental transparency. To ensure that it is liked, the party-state has to justify its positions on domestic policies in front of both the domestic and international audiences. This section will show how international constraints have pressured the party-state to mobilise "double justifications". Under international pressure, the party-state chooses the courses of actions that are likely to increase its international credentials. Loosening information control figures in such actions.

For the autocrats aspiring for global leadership, the rising necessity of undertaking "double justifications" on domestic issues is due to the reduced effectiveness of the "feigned virtues", term coming from Machiavelli's *The Prince* (1532). In this fundamental piece of classic realism, Machiavelli recommended that it sufficed that the Prince *appeared to be* virtuous. The Prince should not hesitate to behave immorally when the defence of national interests required. If Machiavelli's thoughts are to be extrapolated to the American leadership in the post-1989 world, it can be inferred that feigned virtues no longer work so well in an international context characterised by an easy access to information and a non-stop exposure of the "Prince" to the public scrutiny.

After 9/11 terrorist attacks in 2001, the USA invaded Iraq in March 2003. Before doing this, the Bush administration spent one year and half in doing "double justifications" at the national and international levels to "appear virtuous". For instance, the speech of the American Secretary of Defence Donald Rumsfeld on 27 May 2003 at the UN headquarters was intended to seek international support for the American military actions in Iraq. Allegedly instructed by the Vice-President Dick Cheney, president of the American petroleum behemoth Halliburton before 2000, speechwriters used concocted proves to toughen up the arguments. It is important to note that what differentiates today's world from the epoch of Machiavelli concerning leaders' accountability is that operating within a highly connected world, feigned virtues are more likely to be denounced than in a closed world. The failure to corroborate the alleged existence of massive destructive weapons in Iraq, and the denounced atrocities committed to the Afghan soldiers detained in the Guantanamo Bay prison, cast a heavy blow to the American global leadership. As Nye argued, the international outcry against the US invasion of Iraq "undercut soft power, reducing the ability of the United States to achieve its goals without resorting to coercion or payment" (Nye, 2004: 16). The stained American leadership due to the Iraqi war bears witness to the ineffectiveness of "feigned virtues". It also shows that the influence of the virtues on the preservation of the global leadership is substantial, rather than rhetoric.

Since the virtues of a global leader nowadays are put under non-stop international spotlight, it becomes very complicated to preserve international credibility by pretending to be virtuous without supporting the pretence with concrete efforts. China is no exception. Its secretive bureaucratic culture and the increasingly sophisticated Great Firewall do not suffice to shield the country from international scrutiny any more. The international rally intended to make independent investigations into the roots of the covid-19 in China due to lack of transparency from Chinese authorities shows how feigned virtues no longer work. The reduced effectiveness of "feigned virtues" entails important implications for the politicking of the party-state. Instead of pretending to be accountable, the party-state must make visible efforts to show that as a would-be global leader, it is genuinely rather than apparently virtuous. To live up to the "virtues", it must show responsiveness to the international audience on foreign policies as well as on domestic policies.

In the contemporary context of "open authoritarianism", the dividing line between the domestic and international spheres is blurred. The possibilities where domestic affairs backfire into the international ambit, and *vice versa*, have increased. In this context, autocracies mobilising systematic censorship or violence run greater risks than before to expose themselves to the international spotlight and to lose international support (Guriev and Treisman, 2019). As the largest autocracy in the world, China sees the interpenetration between its domestic and foreign policies increase. However, it is noteworthy to point out that foreign countries are not equally capable to wield pressure over China. Countries that economically depend on China (e.g., Greece, Venezuela, Ethiopia, Uganda) or politically/ideologically align with it (e.g., Russia, Cambodia) wield less pressure than the countries having no dependence on Beijing. Countries maintaining more competitive relations than dependent relations with

China are also the countries that the party-state needs to persuade in priority by deploying tangible efforts. To illustrate how the international pressure has pressured China to behave in a less authoritarian manner, no example is more appropriate than the international outcry provoked by the alleged detention of Uighurs.

The impressive 403 pages of leaked documents on the detention of Uighurs and Kazakhs were made public by *The New York Times* on 16 November 2019.[7] They showed that the internment campaign was initiated after the killing of 31 people in the Kunming train station by Uighur people on 1 March 2014. From 27 to 30 April 2014, Xi Jinping visited Urumqi and Kashgar, two cities in Xinjiang. This visit denoted the high importance that the Chinese leadership granted to the killing in Kunming in the previous month. The leaked documents showed that Chinese leaders planned to adopt the "war on terror" approach that the Bush administration adopted for tackling extremism and terrorism in the wake of the 9/11 terrorist attacks. The nomination of the hawk, Chen Quanguo, on 29 August 2016 as the Party Secretary of Xinjiang and the First Political Commissar of Xinjiang Production and Construction Corps, was inscribed into the Chinese version of the "war on terror". In contrast with his predecessor Zhang Chunxian, known for his gentle approach to the Xinjiang issue, Chen used highly coercive measures to ensure stability. Since 2018, around 1 million Uighurs of the ethnic region have been detained under the cover of receiving vocational training.[8] The authorities of Turpan, a prefecture-level city in Xinjiang, prepared detailed Q&A documents from which local officials learnt standard answers. This allowed them to answer the questions that college students, who returned to Turpan for Summer holidays, might ask concerning their lost parents (*Ibid.*).

The facts revealed by *The New York Times* and subsequently by other leading international media (e.g., based on the surveys of the Australian Strategic Policy Institute, *Le Monde* revealed that there existed 380 internment centres in Xinjiang)[9] caused wide international attention. After a highly detailed and poignant reporting of the BBC on the rape and

[7] Ramzy, Austin & Buckley, Chris, "Absolutely No Mercy: Leaked Files Expose How China Organised Mass Detentions of Muslims," *The New York Times*, 16 November 2019.

[8] Ibid.

[9] "Chine: une enquête fait état de 380 centres de détention de Ouïgours au Xinjiang," *Le Monde*, 24 September 2020.

torture of women (BBC covered in particular the story of Ms Tursunay Ziawudun in Kunes county) in the detention centres on 2 February 2021,[10] the Chinese Ministry of Foreign Affairs was obliged to reply to these allegations. Although at the press conference on 24 February, Wang Wenbin, spokesperson of the Ministry of Foreign Affairs, argued that Ms Ziawudun was used by foreign forces to denigrate China, the reporting of the foreign media on the issue tainted the image of China. In March 2021, the European Union, USA, Canada, and the UK applied travel bans and asset freezes to four Chinese officials and to the Xinjiang Production and Construction Corps, a Han-dominated governmental organism enjoying the same administrative rank than the provincial government of Xinjiang autonomous region. Multinationals, including H&M, Hugo Boss, Inditex (parent company of Zara), Adidas, Burberry, Nike, and Gap, expressed their concerns towards the human rights conditions in Xinjiang. Better Cotton initiative, a non-for-profit organisation created in 2009, stopped approving products sourced from Xinjiang in 2020 for human rights concerns.

Nowadays, China has become much less vulnerable to international sanctions than during the sanctions for the 1989 crackdown. For instance, the Swedish retailer H&M had to soften its human rights criticisms against China in March 2021 after Chinese consumers' boycott. However, economic power alone does not always provide effective immunity against international criticisms. The reason resides in the entering of new variables into the party-state's equations. While in 1989, China represented merely 1.2% of the global GDP, the percentage rose to 16.28% in 2019. The differences of numbers denote not only the rise of the economic power of China, but also the increase of its political ambitions. Being the second largest economy in the world since 2010, China increasingly aspires to obtain global leadership that matches better its economic status. That is why the concept of "Chinese Dream" of Xi Jinping has been so widely mediatised by the state media since 2013. In contrast with the low-key of Deng Xiaoping who taught officials to "hide your brightness and bide your time", the current Chinese leadership does not hesitate to talk about the "four confidences" and the superiority of socialism over capitalism (Ekman, 2020: 32). Since China has been "coming closer to the

[10] Hill, Matthew, Campagnale, David & Gunter, Joel, "Their Goal Is to Destroy Everyone: Uighur Camp Detainees Allege Systematic Rape," *BBC News*, 2 February 2021.

centre of the global stage day after day",[11] how the country is perceived internationally becomes important for the party-state, because the latter would difficultly assume global leadership without acquiring international respect. Therefore, it is expected that the need for establishing a good international reputation will limit the party-state's freedom of action on domestic matters, including how to conduct information control on ISM platforms.

The overwhelmingly negative reporting in the international media pressured the party-state to provide "double justifications" on the Xinjiang issue. While the justification of domestic policies to the domestic audience has nothing special, the justification of domestic policies to the international audience distinguishes the pre-2013 China from the post-2013 China. A comparison of the reactions of the Chinese Ministry of Foreign Affairs to the international criticisms related, respectively, to the Xinjiang riots in 2009 and the alleged systematic rape in the internment camps in 2021 helps understand the increasing attention that the Chinese government pays to the international audience. After the Xinjiang ethnic unrest on 5 July 2009, the Chinese government did not take effective measures to calm down international criticisms. Invoking the principle of non-interference, Chinese authorities disdained clarifying to foreign observers and handled the issue behind closed doors. On 9 July, President Hu Jintao convened a meeting of the standing committee of the Politburo to discuss the situation in Xinjiang. During the press conference on the same day by the Ministry of Foreign Affairs, the spokesperson, Qin Gang, explained that what happened in Xinjiang "totally belonged to Chinese internal affairs" and that China rejected the proposal of any country to submit the Xinjiang riots for a discussion at the UN Security Council.

The reaction of the Chinese Ministry of Foreign Affairs to the negative reporting on the systematic rape in 2021 was different. Although the emphasis on the principle of "non-interference" remained unchanged, the Ministry showed higher openness to the enquiries of foreign media. Since the beginning of February 2021 to July 27, the provincial government of Xinjiang was allowed to organise 13 press conferences at the

[11] Xi, Jinping, "Speech at the 30th Collective Study Session of the Politburo of the CCCCP," cnr.cn, 31 May 2021, http://news.cnr.cn/native/gd/20210601/t20210601_525501556.shtml.

Ministry of Foreign Affairs to discuss with Chinese and foreign journalists, an initiative that had never been mobilised in the past. Some of these press conferences were organised around a pre-selected topic. For instance, the fourth Press Conference on 10 February 2021 focused on the living conditions within the vocational training schools. On this occasion, 15 graduated trainees presented their learning experiences of computer, cooking, and constructions at the vocational training centres.[12] When these lines were written in August 2021, the ministry continued to organise these press conferences. The transcription for each of them was available online.

Another change is that the Ministry of Foreign Affairs has become more patient and less touchy when discussing human rights conditions in Xinjiang. In contrast with the traditionally laconic answers reminding of the hasty words of an offended person, the Ministry of Foreign Affairs has been giving more detailed explanations to appear persuasive and to calm down international accusations. At the press conference on 23 February 2021, the spokesperson of the Ministry replied to harsh questions from foreign journalists by providing detailed information and facts.

Journalist of *The Globe and Mail* (a Canadian newspaper): The Canadian House of Representatives unanimously voted to approve a motion which identified that China practiced "genocide" in Xinjiang. The motion also stated that if China continued to implement "genocide", the 2022 Winter Olympics should be held in a different place. The Chinese Embassy in Canada responded that Canada should deeply reflect on the tragic experiences of its indigenous people. How should I understand this statement from the Chinese Embassy? Is it true that countries that committed bad practices in the past have no right to criticise China?

Spokesperson: The essence of Xinjiang-related issues is anti-violence, anti-terrorism and de-radicalisation. China has been carefully implementing in Xinjiang the Action Plan to Prevent Violent Extremism elaborated by the United Nations. It takes inspiration from the practices used in other countries and effectuates the deradicalisation work in accordance with laws. China completely abides by the principles and the spirit of the UN Global Counter-Terrorism Strategy. With the efforts of all the ethnic

[12] "Transcript of the Fourth Xinjiang-related Press Conference Organised by Xinjiang Autonomous Region in Beijing," Foreign Ministry of China, 10 February 2021, https://www.fmprc.gov.cn/ce/cedk/chn/zgwj/t1856252.htm.

> groups, Xinjiang has not seen any violent terrorist cases for four consecutive years...In recent years, the population size of the Uighurs keeps rising. From 2010 to 2018, the number of the Uighurs in Xinjiang had increased from 101 715 to 127 184, an increase of 25.04%. The population increase of the Uighurs not only outnumbers the average increase of 13.99% in Xinjiang, but also the 22.14% of all the minorities, let alone the 2% of the Han people...The facts show that there has never been any "genocide" in Xinjiang...[13]

The patience of the Chinese Ministry of Foreign Affairs in replying to the charges of genocide in 2021 interestingly contrasts with its touchiness on the Xinjiang ethnic unrest in 2009. The more China moves closer to the centre of the global stage, the more it becomes patient to discuss the Xinjiang issue the international reporting of which has been tainting the Chinese image since the denunciation of internment camps in 2019. The increasing patience and moderation do not originate from the pure goodwill of the party-state. Instead, it closely relates to the payoffs coming along with the nurturing of a Samaritan image on the international scene. Exposed to international spotlight, the party-state must convince other countries of its adherence to moral principles and live up to the reputation with tangible efforts. The failure to do so will hinder the global ambitions of the party-state, unless the latter deploys efficient strategies allowing it to mitigate or to shake off international pressure. In fact, Xinjiang is only one of the many sensitive issues the handling of which puts the reputation of party-state at serious stake. The aspirations for global leadership imply that China has to prepare itself to resist the criticisms on these sensitive issues and to peddle its own storytelling to international audiences. The next section, to which we now turn, analyses the crafting of the storytelling by the party-state and discusses their efficiency in mitigating international constraints.

[13] "Mr Wang Wenbin, Spokesperson of Foreign Ministry of China, Presides the Routine Press Conference on 23 February 2021," Foreign Ministry of China, 23 February 2021, https://www.fmprc.gov.cn/web/fyrbt_673021/t1856054.shtml.

Limited Efficiency of the Strategies for Shaking off International Constraints

The party-state has been deploying two strategies for mitigating international pressure: at the international level, it supports the doctrine of cyberspace sovereignty; at the national level, it mobilises a jingoistic rhetoric, delegitimising the international public opinion by portraying it as deliberately anti-Chinese. At the international level, the administration of Xi Jinping has developed the doctrine of cyberspace sovereignty as a new philosophy for internet governance. It has promoted the concept at the international conferences under the Chinese patronage. At the 6th World Internet Conference in October 2019 in Wuzhen, cyber-sovereignty is conceptualised in the document titled "*Sovereignty in Cyberspace: Theories and Practices*", project co-launched by Wuhan University, Shanghai Academy of Social Sciences, and China Institute of Contemporary International Relations. The document defines "cyber-sovereignty" in the following terms: "sovereignty in cyberspace is the extension of national sovereignty to cyberspace. It is the supremacy and independence that a state enjoys, on the basis of its national sovereignty, over cyber infrastructure, entities, behaviour, as well as relevant data and information in its territory". The respect of cyber sovereignty entails the respect for a country's choice of the internet development paths, the regulatory frameworks, internet-related policies, and the right to participate in global internet governance.

The implications of the cyber-sovereignty doctrine are twofold. First, claiming the right of each state to govern cyberspace as it sees fit is "about legitimising China's internet controls, and gaining tacit approval and cooperation by foreign countries and foreign internet companies" (Bandurski, 2018: 383). It could reduce international pressure on China by *de facto* introducing the non-interference principle into the governance of cyberspace, thus delegitimising international criticisms towards Chinese internet censorship. The cyber-sovereignty doctrine actually nullifies the moral pressure that the international community could wield on China. It is no surprise that the term "internet sovereignty", interchangeably used with "cyber-sovereignty" in Chinese language, has significantly risen in recent years. As shown in Fig. 7.1, from Q3 2010 to Q3 2015, the frequency with which "internet sovereignty" is mentioned in Chinese media increases. This tendency is particularly obvious since Q4 2012, a moment corresponding to Xi's arrival to power. At the peak moment,

Q4 2014, "internet sovereignty" appears more than 150 times. The momentum in the promotion of the doctrine comes from the CAC. Since its creation in March 2014 as the tsar of content regulation, it has been deploying continuous endeavours to make "internet sovereignty" known to all.[14]

Second, that the CAC has painstakingly elaborated and spread cyber-sovereignty serves not only to mitigate international pressure on China, but also to justify Chinese internet controls. What is inherent to the cyber-sovereignty doctrine is the idea that each national government is sovereign in choosing the regulatory policies without external interference, even though these policies imply restrictions to the freedom of expression of domestic citizens. The party-state's justification of the domestic internet censorship to the international community illustrates that the international context does create constraints to the party-state's freedom of action. Should there be no constraints, the party-state would not have spent the past 9 years on elaborating the "cyber-sovereignty" doctrine.

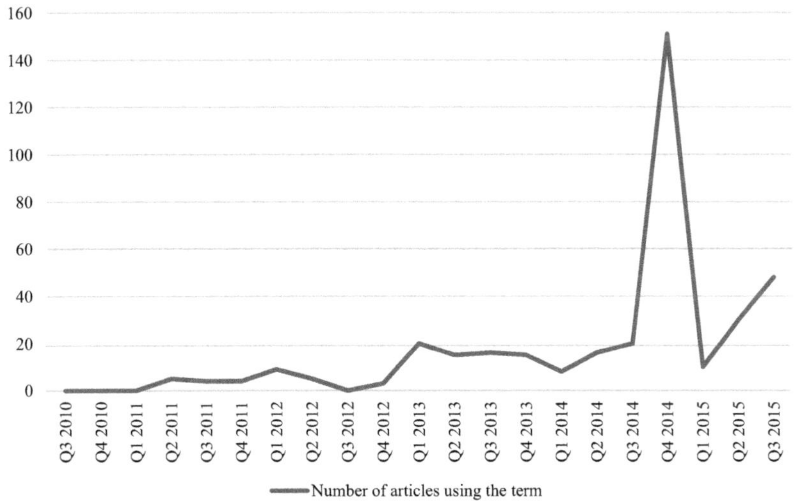

Fig. 7.1 Frequency with which "internet sovereignty" appeared in Chinese media (2010–2015) (*Source* China Media Project)

[14] Bandurski, David, "Re-defining Cyberspace," *China Media Project*, 7 October 2015.

It would not have created the World Internet Conference (WIC) as an international venue to sell the doctrine to the external world.

However, the cyber-sovereignty doctrine has not travelled far. One article of Lu Wei, dubbed as the "gatekeeper of the Chinese internet", in *Huffington Post*, expounded the Chinese multigovernmental and the state-centric approach to internet governance (Bandurski, 2018). This approach stood opposite to the multistakeholderism and human-centric approach (Arsène, 2012: 298–299) championed by the USA. This article only received 49 Facebook likes and 69 Twitter shares in nine months (Bandurski, 2018). The weak feedback to the article shows lack of international interest in "cyber-sovereignty".

The declining participation into the WIC also weakens the global diffusion of the doctrine. At the end of the 2014 WIC, Lu Wei failed to persuade the attendees, including the representatives from Microsoft, Amazon, and Apple, to sign up a declaration which included a provision on cyber-sovereignty (Economy, 2018: 80–81). The participation of high-profile attendees to the China-championed WIC has been in decline. In 2017, during the 4th WIC, high-profile foreign speakers included Tim Cook of Apple and Sundar Pichai of Google. Bob Khan, known as one of the fathers of the internet, addressed the conference. In 2018, the sole Western executive who gave a keynote speech was Steve Mollenkopf of the chipmaker Qualcomm (Table 7.1). In 2019, the only representative of foreign firms, Kang Rong, was the vice-CEO of Microsoft in the Greater China Region. Foreign speakers with the standing of Tim Cook or Sundar Pichai were absent. The loss of international attractiveness of the WIC does not help the global diffusion of the cyber-sovereignty doctrine.

The second strategy for reducing international pressure is to play the card of nationalism and to mobilise the jingoistic rhetoric in cyberspace. In so doing, the party-state allies itself with Chinese firms and netizens to incapacitate foreign critics. The triangular alliance builds a comfortable situation for the party-state. Perceiving the negative international reporting as an offense to national pride more than anything else, some radical netizens lend unconditional support to the Chinese government. Under the circumstances where China as a nation is targeted by foreign media, some netizens concentrate their energy on counterattacking foreign critics, forgetting temporarily their gripes caused by internet censorship. Government-paid trolls are particularly active in such circumstances (Shirk, 2011). The mobilisation of the jingoistic rhetoric intensified in the 2010s. From 2011 to 2016, the previously dominant

Table 7.1 Program of the opening ceremony of the 2018 World Internet Conference

Opening Ceremony of the 5th World Internet Conference	
Message of Congratulations from H. E. President Xi Jinping, read by	
HUANG Kunming	Member of the Political Bureau of the Communist Party of China (CPC) Central Committee Member of the CPC Central Committee Secretariat Head of the Publicity Department of the CPC Central Committee
Keynote Speech	
HUANG Kunming	Member of the Political Bureau of the Communist Party of China (CPC) Central Committee Member of the CPC Central Committee Secretariat Head of the Publicity Department of the CPC Central Committee
Moderator	
ZHUANG Rongwen	Minister of Cyberspace Administration of China
Welcome Address	
CHE Jun	Secretary of the CPC Zhejiang Provincial Committee Chairman of the Standing Committee of Zhejiang Provincial People's Congress
Speakers	
Francis Gurry	Director General of World Intellectual Property Organisation
John Hoffman	CEO of GSMA Ltd
Steve Mollenkopf	CEO of Qualcomm Technologies, Inc
Pony Ma	Chairman and CEO of Tencent
Whitfield Diffie	2015 Turing Award Winner
YANG Jie	Chairman of China Telecom Corporation
Shaukat Aziz	Former Prime Minister of Pakistan

Source Collected from Twitter on 17 June 2021

legitimacy-challenging imaginary grounded in Western democratic references had been absorbed by a nationalist, "civilisational competition" discourse that enhanced the regime's legitimacy (Wu, 2020). Jingoistic discourses have shown to be efficient on several occasions to shield the party-state from international criticisms.

On 14 March 2008, a violent confrontation between Han and Muslim residents broke up in Lhasa, capital city of the Tibet autonomous region in the west of China. Similar conflicts happened between Tibetan monks and laypeople. Tibetans set fire to the vehicles and looted businesses owned by Han and Muslim. After the riot, the videos in which

Tibetan monks were beating Chinese shopkeepers went viral online and sparked strong sense of nationalism. Chinese netizens condemned not only Chinese nationals who showed sympathy towards Tibetans, but also foreign media that criticised China. Wang Qianyuan, a Chinese student studying at Duke University in the USA in 2008, was labelled as a "traitor" by her Chinese classmates for having tried to reconciliate supporters of Tibetan separatism and supporters of Beijing. Nationalists launched a large-scale human flesh search against Wang. They publicised her telephone number, home address in China, number of identity card, the name of her parents, and their work units. They even threatened to kill her. Foreign media were the object of intensive condemnations in Chinese cyberspace. Shirk described how CNN was targeted by Chinese netizens:

> The public opinion furore ratcheted up when Chinese netizens criticised the Western media for biased coverage of the Lhasa events and established an anti-CNN Website. CNN became the target because it supposedly cropped a photograph to remove Tibetans throwing stones at a Chinese tank, and subsequently an intemperate CNN commentator, Jack Rafferty, described the Chinese as 'a bunch of goons and thugs'. (Shirk, 2011: 246)

The explosion of nationalist feelings around the Lhasa riot somehow protected the Chinese government from international pressure, because the backup of the domestic public opinion allowed Beijing to be hawkish in handling international criticisms. Below is an excerpt of the Regular Press Conference on 27 March 2008 by Qin Gang, then Spokesperson of the Chinese Ministry of Foreign Affairs. It showed how Beijing actively positioned itself with the domestic public opinion when the latter allowed it to invalidate international criticisms:

> Question: You may be aware that there's now a non-governmental website, ANTI-CNN.COM, where people criticise the untrue reports on the Tibet issue published by some foreign news agencies like the CNN. Do you appreciate or support this website? Is the Chinese government providing financial or material support to this website?...
>
> Answer:... You ask whether the Chinese government is involved. You won't ask that if you take a look at the reports by the western media. It is these irresponsible and unethical reports that infuriate our people to voluntarily voice their condemnation and criticism.

> The Lhasa incident is over. Our country will have a better future, and so will Tibet. But it leaves us with a legacy, a mirror where we can see the true face of some people out there. It is also a textbook of bad examples, and it helps our people discern clearly the essence of the much-vaunted 'justice' and 'objectiveness' by some western media. What happened in Lhasa is bad and unfortunate; but if there's anything good about the Lhasa incident, this will be one.[15]

On the Lhasa riot, the unity between the party-state and netizens enables the former to delegitimate international criticisms. For netizens, circumstances where China suffers from unjustified international criticisms lead them to defend national honour more than to complain about censorship. However, the efficiency of jingoistic discourses is not always guaranteed. Chinese diplomats and decision-makers are actually very cautious about the expansion of nationalist feelings in the country. They are afraid of not being able to satisfy the extreme nationalist claims of Chinese people if they let nationalism freely grow in China (Shirk 2011b). Since jingoistic discourses constitute a double-edged sword, the party-state does not afford to systematically use them to shake off international pressure. In addition, the effectiveness of jingoistic discourses is also constrained by its weak resonance among Chinese elite who demonstrates a high level of immunity to nationalist discourses.

My fieldwork shows that Chinese netizens develop differentiated receptivity towards nationalist discourses depending on their socio-economic status. To evaluate the prevalence of nationalism among netizens, one question that I asked to my interviewees was whether they would feel proud, anxious, or indifferent if they lived in a China-led global order. I find that the level of education of interviewees partly explains their reactions. The more an interviewee is educated, the less she is likely to feel proud. This is not because they don't like China. Instead, it is often due to the fact that they feel indifferent to whether China becomes the global leader or not, for this can seldom change their personal life.

Interviewees' multicultural experiences count, too. Compared with the interviewees having no overseas experiences, those who had been abroad for a relatively extended period (at least one year in my sample) granted

[15] "Foreign Ministry Spokesperson Qin Gang's Regular Press Conference on 27 March 2008," Foreign Ministry of China, 27 March 2008, https://www.mfa.gov.cn/ce/ceus//eng/fyrth/t419160.htm.

less importance to whether China became a global leader or not. What is interesting is that among some interviewees who attended college and who had multicultural experiences, they took a satirical tone when talking about patriotism and squarely considered it as propaganda. They also showed some sense of sarcasm towards the people claiming to be patriotic, arguing that the explicit display of love for the motherland was seldom sincere. Patriotism for this group of interviewees was almost a pejorative term reminiscent of cowardice and lack of courage in pushing social changes in China. When asked whether he thought a China-led world order would be better than an USA-led one, one journalist I interviewed deliberately imitated the tone of propaganda officials and declared "Oh, I love President Xi Jinping! I love China!". After this, he laughed and explained how he felt anxious about the diffusion of authoritarianism in the world.[16]

I did not encounter many interviewees who expressed explicit nationalist inclinations. Exceptions existed. One interviewee immediately started her professional life after obtaining a master's degree in medicine. She had no opportunity of travelling abroad. Among the group of users I interviewed, her opinion was the most representative of the nationalist feelings among Chinese people.

Researcher: Do you think it is important for China to replace the USA to become a global leader?

Interviewee: Very important, I will be very proud, very excited…The fundamental reason for which Chinese people now enjoy a high social status [at the international level], it is because the overall strength of China has improved. It is like two families, if your family is very poor, the other family will bully you, like Africa. Since China has become powerful, the USA and Japan don't dare to bully us, yes, it is true. So I will be very proud…Frankly speaking, whether the Chinese GDP becomes No.1 in the world has nothing to do with my salary, but when it comes to the overall strength of the country, it does matter a lot for me."[17]

Except a few cases of exceptions, the great majority of the interviewees separated the party-state's interests from their own interests, considering the pursuit of global leadership as the concern of top policymakers only.

[16] Interview O-06/07/2020.

[17] Interview B2-19/12/2019.

Interviewees did not automatically adhere to the party-state's concern of becoming a global leader. One interviewee studied in Italy and was working in a Chamber of Commerce in Beijing at the moment of interview. For her,

> [That China becomes No.1 in the world] may make you feel proud and superior to the people in other countries, but I don't think that it is the most important. What matters for me is the tangible improvements in the living conditions of individuals, improvements in education quality, or significant progress in healthcare, or high level of freedom of speech, or high level of social security. I care about these very mundane aspects.[18]

One comment posted on pincong.rocks on 18 March 2020, a Chinese language forum where netizens discussed current affairs and sensitive topics, revealed how, for many Chinese people, patriotism was a window-dressing gesture intended to obtain real gains from the system:

> In my life experiences, most adult people don't love the country that much. When it comes to their personal interests, their first choice can be to betray the country if betrayal brings them money. On the contrary, if patriotism brings money, they will love the country to the fullest...people who sincerely love the country are often very young people or vulnerable groups of people. Patriotism allows them to temporarily forget their hardship and insignificant life. It allows them to feel like the masters of the country at an unknown moment.[19]

Lack of patriotism among the elites partly comes from their questioning of whether the regime truly represents citizens' interests as it claims. Some reckon that China is *de facto* a regime serving the interests of only a small group of people. One interviewee declared:

> When I hear people say how much they love the country, I just want to laugh. I want to ask: 'Is there any relationship between you and the country you talk about? Do you really consider yourself as the master of the country? The country has nothing to do with you (*guojia he ni meiyou banmaoqian de guanxi*)!' Our country is controlled by a small group of

[18] Interview B1-19/12/2019.

[19] Retrieved on 28 January 2022 on https://pincong.rocks/video/1513.

> people. Although we have not voted for them, they make decisions for us.[20]

For this interviewee and others like him, since the party-state does not sincerely take citizens as masters of the country, it does not have the legitimacy for requiring them to be patriotic. Their understanding of patriotism is from a legal and political perspective in which the love for the country is equal to the support of the political regime in place. This is also the perspective adopted by the Chinese propaganda apparatus. At one conference, Zi Zhongjun, former Director of the Institute of American Studies at Chinese Academy of Social Sciences, distinguishes patriotism at three levels: love for the country, love for the nation, and love for the state. For her, while everybody loves the country (e.g., natural environment, place of birth) and the nation (e.g., culture, tradition, language), love for the state is not unconditional. In her words:

> When it comes to the third level, it is very different, you can choose not to love it, you can hate it, you can love, you can defend it, you can overthrow it...It is not innate, we are not obliged to do this or that. In a globalised world, you can also stop being the national of a country and take up the nationality of another country.[21]

Among the three levels of patriotism, the third level is the least developed among the elites, implying that the regime can difficultly rely on them to sing its praises outside China. On the contrary, it must fight back unfavourable international opinions on its own, leaving government-paid internet commentators (*Fifty Cents Party*) almost the only army in town. However, the voices of the Fifty Cents Party outside China have been increasingly identified and rejected by foreign ISM platforms. The difficulties of internet commentators in convincing the international audience do not help the regime's construction of a friendly international image.

Mobilising nationalist feelings shrinks China's circle of friends. At the beginning of June 2021, a report of the Japanese Ministry of Foreign Affairs showed that from 2008 to 2016, 73 Chinese celebrities and

[20] Interview B-18/12/2019.

[21] "Zi Zhongjun: When 'ism' Adds to 'Patriot', Patriotism Has a Coercive and Utilitarian Nature," Xinhuanet, 23 May 2014, http://www.xinhuanet.com/book/2014-05/23/c_126537512.htm.

71 Chinese organisms had been invited to visit Japan. A high-profile Chinese writer Ms Jiang Fangzhou figured on the list of the invitees. This information quickly became the fodder of nationalist bloggers on Weibo represented by Guyan Muchan and Diguaxiong Laoliu, who had, respectively, 6.4 million and 6 million followers. On the post of 10 June, Guyan Muchan uploaded a video clip where she criticised that Jiang accepted the Japanese financial help, glorified Japan as a free society, and denigrated the Chinese life as being circumscribed within the Great Firewall. Within 12 days (from 10 to 22 June 2021), the post obtained 5279 likes. However, netizens' attention subsided in the following months. When I checked this post again on 24 January 2022, the number of likes reached 5549, showing that this post didn't attract wide attention in the second half of 2021. Another blogger adapted the story into a short video. It featured a Chinese traitor who telephoned to a Japanese official, complaining about the release of the list of the Chinese elites who had received money from Japan. Two weeks after its release on 8 June, the post obtained 17 808 likes.

This time the Chinese government tried to cool down nationalist discourses. At the press conference on 9 June 2021, Wang Wenbin, spokesperson of the Chinese Ministry of Foreign Affairs, explained that "it is common practice in international relations to organise various forms of exchanges and visits. As important close neighbours, China and Japan have a lot of personnel exchanges. Since the normalisation of the Sino-Japanese relations nearly half a century ago, the two sides have supported and facilitated numerous programs of people-to-people exchange, which have positively helped the growth of bilateral relations".[22] The cooperative attitude of Wang contrasted with the aggressivity of his colleague Zhao Lijian who embodied Chinese "Wolf Warrior Diplomacy" on several occasions (e.g., the controversial twit of Zhao Lijian in November 2020 on the killing of Afghan civilians and prisoners by Australian soldiers). The change of attitude of the spokespersons was in line with the spirit of the 30th Collective Study of the Politburo, presided by Xi on 31 May 2021. While he emphasised the importance of increasing the effectiveness of external propaganda, he highlighted that China should befriend foreign countries. This meeting can be considered as a harbinger of future

[22] "Foreign Ministry of China: in International Relation Practices, It Is Very Common that States Organise Reciprocal Visits and Exchanges in Various Forms," Sohu.com, 9 June 2021, https://www.sohu.com/a/471268756_162758.

changes to the party-state's approach to external propaganda. Instead of continuing to practice Wolf Warrior diplomacy, the Chinese leadership is expected to create a friendlier international image of China.

In line with the efforts of the Chinese Ministry of Foreign Affairs to appease netizens' nationalist feelings towards Japan in 2021, content regulators occasionally censor pro-regime voices to prevent them from growing into uncontrollable forms. For instance, the pro-government microblogger Guyan Muchan was censored by Sina Weibo for 15 days in September because her radically patriotic posts were perceived as violating the code of conduct of the platform. Another pro-regime voice Li Guangman was censored awhile after publishing a combative article where he argued that the Chinese society was undergoing "the second revolution". Li called upon "the return of redness, heroes, and blood".[23] The radicalness of Li's article even attracted the criticisms of Hu Xijing, editor-in-chief of the party mouthpiece *Global Times*. Censorship of nationalist speeches reflects a sort of "patriotism without state blessing" (see Han, 2019, about Diba Expedition). Stoking nationalism in cyberspace is a double-edged sword for the party-state (Shirk, 2011), because it can embarrass the latter by awakening among netizens the nationalist requests that the party-state may be unable to satisfy. Under such circumstances, the party-state can become the victim of its own endeavours of promoting cyber nationalism. Since mobilising nationalism is not failsafe, the party-state will continue to be prudent in using it. It implies that the double-bind logic that the international context participates to build in China will persist and continue to constrain the party-state's freedom of action.

Firm-Netizen Alliance in the Pursuit of Corporate Autonomy

That the party-state aspires to take global leadership deepens its predicament by increasing the payoffs for it to be responsive and accountable to the international public opinion. However, increasing responsiveness and accountability of autocrats is not always compatible with their objective to ensure the regime's survival, because to be responsive implies that

[23] "Li Guangman: Why Do the Articles of Chinese Self-Media Writers Cause Anxiety of a 'Cultural Revolution 2.0'," BBC, 13 September 2021, https://www.bbc.com/zhongwen/simp/chinese-news-58500007.

autocrats must accept to reduce arbitrariness in their exercise of power. In other words, it requires autocrats to be less authoritarian. When it comes to the information control, to be less authoritarian implies that autocrats accept to leave the interstices open through which certain boundary-spanning expressions can go online. As shown above, the party-state has made attempts to attenuate the dilemmas facing it, but in vain. This section empirically demonstrates how the party-state's predicament on the political axis turns out to be a structural opportunity for private ISM firms.

Alignment with the Domestic Public Opinion to Lower the Risk of Punishments

On the political axis of the ISM regulation, the party-state's predicament creates opportunities for ISM firms to seek autonomy. On the one hand, the party-state is in need of pursuing global leadership, implying that it must improve its image on the global stage. Allowing ISM firms to carry a larger diversity of contents than government-sanctioned ones is a sign of greater openness, helping increase the global acceptance of China. On the other hand, the imperative of preserving authoritarian reign makes it dangerous, even suicidal, for the party-state to do so, because it is not in its interests to let the governed express critical or subversive opinions. The predicament between pursuing global leadership and preserving domestic authoritarianism complicates the political regulation of the ISM. It obliges the party-state to navigate between the two objectives all the time, loosening information control at one moment, tightening it immediately after. The constant shift between more control and less control implies that the party-state cannot choose a definitive regulatory approach to the ISM industry. Since it does not afford to permanently resort to one option or another, the alternance between the two becomes an iterative process.

To make the party-state concede, the international public opinion can amplify its influence if sustained by the domestic public opinion. In case of disagreement between the two, the international audience tends to have a limited influence on the content regulation: either the party-state allies with the domestic opinion to resist international pressure; or it calls upon the domestic audience to calm down, keeping the overly nationalistic remarks at bay. In both situations, the opinion of the domestic audience is a game-changer for shifting the power relations between firms and the party-state. That is why even without international support, the coalition with the domestic audience alone can also help private ISM firms to push the boundaries of the "speakability" in cyberspace. An extensive literature has been done on the strength of the domestic public opinion in China (Teng, 2012; Negro, 2017; Tai, 2006; Yang, 2009; Lei, 2018). The research by Teng (2012) on how Chinese rights activists have used Sina Weibo to gather larger-scale support and seek favourable rulings from courts shows the importance of the domestic public opinion.

The domestic audience weighs high to the party-state for two reasons. First, citizens are perceived as a potential source of instability that the party-state should bring under control. As a sign of the party-state's vigilance vis-à-vis citizens, the latter are not allowed to have the same level of information as government officials. As one interviewee pointed out,

> [Chinese internet] is not an internet, but an intranet. It is because this intranet is sufficiently large that we mistakenly consider it as an internet, but it is an intranet. Therefore, information exchanges with the outside world are not smooth. This is a system of management specifically created for domestic politics... However, due to the regime itself, the nature of the regime decides that it does not allow information sharing. The reason is that information sharing will destabilize power relations in China, information sharing will damage the distribution of power. Access to information in China depends on how much power you have. Of course, this phenomenon exists in all countries of the world, but it is particularly prevalent in China. Power-holders and ordinary people have asymmetric access to information. In particular, some basic information, the access to which is part of citizens' rights, is not accessible.[24]

Second, citizens are perceived as the source of legitimacy that the party-state cannot lose. In autocracies like China where "preferences falsifications" prevail (Kuran, 1997), ISM platforms constitute an important window through which the party-state takes the pulse of the public opinion. Strict information control which deepens the falsification of citizens' preferences harms the party-state' interests, because it goads citizens to lie. Moreover, pushing information control too far contradicts the party-state's discourses according to which it takes people's interests to

[24] Interview B1-21/12/2019.

heart. In his widely quoted speech at the celebration of the 100th anniversary of the CCP, Xi Jinping described the altruism of the CCP in its relationship to the people:

> The Chinese Communist Party has always represented the fundamental interests of the overwhelming majority of the people. It has no special interests of its own. It has never represented the interests of any interest group, any powerful group, or any privileged class. Any attempt to separate and antagonise the Chinese Communist Party from the Chinese people will fail! More than 95 million Chinese Communists do not agree! More than 1.4 billion Chinese people do not agree![25]

Xi's speech on the altruism of the CCP contains some truth. Although authoritarian regimes often justify themselves by a performance-based legitimacy, rather than a norm-based legitimacy (Gerschewski, 2018), it is important that autocrats have both for being perceived as legitimate by the general public. To achieve this objective, the party-state needs public support. Therefore, if the content regulation goes as far as to provoke public dissatisfaction, or if content regulators harass ISM firms that enjoy a high level of popularity, regulators tend to seek compromise by tolerating explosive news reporting on ISM platforms. The dynamics in which internet firms ally themselves with netizens to fight back regulators also exist in democracies like the USA. In their insightful research on American platform firms, Culpepper and Thelen (2020) argue that the source of power of these firms is that "unlike the monopolies of the past, today's tech firms enjoy a direct, indeed virtually unmediated, link to their users, most of whom connect to these firms through devices they carry in their pockets every day" (2020: 294). Similarly, the dependence of Chinese netizens on private ISM firms for satisfying their needs for information, entertainment, consumption, and even public services allows firms to coalesce with netizens to deter business-unfriendly policies. However, although platform firms in both China and the USA can mobilise their users to fight economic regulators, the particularity of Chinese firms is that they also need netizens to deal with content regulators, the latter being non-existent in the USA.

[25] "Speech of Xi Jinping at the celebration ceremony of the 100th anniversary of the Chinese Communist Party," Xinhuanet, 1 July 2021, https://www.gov.cn/xinwen/2021-07/01/content_5621847.htm.

That content regulators are constrained by netizens' preferences creates an important opportunity for private ISM firms. More concretely, if the latter ally themselves with netizens, they are much less likely to be penalised by content regulators if they carry politically unpleasant speeches. One interviewee, working as a content moderator in Sina Weibo since 2012, explained how the domestic public opinion counts for regulators:

> Obviously, Sina Weibo is not happy to censor its users, it is not happy to shut down users' accounts either, in particular for the users having tens of thousands of followers. They drive up the traffic on our platform. Some colleagues secretly deblocked the account of some blocked users, and they did not get in trouble with managers...In August 2013, Lu Wei [Director of the CAC] announced the seven baselines. Netizens complained so much about them and ironic comments on the "socialism" baseline were intensive. Like Tencent Weibo, we were under pressure. However, there were so many comments on Weibo that the human checking became gradually laxist, although these comments were still checked by the machine. The CAC seemed to concede too, because we did not receive further instructions from it later on.[26]

Large-scale public outcry against the Seven Baselines created pressure over content regulators. Instructions from the CAC for deleting critical and satirical comments, although categorical in the beginning, attenuated over time. The softening of regulators' attitude was explained by the concern that pushing censorship too far against the expressed public preferences risked backfiring against censors and lowering public support towards the regime. Therefore, when the alliance between firms and netizens is explicitly signalled, it becomes politically risky for regulators to continue to coerce firms. Private firms that find themselves in such situations tend to reduce the intensity of their content moderation and to compete for looking freer than their competitors (Miller, 2018). That content regulators cannot afford to ignite the public anger lowers the costs of committing disobedience for firms. What equally helps ISM firms is that in China netizens show greater sympathy to firms than to regulators. More than one interviewee explained that regulators, not firms, are

[26] Interview O-24/09/2020.

the real culprit of the shrinking freedom of speech in China. One interviewee participated into the deployment of the earliest internet networks in China in the 1990s. When asked to make an overall assessment on the contributions of private ISM firms to the Chinese society, he said:

> Caught between information control and market pressure, I think they [private ISM firms] have already performed pretty well. They have grown into international stars, become a name card of China. They obviously have the potential to grow even bigger...Information control is linked to the political reality of China and it is certainly not fair to take firms as the scapegoat of internet censorship.[27]

The alignment with the public opinion is facilitated by the fact that platforms decide on their own what contents to censor, what policies to apply, and what technical methods to deploy. Firms' discretion in the day-to-day content management is explained by the content regulation regime in China. Since the release of the earliest content regulation rules in 2000 by the SCIO and the MII, private ISM firms have been delegated the task of overseeing their users. The outcome of the delegation is that firms bear both the costs and the moral hazard issues when implementing the content regulation (Jiang, 2012). Although content regulators send daily instructions to leading ISM platforms and that more than 50 governmental organs can instruct content providers to take down contents, firms have high discretion for deciding which measures to take for which types of contents. For instance, platforms rather than content regulators elaborate the lists of sensitive words to censor. They decide whether to put one post into the "small black room", or whether to send a warning message to a concerned microblogger. In addition, technical superiority of ISM firms over regulators makes corporate noncompliance difficult to detect. It has been found that private internet companies in China hide "information about their technical capabilities and budgets to invest as little as possible in developing high-performance censorship systems" (Miller, 2018: 6).

When pursuing margins of manoeuvring, private ISM firms can ally themselves with netizens to lower the costs of committing noncompliance. Netizens' dependence on ISM firms facilitates the creation of the firm-netizen alliance. What equally goes to the interest of private firms

[27] Interview B-23/12/2019.

is content regulators' consistent fear of backfiring that could result from overly coercive regulation rules. Therefore, when the firm-netizen alliance is clearly signalled under the form of netizens' massively explicit disapproval of the content regulation rules and ISM firms' relay of such messages, regulators tend to make concessions, leaving firms certain autonomy in their content management. In regulator-firm battles, the position of netizens often acts as a game-changer. Whoever succeeds in taking netizens on board wins an advantage against the other party. However, it is not wise to be overly optimistic, for the potential of the firms-netizen alliance to make regulators concede is not unlimited.

Limits of the Firm-Netizen Coalition in Firms' Pursuit of Autonomy

It is noteworthy to point out that compared with the post-2013 period, it is in the pre-2013 period that private ISM firms can more easily take content regulators hostage by allying with netizens. The cases where public mobilisations on ISM platforms succeed in making public authorities surrender (e.g., Sun Zhigang case in 2003, Deng Yujiao case in 2009, "Watch Brother" case in 2012) have reduced after 2013. The nomination of Lu Wei in 2013 as the head of the CAC partially explains the strengthening of internet censorship (Economy, 2018). After rolling out the Seven Baselines in August 2013, he initiated the anti-rumour crackdowns against online celebrities in subsequent months. This move reduced the number of Sina Weibo users by 70% from 2011 to 2013 (Economy, 2018: 79). Other variables also contribute to shrink the room to manoeuvre of private ISM firms. For instance, ISM firms have limited autonomy in recruiting key editorial staff. Miller explains that:

> Employees working in editorial functions at internet content providers in China can no longer remain on private payrolls. This means that according to Chinese law, the employees who censor content can no longer be employees of Sina Weibo, but now must be employees of a government agency such as the Network Management Office or the Supervision Department. Whether or not this solves principal-agent problems identified in the censorship system has yet to be seen, but control over the internet has been tightening at a rapid pace over the last few years. (Miller, 2018: 22–23)

Increasing presence of official accounts on ISM platforms also risks shrinking firms' leeway. Liu Lipeng had worked as a content moderator for over 10 years in China. He worked in several leading private ISM firms, including Sina Weibo and Le.com, a Chinese video-sharing platform. Fearing to get into trouble for having illegally collected the logbooks of censorship practices in Sina, he moved to the USA in March 2020. Interviewed by the *Voice of America*, he recalled how the party-state had increased its presence on private platforms since 2013:

> In 2013, after the government learnt to control the internet with greater easiness, not only were platforms moderated by moderators, but also official accounts came to platforms, including internet commentators, public security *xi tong (system)*, and the communist youth league, etc. In the past, there existed internet commentators too. They were the so-called Wumao [50 cents], but most of the time their job was to make the water trouble. However, at the current times, internet commentators rule above all. In contrast, people are less impressed with the suppression of posts and the blockage of accounts. What leaves a clear impression on you is that now each time you click a piece of news, all comments below are pro-CCP.[28]

The party-state's infiltration in cyberspace risks hollowing out the very notion of "public opinion". In one famous article, Pierre Bourdieu declares that "public opinion does not exist" (1979). For him, since the design of questionnaires in opinion polls mainly reflects the interests of the dominating class and that not every surveyed individual is able to produce an opinion, the public opinion produced through opinion polls is merely an artefact. Bourdieu's warning can also be applied to China, although the reason behind the non-existence of the public opinion is different. By pouring pro-CCP comments in cyberspace, internet commentators twist the reality and create a false impression in which only pro-regime voices represent the public opinion. They demoralise and marginalise anti-CCP opinions as not representative of the thoughts of the Chinese people.

Increasing presence of the Wumao party hinders the capacity of ISM firms to fight back unfavourable regulations, because even though the

[28] "Special Interview with the Former Content Moderator at Sina Weibo: How China Is Building an Online 'Ministry of Truth'," VOA, 12 August 2020, https://www.voachinese.com/a/internet-censorship-20200812/5540475.html.

discourses of Wumao are easily identifiable due to their cheerleading characteristics (King et al., 2017) and their lack of engagement with other netizens (Han, 2018), they still signal the strength of the propaganda apparatus (Huang, 2015, 2018). By stint of putting the power of propaganda on display, the increasing online presence of the party-state deters netizens from expressing their genuine opinion. In return, if the Wumao party's pro-regime discourses signal the strength of the regime more than netizens' mockery of regulators signals the netizen-firm coalition, firms will have difficulties in reclaiming their room to manoeuvre. The activism of internet commentators in releasing pro-regime comments creates a *prima facie* coalition between regulators and "netizens". In these circumstances, firms can fail the battle with regulators, for the latter can falsely pretend that they, not firms, succeed in taking "netizens" on board.

Increasing Relevance of the International Public Opinion

If the domestic public opinion can help private firms to resist regulators, the international public opinion has developed a similar potential in recent years. The increasing relevance of the international public opinion to the party-state's pursuit of global leadership makes it possible for firms to resist national regulators by allying themselves with the international public opinion. As explained above, the effectiveness of the international public opinion tends to be limited if it is not echoed by the domestic public opinion. This section dives into how international public opinions contribute to shape the opportunity structure for private firms and how firms exploit this structure to amplify their margins of manoeuvring.

The increasing relevance of international perceptions to the party-state is related to two factors: the party-state's aspirations for global leadership, on the one hand, and the ineffective strategies to shake off international pressure, on the other hand. As for the party-state's aspirations for global leadership, the more it aspires for it, the more it becomes vulnerable to international perceptions. The vulnerability is due to the fact that to become a global leader, the party-state needs greater legitimacy: legitimisation in front of both the domestic and international audiences. The American experiences in claiming the championship of democratic values and the stint to its global leadership for having failed to live up to such values in the Iraqi war show that it is not because a country manages to impose itself militarily that it can become a global leader. It is difficult for any country to establish global leadership merely by dint of sheer heft. Global supremacy established by deterrence only is better called global hegemony (like imperialism), rather than global leadership, because the latter entails wilful acceptance and recognition by other countries. Therefore, on the rugged path leading to global leadership, international acceptance is as important as the economic and military clout. It is particularly true for China whose international reputation suffered and continues to suffer from the Tiananmen Massacre, the allegedly unfair treatment of Uighurs, and lack of transparency on the management of the coronavirus crisis.

The relevance of the international public opinion is likely to increase further, for, as shown above, the two strategies that the party-state has mobilised to mitigate international pressure—advocating the cyber-sovereignty doctrine and mobilising a jingoistic rhetoric—have not proven efficient so far. To illustrate how the pursuit of global leadership constrains the freedom of action of the party-state in the ISM regulation, I focus on the ISM regulation under Xi Jinping (2012–2020). The rationale is twofold. First, compared with its predecessors, the administration of Xi is the most vocal in pursuing global leadership. While his predecessor Hu Jintao talks about the "peaceful rise", Xi Jinping is centred around "building a new type of major power relations", objective reiterated during a confab with Obama in a ranch in California. The pursuit of global leadership under Xi is not purely discursive. It has been accompanied by the rollout of signature projects epitomised by the Belt and Road Initiative (BRI). By 30 January 2021, a total of 171 countries and international organisations join the BRI.[29] No project initiated by Xi's predecessors has reached a similar level of international coverage. In this sense, it can be said that China under Xi is closer to the centre of the global stage than ever before. If the pursuit of global leadership limits the party-state's capacity to mobilise coercion in the political regulation of the ISM, it is expected that Xi's administration faces the highest level of constraints.

Second, Xi's presidency is recognised as exercising the strictest internet censorship, called an "all-dimensional" control (Bandurski, 2018: 383) or a "near-total control over information" (Shirk, 2018: 26). The current tightening internet control seems to enfeeble the above hypothesis

[29] "China Has Signed 205 Cooperation Documents on the Joint Construction of One Belt One Road," Ministry of Commerce, 30 January 2021, https://www.yidaiyilu.gov.cn/xwzx/gnxw/163241.htm.

according to which pursuing global leadership limits the party-state's freedom of action in applying domestic internet censorship. Therefore, if it can be shown that the strict internet control under Xi is not an exception to the rule, then it can be established that simultaneously implementing domestic information control and pursuing global leadership does throw the party-state into a double-bind predicament.

The reason for which the party-state's aspirations for global leadership hinder the implementation of internet censorship is that the two goals are hardly compatible. To be a global leader, China must be respected globally; however, practising internet censorship hardly wins any respect. Krastev (2011) advances the "Lipset hypothesis", the "effect of openness" and the "imitation argument" to explain the instability of authoritarian regimes. While the first two arguments explain the instability with the modernisation and the integration of autocracies into the international system, the last argument says that authoritarian rulers pretend to be democratic and use the democratic language to "gain them acceptability and less international pressure for real change" (2011: 10). However, quoting Anton Chekhov, an influential Russian playwright in the 19th century, Krastev argues that the adoption of democratic institutions and democratic languages by autocrats is similar to a pistol hung on the wall. At some point, these institutions, set up "even in a limited, manipulative way", will fire, pushing autocracies to change into something else (2011: 10). Krastev's brilliant metaphor reminds of the embarrassment that the party-state has created for itself by donning a democratic hood and speaking democratic languages. The embarrassment in return narrows down the party-state's repertoire of actions in handling private ISM firms.

Negative Domestic Public Opinion

The reporting of coronavirus shows how private ISM firms ally themselves with international audiences to push the boundaries of "speakability" and how the joint influence of the domestic and international public opinions can push the party-state to concede. The party-state's management of the coronavirus was subject to netizens' criticisms. Xi's absence in the front pages of *People's Daily* and in the news broadcasting of the CCTV for one whole week until the beginning of February caused wide criticisms and speculations. On social media like Weibo, netizens asked "where is that person?", avoiding directly mentioning Xi Jinping for concerns of censorship. However, to make their message sent, netizens posted the photos of Xi's predecessors who appeared at the moments of crises, thus making Xi's absence clearer.

The propaganda machine was inefficient then. Netizens resisted the two talismans, called "Jiang Shan Jiao" (literally "Beautiful Rivers and Mountains") and "Hong Qi Man" (literally "Hovering Red Flags"), that Chinese Youth League (CYL) posted on Weibo on 17 February 2020. The CYL said that Jiang Shan Jiao was the big sister of Hong Qi Man, and called upon netizens to put a thumb-up like. One blogger posted: "the members of the CYL are Chinese citizens at first, then they are members of the CYL, but by no means are they fans". This post obtained more than 50 000 likes before the CYL suppressed the two talismans 4 hours later.[30] Another blogger posted "Stop this please, it is disgusting" to protest the post of the CYL. To emphasise his(her) anger, the blogger repeated the same sentence seven times in the same post. Netizens also criticised that state-backed media reported the coronavirus in an excessively positive manner. They considered positive reporting of the crisis as against ethical principles. For instance, the CCTV showed a beaming young woman who went outside the Fangcang hospital after recovering her health. She was reported as saying that "after being hospitalized here, I even don't want to leave the hospital". Deng Xueping, a renowned Chinese lawyer, criticised:

> The intentions of certain authorities are quite easy to understand: the emphasis should be put on positive reporting to cheer people up, because it is said that in so doing, we can calm down people and avoid public panic. However, the problem is that the coronavirus is obviously a disaster. If the images show a different atmosphere to the public, don't they cause public suspicion?... Official media like CCTV should use their precious camera to dig and reveal the truth on the pandemics, and dedicate their prime time to the people who are really in need of care or assistance. The world may have many 'negative news'. It is only when we bravely face them that we can solve them.[31]

[30] "After Being Scolded for 4 Hours, the Two Virtual Idols Created by the Communist Youth League Were the Most Short-Lived", DW news,18 February 2020, https://blog.dwnews.com/post-1308899.html.

[31] Weibo of Deng Xueping, posted on 15 February 2020, retrieved on 9 February 2022 on https://m.weibo.cn/status/4472210771011559?display=0&retcode=6102.

Fearing that Chinese young people forgot the coronavirus crisis in the same way that they knew little about the Tiananmen Square massacre, the novelist Yan Lianke explained at one seminar at Hong Kong University of Science and Technology:

> If we cannot be a whistle-blower like Li Wenliang, let's become a person who hears the whistle blow; if we cannot speak loudly, let's become a whisperer. If we cannot be a whisperer, let's become a silent person with a good memory....[I]n the forthcoming collective chorus in memory of the so-called victory, let's stand aside, let's be a person with a grave in the heart.[32]

Negative International Public Opinion

The international reporting of the coronavirus was principally negative from the outset. It criticised Chinese bureaucratic opacity, strict information control, and the ineffective public healthcare system. Professor Cheng Meidong at Peking University researched on how the international media reported coronavirus. He said that the earliest foreign coverage of the Chinese coronavirus crisis was on CNN on 9 January 2020.[33] At the beginning of February 2020, *The New York Times* criticised Chinese officials' mutual blaming. It criticised that Chinese officials cared more about pleasing their superiors than serving citizens in their jurisdiction: "Too many of its officials have become political apparatchiks, fearful of making decisions that anger their superiors and too removed and haughty when dealing with the public to admit mistakes and learn from them".[34] In another article published in late February 2020, *The New York Times* contended that the coronavirus constituted the biggest challenge to the

[32] "Yan Lianke: After This Catastrophe, Let's Get a Good Memory," speech of Yan Lianke at Hong Kong University of Science and Technology, 21 February 2020, https://theinitium.com/article/20200221-mainland-coronavirus-yanlianke/?utm_medium=copy.

[33] Sun, Meijuan, "A Review of Foreign Media Reports on Coronavirus—Interview with Professor Cheng Meidong, Professor at the School of Marxism at Peking University," 1 June 2021, http://ex.cssn.cn/zx/bwyc/202006/t20200601_5137166.shtml.

[34] Yuan, Li, "Coronavirus Crisis Shows China's Governance Failure," *The New York Times*, 4 February 2020.

legitimacy of the CCP and that the propaganda apparatus was inefficient this time for saving the legitimacy of the Chinese ruling party.[35]

The Chinese language edition of BBC news commented that "the unexpected sanitary crisis overnight exposed the true image of China, revealing the rigidity of the bureaucratic system and tight information control. The bloated image of big power suddenly collapses in the public eyes, the early warning mechanism of the public health system seems to be non-existent".[36] The British media ironically criticised the absence of the Chinese leadership in handling the crisis, arguing that Xi, who emphasised several times that he was the commander-in-chief for combating coronavirus, seldom appeared in public. It was only in late March, two months after the confinement in Wuhan, that Xi turned up there.[37] In April, the VOA explained how Chinese practices of information control made it difficult to bring the sanitary crisis under control. It commented that the earliest 8 whistle-blowers, including Li Wenliang, were summoned up for having spread the information on social media. That the CCTV had reiterated the punishment of Li Wenliang more than 10 times deterred Chinese medical staff from speaking up, which did not help address the crisis in a timely manner. However, when the rumours turned out to be true, the responsibility was assumed by the Deputy Director of a Wuhan police station and one ordinary policeman, "as if the Deputy Director of the police station in Wuhan was able to mobilise the CCTV".[38]

Caught between domestic and international criticisms, the party-state softened its stance on internet control. To understand the evolution of the regulation intensity of private ISM firms, I collected the censorship index recorded by Weiboscope from 1 January to 30 June 2020, period in which the coronavirus occupied the central stage of Chinese political, economic, and social life. Weiboscope is a media visualisation project

[35] Yuan, Li, "Coronavirus Weakens China's Powerful Propaganda Machine," *The New York Times*, 27 February 2020, https://cn.nytimes.com/china/20200227/china-coronavirus-propaganda/

[36] "How Does the Covid-19 Crisis Tarnishe the Chinese Prestige of Being a 'Rising Big Power'," BBC, 20 March 2020, https://www.bbc.com/zhongwen/simp/chinese-news-51942287.

[37] Ibid.

[38] Jin, Zhe, "The CCP Spared No Efforts to Shake off Its Responsibility for the Coronavirus," *VOA*, 8 April 2020, https://www.voachinese.com/a/China-Ongoing-Disinformation-Campaign-20200407/5363459.html.

of the Journalism and Media Studies Centre at the University of Hong Kong. Since January 2011, Weiboscope has been regularly tracking and collecting censored posts of the Chinese microbloggers with more than 1000 followers on Sina Weibo. Since January 2018, Weiboscope has been releasing the censorship index of Sina Weibo every day. A higher censorship index indicates stricter content regulation. The index represents the percentage between the number of censored posts and the total number of the posts released in the day. During these six months, the censorship index of four dates lacks for unknown reasons: 2 February, 17 March, 12 June, and 30 June. For the concern of visualisation, the censorship index of these dates is marked as 0 on Fig. 7.2.

Overall, censorship on Weibo from 1 January to 30 June 2020 goes upward, meaning that the content regulation in June is much harsher than in the preceding months. That censorship intensifies at the moments when the international attention to the Chinese management of the covid-19 weakens shows the linkage between international pressure and censorship intensity in China: the more the party-state is put under international pressure, the more it shows a sense of self-restraint in information control. The censorship in January 2020 is divided in two periods: the first 20 days and the last 10 days. While censorship keeps intensifying in the first 20 days, it declines in the last 10 days. The explanation for this contrast is that before the suddenly declared lockdown by the Wuhan government on 23 January, coronavirus-related information was strictly controlled in China. Poorly informed, foreign media could not report the case to the international audience, let alone to create pressure on Chinese regulators. The period before the Wuhan lockdown was, therefore, characterised by weak international attention. This partly explained the tranquillity with which censorship was conducted on Weibo. From 1 February to 30 April, despite occasional peaks, the regulation intensity decreased to stabilise in March and in April. The distribution of the number points from February to April confirms the effect of international pressure, too.

The two peak moments on Fig. 7.2 deserve special attention. The first peak appears on 9 February (censorship index: 6.115792). Throughout the whole period covered in Fig. 7.2, it is the first time that the censorship index exceeds 6 and it is only in mid-June that the record is broken. On 15 June, the censorship index exceeds 6 for the second time, reaching 6.28963. One week later, on 22 June, the censorship intensity reaches the highest peak (7.60461) of the first half of 2020. To a large extent,

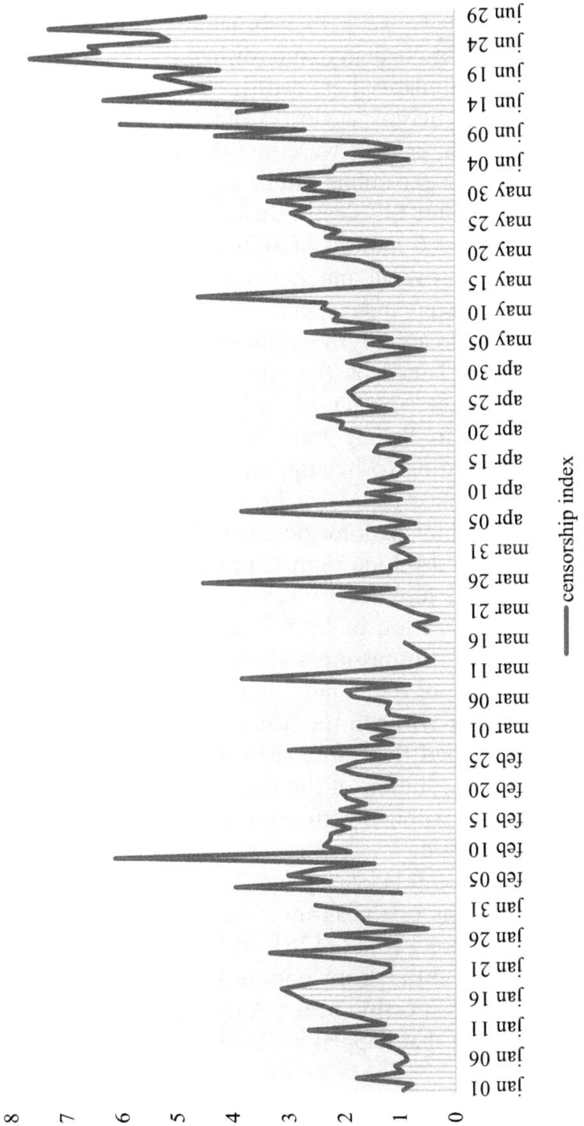

Fig. 7.2 Censorship index of Weibo from 1 January to 30 June 2020 (*Source* Compiled by the author based on Weiboscope)

the peak on 9 February is due the death of Doctor Li Wenliang on 7 February, the event having the highest potential to cause collective mobilisation against the party-state in February. The injustice inflicted to Li Wenliang, reflected in his being admonished by Wuhan local police for having reported the similarity between the SARS and the new virus, caused netizens' wide criticisms against Chinese information control. Therefore, after Li's death, censors were mobilised in an all-out manner to clean cyberspace to prevent online public anger from developing into offline unrest. Two days after Li's death, the censorship intensity reached the highest level of the first 5 months of 2020.

Censorship in June is overall much stricter than in the preceding months, with the censorship index being systematically above 4 from 15 to 30 June. The increase of the censorship index in the second half of June is largely explained by the regional outbreak of the covid-19 in the Xinfadi wholesale market of Beijing in mid-June. From 12 to 15 June, 79 persons in Beijing were tested positive. On 15 June, Hebei province, neighbouring to Beijing, reported 3 cases. The regional outbreak in Beijing provoked panic in China, for citizens were afraid of undergoing a new round of nationwide expansion of the coronavirus. To curb public panic, on 14 June, Sun Chunlan, Vice Prime Minister, convened a State Council high-level meeting at which she required to apply the "strictest investigation in Xinfadi and the surrounding areas" to identify the sources of transmission and cut the transmission belt. On 24 June, the spokesperson of the Beijing municipal government declared that the outbreak of coronavirus in the Xinfadi market was well brought under control. He explained that more than 95% of the identified cases were not serious.[39] The evolution of the censorship intensity in June, as reported in Fig. 7.2, corresponds with the evolution of the coronavirus in Beijing in the same period.

Besides the two peaks on 9 February and 22 June, Fig. 7.2 shows that for each month there exists one peak moment: January 24 (3.329754), March 26 (4.514343), April 7 (3.832478), and May 13 (4.593341). The reasons behind the peaks differ from one month to another. To clarify the reasons behind these censorship peaks, Annex 1 resumes the principal coronavirus-related events that happen in the first half of 2020.

[39] "Spokesperson of the Beijing Municipal Government: The Coronavirus Outbreak Directly Related to Xinfadi Is Brought under Control," Xinhuanet, 25 June 2020, http://www.gov.cn/xinwen/2020-06/25/content_5521705.htm.

Although it is true that international pressure can constrain the party-state to loosen internet censorship, it is important to note that upon the occurrence of major international events, the party-state can tighten information control, too. Whether the domestic public opinion is supportive or not towards the party-state is a game-changer for predicting whether internet control will be loosened or not. The contrast between the party-state's compromise on the coverage of the SARS crisis in 2003 and its intransigence on the reporting of the Hong Kong protests in 2019 due to the extradition bill is in large part explained by whether the domestic public opinion is supportive or not towards Beijing. If the domestic public opinion disapproves the practices of the party-state, represented by public outcry against lack of governmental transparency in the reporting of the SARS crisis and the covid-19, the party-state yields to the pressure from the international community. In contrast, if the domestic public opinion sides with the party-state, as in the case of the 2019 Hong Kong protests, international criticisms seldom pressure the party-state into making compromises.

Conclusion

Against the backdrop of the interplay between national and international politics, the dilemma facing the party-state is that between the pursuit of a global leadership and the continuity of domestic information control. These two sets of interests are seldom compatible because countries applying stringent information control hardly obtain international respect. To make things worse, in recent years, the international image of China has been tainted by several political events: the alleged internment of Uighurs in Xinjiang, lack of governmental transparency on the origins of the covid-19, the adoption of the controversial national security law in Hong Kong in June 2020, the expansion of "Wolf Warrior diplomacy", to name only a few. China seems to be plagued by a crisis of public relations at a moment when it needs international support more than ever before. Since the growth of the Chinese economic power nurtures the political ambition of the country to take up global leadership, a bad international reputation could be an obstacle to the realisation of this objective. In this sense, the pressure of the international public opinion will matter more in the future for pushing the authoritarian regime to behave in a less authoritarian manner, including in the domain of information control.

The limited efficiency of the deployed strategies for shaking off international pressure—advocacy of the cyber-sovereignty doctrine and the mobilisation of jingoistic discourses—does not allow the party-state to shake off international pressure. Both strategies contain weaknesses. The party-state has shown to be unable to increase the international acceptance of the cyber-sovereignty doctrine, for the latter has been internationally viewed as a sophist justification of domestic internet censorship. In the struggles between the US-led multistakeholderism and the China-led cyber-sovereignty, neither of them has shown a clear potential to win over the other at the current time. With the Sino-American geopolitical competition scaling up, the competition of norms will intensify, too. Under such a context, whether the cyber-sovereignty approach to internet governance will take precedence over the alternative approaches becomes even more uncertain.

The efficiency of jingoistic discourses in mitigating the party-state's international pressure is not guaranteed. On the one hand, Chinese young generations have become less receptive to nationalist speeches. Therefore, that government-paid commentators disseminate pro-regime discourses in cyberspace does not automatically increase netizens' adherence to the party-state. More often than not, even though such discourses do signal the strength of the regime, they produce indifference among netizens. On the other hand, as Susan Shirk (2011) argued, the party-state has to be cautious when stoking nationalism among Chinese citizens. Since the Chinese government itself cannot satisfy extreme nationalist demands of netizens, it cannot allow online expressions of nationalism to grow indefinitely. The party-state's difficulty in fixing the equilibrium position in the online expression of nationalism, a position at which the Chinese government can safely incite nationalism among netizens without getting itself into trouble, predicts that the party-state will deploy the greatest prudence when using the trump of "patriotism". Since the strategies for shaking off international pressure have limited efficiency, the party-state is expected to continue to coexist with the double-bind logic. The autonomy-enabling potential of the latter for private ISM firms will persist, too.

The party-state's vulnerability to international perceptions enables private ISM firms to seek autonomy by allying themselves with international audiences. For instance, on the reporting of the SARS in 2003 (Tai and Sun, 2007) and the covid-19 in 2020, under significant domestic and international pressures, the party-state changed its attitude from total censorship to partial openness. However, this chapter has shown that the international public opinion alone seldom works as a game-changer. It can only wield pressure on the party-state when it is aligned with the domestic public opinion. It does not suffice that the international public opinion is hostile to China that the party-state concedes. This chapter has shown that the concessions come when a tripartite coalition takes form: the domestic public opinion, the international public opinion, and private ISM firms, with the first condition being the utmost one among the three. Scholars are encouraged to test whether in the future the international public opinion can produce its influence in a relatively autonomous manner. The rationale for exploring this research path is that the more the party-state aspires for global leadership, the more international perceptions matter for it, unless it succeeds in imposing its own discursive system as a universally acceptable one in the world.

References

Arsène, Séverine. 2012. "Chine: Internet, levier de puissance nationale." *Politique Etrangère* 2: 291–303.

Bandurski, David. 2018. "Can the Internet and Social Media Change the Party?" In *Routledge Handbook of the Chinese Communist Party*, edited by W.W.-L. Lam, 372–390. Oxon, UK: Routledge.

Bourdieu, Pierre. 1979. "Public Opinion Does Not Exist." In *Communication and Class Struggle 1*, edited by A. Mattelart and S. Siegerlaub, 124–130. New York: International General.

Brady, Anne-Marie. 2018. "'The New (Old) Normal': The Party's Propaganda System Under Jiang, Hu, and Xi." In *Routledge Handbook of the Chinese Communist Party*, edited by W.W.-L. Lam, 165–181. Oxon, UK: Routledge.

Cheng, Manli. 2017. "Zhong Guo Dui Wai Chuan Bo De Li Shi Hui Gu Yu Zhan Wang 2009–2017 Nian" [History and Perspectives of Chinese External Propaganda]. *Xin Wen Yu Xie Zuo* [Journalism and Writing] 8: 5–9

Culpepper, Pepper D., and Kathleen Thelen. 2020. "Are We All Amazon Primed? Consumers and the Politics of Platform Power." *Comparative Political Studies* 53(2): 288–318.

Economy, Elisabeth. 2018. *The Third Revolution: Xi Jinping and the New Chinese State*. Oxford: Oxford University Press.

Ekman, Alice. 2020. *Rouge Vif: l'Idéal Communiste Chinois*. Paris: Editions de l'Observatoire.

Gerschewski, Johannes. 2018. "Legitimacy in Autocracies: Oxymoron or Essential Feature." *Perspectives on Politics* 16(3): 652–665.

Guriev, Sergei, and Daniel Treisman. 2019. "Informational Autocrats." *Journal of Economic Perspectives* 33(4): 100–127.

Han, Rongbin. 2018. *Contesting Cyberspace in China: Online Expression and Authoritarian Resilience*. New York: Columbia University Press.

Han, Rongbin. 2019. "Patriotism Without State Blessing: Chinese Cyber Nationalists in a Predicament." In *Handbook of Protest and Resistance in China*, edited by T. Wright, 346–360. Cheltenham: Edward Elgar Publishing.

Harwit, Eric, and Duncan Clark. 2001. "Shaping the Internet in China: Evolution of Political Control over Network Infrastructure and Content." *Asian Survey* 41(3): 377–408.

Hay, Colin. 2009. "King Canute and the 'Problem' of Structure and Agency: On Times, Tides and Heresthetics." *Political Studies* 57: 260–279.

Hay, Colin. 2011. "Ideas and the construction of interests." In *Ideas and Politics in Social Science Research*, edited by D. Béland and R. H. Cox, 65–82, Oxford: Oxford University Press.

Hay, Colin, and Daniel Wincott. 1998. "Structure, Agency and Historical Institutionalism." *Political Studies* 46: 951–957.

Hsueh, Roselyn. 2011. *China's Regulatory State: A New Strategy for Globalisation*. Ithaca, NY: Cornell University.

Huang, Haifeng. 2015. "Propaganda as Signalling." *Comparative Politics* 47(4): 419–437.

Huang, Haifeng. 2018. "The Pathology of Hard Propaganda." *The Journal of Politics* 80(3): 1034–1038.

Huntington, Samuel P. 1991. *The Third Wave: Democratisation in the Late Twentieth Century*. Norman, OK: University of Oklahoma Press.

Jiang, Min. 2012. "Internet Companies in China: Dancing Between the Party Line and the Bottom Line." *Asie Vision* 47: 1–50.

Jiang, Min. 2015. "Managing the Micro-Self: The Governmentality of Real Name Registration Policy in Chinese Microblogosphere." *Information, Communication, and Society* 19(2): 203–220.

Jiang, Fei, and Nan Zhang. 2019. "Zhong Guo Dui Wai Chuan Bo De San Ci Lang Chao (1978–2019)" [Three Waves of International Communication in China (1978–2019)]. *Quan Qiu Chuan Mei Xue Kan* [Global Media Journal] 2: 39–58.

King, Gary, Jennifer Pan, and Margaret E. Roberts. 2017. "How the Chinese Government Fabricates Social Media Posts for Strategic Distraction, Not Engaged Argument." *American Political Science Review* 111(3): 484–501.

Krastev, Ivan. 2011. "Paradoxes of the New Authoritarianism." *Journal of Democracy* 22(2): 5–16.

Kuran, Timur. 1997. *Private Truths, Public Lies: The Social Consequences of Preference Falsification*. Cambridge, MA: Harvard University Press.

Lagerkvist, Johan. 2012. "Principal-Agent Dilemma in China's Social Media Sector? The Party-State and Industry Real-Name Registration Waltz." *International Journal of Communication* 6: 2628–2646.

Lei, Ya-wen. 2018. *The Contentious Public Sphere: Law, Media and Authoritarian Rule in China*. Princeton, NJ: Princeton University Press.

Légaré, Kathia. 2007. "L'influence du système international sur l'évolution des régimes semi- autoritaires: quatre approches des révolutions de couleur (2003–2005)." *Etudes Internationales* 38(4): 501–522.

McGregor, Richard. 2010. *The Party: The Secret World of China's Communist Rulers*. New York: HarperCollins Publishers.

Miller, Blake. 2018. "The Limits of Commercialized Censorship in China." Working Paper, http://eprints.lse.ac.uk/101417/1/miller_limits_of_commercialized_censorship_in_china.pdf.

Negro, Gianluigi. 2017. *The Internet in China: From Infrastructure to a Nascent Civil Society*. Cham: Palgrave Macmillan.

North, Douglass. 1990. *Institutions, Institutional Change and Economic Performance*. New York: Cambridge University Press.

Nye, Joseph. 2004. "The Decline of America's Soft Power: Why Washington Should Worry." *Foreign Affairs* 83(3): 16–20.

Oberschall, Anthony. 1996. "Opportunities and Framing in the Eastern European Revolts of 1989." In *Comparative Perspectives on Social Movements. Political Opportunities, Mobilizing Structures, and Cultural Framings*, edited by D. McAdam, J.D. McCarthy, and M.N. Zald, 93–121. Cambridge: Cambridge University Press.

Polanyi, Karl. 2001. *The Great Transformation of Our Time: The Political and Economic Origins of Our Time*. Boston: Beacon Press.

Shirk, Susan. 2011. "Changing Media, Changing Foreign Policy." In *Changing Media, Changing China*, edited by S. Shirk, 225–252. Oxford: Oxford University Press.

Shirk, Susan. 2018. "China in Xi's 'New Era': The Return to Personalistic Rule". *Journal of Democracy* 29(2): 22–36.

Strittmatter, Kai. 2020. *Dictature 2.0: Quand la Chine surveille son people (et demain le monde)*. Translation by Olivier Mannoni. Paris: Tallandier.

Tai, Zixue. 2006. *The Internet in China: Cyberspace and Civil Society*. New York: Routledge.

Tai, Zixue, and Tao Sun. 2007. "Media Dependencies in a Changing Media Environment: The Case of the 2003 SARS Epidemics in China." *New Media & Society* 9(6): 987–1009.

Teng, Biao. 2012. "Rights Defence (weiquan), Microblogs (weibo) and the Surrounding Gaze (weiguan): The Rights Defence Movement Online and Offline." *China Perspectives* 3: 29–39.

Wolchok, Scott, Randy Yao, and J. Alex Halderman. 2009. "Analysis of the Green Dam Censorware System." https://jhalderm.com/pub/gd/.
Wu, Angela Xiao. 2020. "The Evolution of Regime Imaginaries on the Chinese Internet." *Journal of Political Ideologies* 25(2): 139–161.
Yang, Guobin. 2009. *The Power of the Internet in China: Citizen Activism Online*. New York: Columbia University Press.

CHAPTER 8

The Double-Bind Regulation Beyond China

Arguments of the Book

This book has conceptualised China as a "double-bind regulatory state" and the Chinese way of regulating private ISM firms as "double-bind regulation" (DBR). DBR denotes a two-step autonomy-enabling process in which regulatees exploit regulators' predicament to seek margins of manoeuvring. In the first step, the party-state's pursuit of competing objectives, RFOs and SFOs, puts it in a predicament. In the second step, private ISM firms consciously exploit regulators' predicament to enlarge their room to manoeuvre. Given that private firms in China have few institutionalised channels to contest regulation rules (e.g., they can hardly challenge a regulation in front of courts or through media), their exploitation of the party-state's predicament represents an important opportunity to repair imbalanced state-firm relations in the Chinese politico-economic setting.

The core issue addressed in DBR is the extent to which Chinese private ISM firms can succeed in circumventing unfavourable regulations without being punished. This sphere of immunity constitutes margins of manoeuvring of private ISM firms. DBR adds several nuances to the conventional polarised representation of state-firm interactions in China. The "liberalisation-control" paradigm captures pretty well the picture of polarisation: either the party-state wins over firms and brings them under control, or private ISM firms overwhelmingly outsmart the party-state

A. Ma, *China as a Double-Bind Regulatory State*,
https://doi.org/10.1007/978-981-99-8857-0_8

and change themselves into agents of liberalisation (or even democratisation) of China. Going one step further than the liberalisation-control paradigm, DBR does not preordain the party-state and private ISM firms into the roles of the predator and the predated, the capturer and the captured. Instead, it has unveiled, at a more structural level, why private businesses are not powerless despite political authoritarianism in China.

The three pairs of dilemmas discussed in this book lay the foundations of the double-bind regulation of private ISM firms in China. The pursuit of contradictory objectives in the past 30 years leads to one Chinese particularity: the largest authoritarian state in the world is also the most connected country in the world. This outcome stems from the fact that the party-state must simultaneously pursue the two ends of each equation. This task has been difficult but rewarding. It is difficult because the party-state must arbitrate between competing goals all the time. It is rewarding because pursuing the two sides of objectives rather than one side only has brought China economic success and international prestige. The sense of self-restraint demonstrated by the party-state ensures that the party-state's level of legitimacy is not dangerously low and that internet censorship currently applied in China does not backfire against the regime.

On the regulation of private ISM firms, China is a "double-bind regulatory state", with the party-state being driven by contradictory objectives the achievement of which is nevertheless indispensable to underpin and legitimise its power. The double-bind logic of the party-state fundamentally originates from the contradiction between the economic and political regulations of the ISM industry. Since the party-state needs both and is unable to achieve them simultaneously (although it does simultaneously pursue both), a grey zone emerges. In this grey zone, private ISM firms can commit noncompliance at low cost by taking netizens on board when it comes to implement the political regulation, or by taking local regulators on board when it comes to implement the economic regulation.

Fragmented authoritarianism is the second source of the party-state's double-bind logic. It unfolds at two levels: vertical/territorial fragmentation and horizontal/bureaucratic fragmentation. Vertical and horizontal fragmentation, respectively, generate the predicament for the party-state within the economic regulation and the political regulation of the ISM industry. When it comes to the contradiction between the party-state's pursuit of global leadership and the continuity of domestic information control, it generates the party-state's dilemmas mainly in the realm of the political regulation of the ISM industry. The three pairs of contradictions have been unfolding in parallel to Chinese economic liberalisation and globalisation since the 1990s. They are part of the structural constraints inherent in the Chinese governance logic itself. The built-in nature of the party-state's predicament implies that the manoeuvring opportunities of private ISM firms in China are embedded into the Chinese political system itself.

The conceptualisation of DBR in this book is based on a detailed study of nearly 40 years of state-firm wrestling in China from 1987 to 2022. The long period covered by this research increases the systematicity of the research findings. Moreover, that DBR holds explanatory power for the regulation of both ICPs and non-ICPs illustrates its theoretical robustness. At this stage, this book has principally demonstrated a sort of "internal robustness" of DBR, namely its theoretical power in explaining the regulation of one sector in one country. To increase the reliability of the findings, it is necessary to ensure two other levels of robustness: external robustness and sectoral robustness. First, concerning external robustness, this chapter tests whether DBR holds explanatory power for the state-firm wrestling in other regulatory settings than China. This chapter extrapolates the double-bind regulatory model to Singapore to check the generalisability of DBR. Second, concerning "sectoral robustness", this chapter extends DBR to other economic sectors than the ISM, the objective being to clarify on whether the explanatory power of DBR goes beyond the ISM industry. This chapter will test the validity of DBR by applying it to the regulation of the cosmetic industry.

The methodology behind the robustness test in this chapter is inspired by Yeo (2020). Yeo insightfully advanced the "market-institutional" approach to capture the particularity of the market regulation in China: the leadership role of socialist politico-economic institutions. For the robustness test, she adopted a two-layered testing structure. On the one hand, she applied the market-institutional approach to Vietnam, because, in a similar way to its Chinese counterpart, the Vietnamese Communist Party also played a primordial role in planning, deciding, and implementing FYPs. On the other hand, she tested the market-institutional approach in other sectors (including, among others, finances and civil aviation) than the automobile and telecommunication sectors. The enlargement of the sample of the economic sectors amplifies the testing ground. It contributes to increase the validity of the concerned theories.

Robustness Test I: DBR in Singapore

In their seminal book *Rightful Resistance in Rural China*, O'Brien and Li pointed out that "a truly useful construct has to survive applications beyond its original context" (2006: 15). Gerring also recognised the importance of extending the application of one concept from its original context to other ones. In his words, "A concept that applies broadly is more useful than a concept with only a narrow range of application. A good concept stretches comfortably over many contexts; a poor concept, by contrast, is parochial – limited to a small linguistic turf" (2001: 54). After conceptualising DBR on the basis of the ISM regulation in China, it now becomes important to examine how far DBR can travel. The fundamental problem to be solved here is to clarify the extent to which this peculiar phenomenon, in which the predicament of the regulators enables the regulatees to manoeuvre, distinguishes China from other countries. Readers can legitimately raise the following questions: Does DBR only exist in China? If so, how to explain this Chinese particularity? If not, does the emergence of DBR need some specific conditions?

The Absence of DBR in Democratic Regulatory Settings

Before moving on to the Singaporean case, it matters to clarify that established democracies are not double-bind regulatory states. The fundamental reason is that, in democracies, RFOs and SFOs serve, rather than contradict, each other. Therefore, regulators are not pulled into opposite directions. Democratic countries are characterised by the rule of law. It is anti-constitutional and illegal for any political party or candidate to stay in power without regularly renewing citizens' consent through fair and open elections. Therefore, regulatory activities in democracies do not serve politicians' parochial interests of preserving unchecked power indefinitely. Instead, since political parties can only gain power by defeating competing parties in elections, serving voters augments their chances of taking power. Therefore, RFOs and SFOs do not clash in democracies.

Another reason which explains the absence of DBR in democracies is the high political costs for political parties to deviate from regulatees' interests. As it has been demonstrated in this book, the firm-citizen coalition is an important condition for lowering the costs of the ISM firms for committing noncompliance. Without the backup from citizens, firms may not be able to challenge regulators, for the latter would face no public pressure in punishing firms. Therefore, the capacity of the regulators to break up the firm-citizen coalition and to take citizens on board largely influences whether regulators can narrow down the room to manoeuvre of private ISM firms without causing backfiring against the regime. Wooing the support of citizens takes on importance for regulators in China.

Democracies differ from autocracies on regulator-citizen power relations. Citizens in democracies enjoy institutionalised channels to hold regulators accountable. These channels range from regular elections, media denunciations, inter-party supervisions to street demonstrations and petitions. The consequence is that the costs of being caught going against voters' preferences are particularly high for regulators. The latter have greater interest in aligning their preferences with those of citizens. Regulator-citizen power relations in democracies are, therefore, characterised by the relatively strong position of citizens. In contrast, in autocracies, regulators are more accountable to autocrats than to citizens. Their objectives converge with those of autocrats more than with those of citizens. Therefore, regulator-citizen interest alignment is much tenuous in autocracies. That regulators in authoritarian regimes lack independence deepens interest misalignment, for regulation is, more often than not, an extension of the policy instruments enabling autocrats to self-serve. Since citizens have limited channels to hold autocrats accountable, the costs for regulators to deviate from the interests of citizens are relatively low.

Given the independence of regulatory agencies in democratic settings, they are not held to serve the incumbent political party. In the USA, for instance, although the administrative chief of regulatory agencies does have political affiliations, she serves the interests of her agency before serving her party. As Carpenter argues, that a bureaucratic official in the USA is often affiliated to a political party is "an asset to his bureaucratic entrepreneurship, not a liability" (Carpenter, 2001: 117). Regulators in democracies are thus not thrusted into opposite directions of serving citizens, on the one hand, and serving the incumbent government, on the other, for the latter does not afford to contradict citizens' preferences. The regulator-citizen interest alignment is the fundamental reason behind the absence of DBR in democratic regulatory settings.

Comparing the ISM Regulation Between China and Singapore

The double-bind regulation of the ISM industry is rather circumscribed to autocracies the great majority of which live with the clash between RFOs and SFOs. However, it is not sufficient to simply conclude that DBR is specific to autocracies. To increase conceptual granularity, it is necessary to further pinpoint the scope of applicability of DBR. In which types of authoritarian regimes does DBR turn up? To answer this question, this section adopts the "most similar" comparative methods (Mill, 1843/ 2011). It tests the robustness of DBR by examining the ISM regulation patterns in Singapore.

The rationale of case selection lies in four similarities between Singapore and China. First, they have similar cultural traditions. Intersubjectively shared ideologies, cultures, traditions, and beliefs deeply influence the making of institutions in a country (North, 1990; Hay, 1999). China and Singapore are both influenced by Confucianism, which is the cornerstone of the "Asian values" championed by the former PAP leader Lee Kuan Yew in the 1990s. In both countries, collective interests are given greater importance than individual well-being.

Second, they are both characterised by the one-party rule and the domination of the state in state-society relations. The CCP and the PAP, ruling parties in China and Singapore, have been in power respectively for 74 years and 55 years by 2023. Both parties monopolise traditional media and actively compete with alternative voices in cyberspace. They monopolise the agenda-setting power and hold strong capacities to resist business lobbying. Based on their comparative research on the governance of Airbnb in Singapore and Amsterdam, Li and Canelles (2021) found that the Singaporean government adopted a hierarchical governance style, in contrast with the collaborative approach in Amsterdam. The PAP-led government elaborated rental policies on its own, without engaging platforms as stakeholders. The command-and-control governance style also characterises state-firm relations in China. The domination of the party-state in policymaking processes has caused several outcomes, including, among others, the "generally submissive nature" of private ISM firms (Lagerkvist, 2011: 171).

Third, both countries enjoy high internet penetration rates. By mid-2022, the internet penetration rate in China and Singapore, respectively, achieves 74.4% and 92%. Fourth, despite their authoritarian stances, political leaders in both countries enjoy high popularity. From 2003 to 2020, more than 80% of the Chinese citizens declare themselves supportive of the Chinese central government (Cunningham et al., 2020). In Singapore, that the PAP has continuously won general elections since the Singaporean independence in 1965 from the Malaysian federation illustrates the popularity of the party.

The robustness test includes two parts. The contradiction of regulatory objectives is the core condition for setting the double-bind logic in motion. To ascertain whether Singapore practices DBR, the first part of the test investigates whether the PAP is caught in the predicament of having to serve contradictory objectives in regulating private ISM firms. The second part of the test examines the implications of DBR for the margins of manoeuvring of private ISM firms in Singapore. The objective of the robustness test is to ascertain whether the double-bind logic facing the PAP has enabled Singaporean ISM firms to enlarge their room to manoeuvre. If these firms seek their autonomy principally by other means than exploiting regulators' predicament, Singapore will not be considered as a double-bind regulatory state. The reason is that the autonomy-enabling opportunities opened by the predicament of the PAP are wasted.

Contradiction Between SFOs and RFOs in Singapore

China and Singapore demonstrate remarkably similar patterns in the regulation of private ISM firms. Like China, Singapore "appears to have come up with a strategy by which an authoritarian regime can successfully engage with technology without allowing itself to be overwhelmed politically" (Kalathil and Boas, 2003: 73). The PAP-led Singapore has adopted a two-pronged approach to regulating ISM firms since the mid-1990s: relax control on economic issues and tight control on political issues. Like China, the Singaporean government faces the predicament of having to strike right balances between RFOs and SFOs. It has to engage itself into an iterative process of switching between the two, for exclusively pursuing either side of objectives risks destabilising the one-party-rule.

The contradiction between RFOs and SFOs in Singapore is easily identifiable. For SFOs, the PAP has vigorously pursued public-regarding policies to benefit private ISM firms, in particular from 1991 to 2003. This in large part matches the period in which the Chinese government supports the technological development of the ICTs. Lee Kuan Yew, then Prime Minister of Singapore, said, "the government must learn to use information technologies, rather than to reject it. Otherwise, this government would be doomed to fail". Since 1991, the year in which the National University of Singapore established the country's first internet connection, the PAP has vigorously developed a broadband infrastructure, promoted internet literacy, and endeavoured to turn Singapore into a hub for E-commerce (Kalathil and Boas 2003: 76). In 1992, the Singaporean government advanced the project "Intelligent Island: IT2000", the objective being to cover the whole country with high-speed broadband network within 10 years. The PAP's support to the E-commerce sector was spectacular. It amended several laws and statutes, including Electronic Transactions Act, Sale of Goods Act, Computer Misuse Act, to increase consumers' confidence in online shopping. Thanks to substantial public support, Singapore has developed into the best equipped E-commerce hub in Southeast Asia.

The PAP's flexibility and rule-abiding in pursuing SFOs contrast with its intransigeance and arbitrariness in pursuing RFOs. As in China, internet regulation is one of the instruments enabling the PAP to consolidate its rule. Content regulation in Singapore started in 1996 under the supervision of the SBA (Singapore Broadcasting Authority). In the same year, the Singaporean government started developing legal regulatory means. The Singaporean legal system was stricter than that in China, for many moral rules were inscribed into the legal system. This is what Long Yang (2010) called the "moralisation of the laws" (*falü de fan daode hua*). Successive content regulators, including SBA, MDA (Media Development Authority), and IMDA (Infocomm Media Development Authority), have combined legal and extralegal means to ensure corporate compliance.

The Internet Code of Practice in 1997 required all ISPs and ICPs to apply for an operating licence. ISPs must unsubscribe any newsgroup upon the request of regulators (article 3). They must ensure that a great variety of materials did not appear online. Except pornography, other categories of prohibited content were vaguely defined, thus increasing the discretionary power of regulators in interpreting whether firms complied or not. For instance, article 4(1) of the Internet Code of Practice provided a vague definition of "prohibited material". The latter was the "material that is objectionable on the grounds of public interest, public morality, public security, national harmony, or is otherwise prohibited by applicable Singapore laws". That lawmakers did not clearly explain what they mean by public interest or public security increased the difficulties of ISM firms in implementing content regulation rules. On extralegal measures, regulators mobilised and continue to mobilise pressure, intimidation, and litigation to "enshrine the practice of self-censorship" among domestic and international media outlets (Kalathil and Boas, 2003: 76).

Similar to the Chinese party-state, the PAP-led Singapore faces similar constraints in regulating private ISM firms. The sources of the predicament of the PAP, however, differ from those of the CCP. They centre around the contradiction between political and economic objectives. The other two sources of the double-bind logic in China, "fragmented authoritarianism", on the one hand, and the contradiction between the pursuit of global leadership and the continuity of domestic information control, on the other hand, are barely visible in Singapore.

First, given that Singapore is a city-state small in size, its regulatory framework of the ISM sector is centralised. Mandated by the Info-Communications Media Development Authority Act, the IMDA is set up in 2016. As the regulator-in-chief of the ISM sector in Singapore, the IMDA pursues both political and economic objectives. It enjoys a similar institutional status as its Chinese Counterpart the CAC. The IMDA practices centralised regulation both on economic and political issues. The official website of the IMDA shows that as a statutory body, it regulates economic sectors (e.g., electronic transactions) and political sectors (e.g., films, broadcasting, newspapers, printing, personal data protection). In contrast, the ISM regulation in China is supervised by the regulators operating at different administrative levels: whereas the political regulation of ISM firms is centralised with the CAC, the economic regulation is decentralised to local governments. For instance, the regulation of the E-commerce sector in China is in large part decided by local governments. Tables 8.1 and 8.2, respectively, capture the distribution of regulatory competences in China and Singapore.

Table 8.1 Distribution of the ISM regulation competences in China

	Political regulation	*Economic regulation*
Regulators	CAC, CDP, MPS, NRTA, Ministry of Culture	Local governments, State Council, MIIT, NDRC, MST, SAMR
Regulation levels	Central-level	Local-level and central-level
Regulation approach	Centralisation	Decentralisation

Table 8.2 Distribution of the ISM regulation competences in Singapore

	Political regulation	*Economic regulation*
Regulators	IMDA	IMDA
Regulation levels	Central-level	Central-level
Regulation approach	Centralisation	Centralisation

Second, Singapore differs from China also in the respect of international ambitions. The Chinese government has been endeavouring to become a cyber big power (*wangluo qiangguo*), objective advanced in February 2014 at the first meeting of the Central Cybersecurity and Informatisation Small Leading Group (CCI-SLG). It tries to gain a greater say over the norm-making and standard-making in international internet governance, Singapore has much limited diplomatic and economic influence enabling it to move to the centre of the global stage. In this sense, if the Singaporean government demonstrates a sense of self-restraint in regulating online speeches, this softening stance should not be motivated by the aspirations for global leadership. After comparing the sources of the predicament of the PAP and the CCP, it is safe to say that while Singapore is indeed caught by the double-bind logic, the predicament facing it turns out to be single-sourced, in stark contrast with China.

That the PAP is confronted with less dilemmas than the CCP holds important implications for the configuration of state-firm power relations in Singapore. For one thing, the PAP has less equations to balance in the regulatory process. Besides the contradiction between political and economic objectives, the PAP has no balances to strike between the central and local regulators, or between the pursuit of global fame and the continuation of domestic information control. An important issue here is whether the fact that the PAP faces less dilemmas creates a smaller opportunity structure for private ISM firms in Singapore. If Singaporean regulators are less double-binded than their Chinese counterparts, does this strengthen or weaken the capacities of private ISM firms in Singapore to resist unfavourable regulations? The next section moves to the second half of the robustness test: the exploitation and the exploitability of the DBR-enabled opportunities for ISM firms in Singapore.

The Manoeuvring Room of Private ISM Firms in Singapore

The second step of the robustness test is to investigate the implications of the PAP's double-bind logic for the manoeuvring room of private ISM firms. Does the fact that the PAP pursues contradictory political and economic objectives open a window of opportunities for private ISM firms? Do the latter exploit such opportunities to circumvent unfavourable restrictions or leave the opportunities unexplored? If, in the course of pursuing greater margins to manoeuvre, private ISM firms in Singapore rely on other avenues than exploiting the regulators' predicament, it will be inaccurate to call Singapore a "double-bind regulatory state", because, in this case, autocrats' constraints leave regulators' bargaining power unchanged.

The PAP's predicament in simultaneously preserving the one-party-rule and legitimising it does open manoeuvring opportunities for ISM firms. The PAP has demonstrated much responsiveness to appease the public outcry caused by its arbitrary repression of online criticisms. It balances between liberalisation and control in a similar way to China (Ortman, 2011: 158). That the PAP does not afford to lose its popularity explains a paradox in Singapore: while the coercive means at the PAP's disposal are no less dreadful than those of the CCP, the PAP uses them only as the last resort (George, 2007). To preserve its popularity, the PAP mobilises two approaches to regulating cyberspace: on the one hand, it increases its presence online to build a positive image of itself; on the other hand, it avoids blocking or shutting down critical websites as much as possible. To some extent, the PAP has faithfully sticked to the principle of "strategic self-restraint" in its relations with ISM firms, "strategic self-restraint" defined as a choice of "calibrating its coercion to get the job done with as little force as possible" (George, 2007: 140).

The contrast between China and Singapore in terms of content regulation illustrates the "strategic self-restraint" in Singapore. The Chinese government has been frequently reported to send instructions to ICPs and to sanction their disobedience. Since its inception, the CAC has kept a column on its website called "Law Implementation and Inspection" (*zhifa ducha*), where the sanctions against ICPs are regularly publicised. The sanctions include, among others, temporary suspension of websites or accounts, definitive shutdowns, warnings, and administrative fines. In contrast, the PAP has been relatively rare to shut down websites. In 2007, George wrote that "ten years after the public internet was introduced, no political site has been blocked, even though this would be a technically and administratively straightforward move" (2007: 139).

As a result, it becomes possible for ISM firms to exploit the caution with which the PAP deals with "the growing political activism that resulted from its own liberalising measures" (Ortmann, 2011: 159). Ortmann took the PAP's crackdown on the critical blog *The Online Citizen (TOC)* as an example. In January 2011, the MDA obliged TOC to register itself as a political association. TOC became the first website in Singapore to be gazetted as a political association. The considerations of the content regulator were that after being registered as such, TOC could no longer accept foreign donations and it became easier to control. This measure contradicted the PAP's citizen-friendly image. Exploiting the PAP's fear of losing popularity, TOC continued to disseminate dissenting voices, "as if to prove that Singapore's new openness cannot be closed again" (Ortmann, 2011: 159).

Besides the exploitation of regulators' predicament, private ISM firms in Singapore enjoy, at least in principle, other methods to augment their room to manoeuvre. For instance, that Singapore has an efficient judiciary system can be an effective means for private ISM firms to contest the PAP. However, even though ISM firms can take regulators to justice, their chances of winning the lawsuits are fairly slim. In this respect, the differences between China and Singapore are not significant. It is certainly true that Chinese ISM firms have never been reported to take content regulators to court. Nevertheless, the regulator-firm power relations in both countries are basically the same: since judiciary courts in both countries are largely controlled by the incumbent party, they cannot afford to rule in favour of ISM firms should the latter dare to challenge regulators' decisions.

In October 2019, less than one year before the Singaporean general elections in July 2020, the law POFMA (Protection of Online Falsehoods and Manipulation Bill) took effect, granting the PAP almost unfettered power to censor dissent. With the POFMA, the government could issue "correction directions" (CDs) to individuals or intermediaries if it deemed online contents about a public institution as false or misleading. The law required social media companies and other ICPs to remove contents or display prominent corrections when required by the government. Noncompliance exposed firms to fines of up to \$730 000. Exasperated, ISM firms blamed by the PAP for having carried hostile messages against it contested this decision in front of courts. For instance, after receiving a CD for having disseminated a critical report of Lawyers for Liberty, a Malaysian human rights organisation, TOC challenged the decision and filed a lawsuit in 2019. However, it lost the battle in 2020 in front of the Court of Appeal, the highest court in Singapore.

The fact that large media corporations are formally or informally controlled by the government does not help private ISM firms to effectively challenge the PAP. The *Strait Times*, flagship newspaper in Singapore, was founded in 1845 by Robert Carr Woods. It initially provided commercial information. Today, *Strait Times* is part of the government-controlled Singapore Press Holdings. As the newspaper with the largest number of readers in the city-state, *Strait Times* is known for its promotion of pro-government propaganda. Another important news group Mediacorp is a dominating media conglomerate in Singapore. As the largest content creator and network in Singapore, the activities of Mediacorp cover TV, radio, and digital platforms. Given the importance of Mediacorp, it is unsurprising that it is totally controlled by Temasek, the state fund of Singapore. The close oversight of the PAP over the mainstream media implies that dissenting voices or contestations from private ISM firms are barely audible for the large majority of Singaporean citizens.

Since Singapore practices multipartisan competition, private ISM firms can vote for other political parties than the incumbent PAP as a sign of protest. This trait also distinguishes the Singaporean context from the Chinese one where genuine multiparty competition is non-existent. However, this means of protest has not been effective in making the voice of ISM firms respected by Singaporean regulators. One straightforward reason is that since the PAP became the ruling party in 1959 after winning elections, it has never been voted down thus far. Voting for alternative parties to the PAP has been ineffective for changing imbalanced state-firm power relations in Singapore.

As shown above, the operating conditions of Chinese and Singaporean ISM firms are strikingly similar. State-firm relations in both regulatory settings are highly asymmetric, with the state (or, more precisely, the party-state in both cases) largely taking precedence over firms. Private ISM firms are ultimately left with few effective channels through which to contest arbitrary or overly restrictive decisions of political regulators. The judiciary system in Singapore, known for its efficiency for handling economic issues, becomes merely a window-dressing institution when it comes to adjudicate over the regulator-firm conflicts on political issues. The same is true for the protest via electoral means. That private ISM firms are allowed to vote for other parties than the PAP does not mean that they can improve their operating conditions, because the PAP, at least for the current times, seems unshakable in electoral games.

The diversification of the contestation means at the disposal of private ISM firms in Singapore opens, in principle, an ampler structure of political opportunities than in China. However, these opportunities have been more wasted than exploited, for they are simply unexploitable. These diverse contestation means have done little to render Singaporean ISM firms freer than their Chinese counterparts. At the end of the day, firms in both countries can seldom resort to institutionalised means like courts, media, or elections to defend their legitimate interests in the political regulation of the ISM sector. They have to rely on informal methods to fend off unfavourable regulations. The exploitation of regulators' predicament figures as one of these methods.

The robustness test reveals that Singapore satisfies the criteria of a double-bind regulatory state in two fundamental respects. First, the simultaneous pursuit of political and economic objectives does create governance dilemmas for the PAP. Second, the predicament of the PAP in striking right balances between contradictory regulatory objectives enables ISM firms to seek manoeuvring room. The structure of political opportunities of Singaporean ISM firms resembles that in China. In both countries, firms have to mobilise informal means opened in the interstices of the authoritarian regime to dodge freedom-unfriendly regulation rules.

In China and Singapore, the democratising effect of the internet has not fully played out (Han, 2018: 183). Singapore is a "semi-authoritarian" (Kalathil and Boas, 2003: 74) or an "authoritarian but newly competitive" regime (Ortmann, 2011). Compared with the controlocrat China (Ringen, 2016: 138), the way in which the Singaporean government regulates private ISM firms is more rule-abiding than in China. However, the rule-abiding governance in Singapore has not generated a significantly larger room to manoeuvre to private ISM firms, because it is highly selective: it mostly appears in the regulation of economic issues, not that of political issues. On political issues, the PAP is as ruthless and intransigent as the CCP, narrowing down the margins of manoeuvring of private ISM firms.

Sectoral Robustness: DBR in the Cosmetic Industry?

To further ascertain the robustness of DBR, it is indispensable to examine whether its explanatory power remains valid when it is used to explain state-firm relations in other economic sectors than the ISM sector. Two important questions need to be answered here:

- Is the party-state caught within the double-bind logic when it regulates other economic sectors than the ISM?
- To what extent do private firms in other economic sectors than the ISM draw their autonomy by exploiting the party-state's quandaries in balancing competing regulatory objectives?

These two questions constitute the organising thread of this part. To answer them, the robustness test unfolds in two steps. In the first step, I investigate the regulatory patterns of the cosmetic industry, the objective being to ascertain whether the party-state finds itself caught within governance dilemmas and what these dilemmas are. In the second step, I examine the room to manoeuvre of private cosmetic firms in China, the objective being to understand whether firms seek their autonomy by exploiting the manoeuvring opportunities (if any) opened by the party-state's predicament (if any).

State Regulation of the Cosmetic Industry

To investigate the cross-sectorial validity of DBR, this section compares the regulatory patterns between the ISM sector and the cosmetic sector. The two economic sectors share the commonality of being over-represented by private businesses. I exclude SOEs from this comparison. SOEs operate in priority in the sectors that constitute national economic lifelines (e.g., energy, aviation, telecommunication). They are considered as part of state assets. Therefore, there is no organisational separation between the party-state and SOEs. Mitnick (1980: 7) considers that in a regulation, "there is the requirement that the regulatory entity not be a direct party to or involved in the activity of the regulatee, or subject". Instead of facing genuine competition as private firms do, SOEs operate within a context of "orchestrated competition" (Ma, 2022: 180). In this respect, although SOEs constitute "one highly distinctive characteristics of state capitalism in China" (Lin and Milhaupt, 2013: 699), the interactive patterns between them and the party-state are not those appearing in a regulatory process. Therefore, the robustness test in this part only focuses on private cosmetic firms.

The quandaries of the party-state in the regulation of ISM firms are in large part due to the particularity of such firms. What is particular with ISM firms is that they hold both economic and political importance for the party-state, a characteristic that firms in other sectors don't have. Therefore, to ascertain whether the party-state is caught within regulatory dilemmas when regulating cosmetic firms, it matters to clarify what types of stakes cosmetic firms represent the party-state. The importance of the cosmetic industry for the Chinese government is above all economic. The 2017 version of the National Economy Catalogue (*guomin jinji fenlei*, GB/T4754-2017) defines cosmetics as "chemical-industrial products that reach any part of the human body (skin, hair, fingernail, mouth, lips) by means of application, spray, or other similar methods, in order to clean, dissipate unpleasant smell, protect skin, improve and correct outlooks". *Forward-The Economist*, a leading Chinese think-tank known for producing cutting-edge policy analysis, divides the cosmetics industry in three parts: upstream, midstream, and downstream. The total value of the Chinese cosmetic industry reaches $73.2 billion in 2020, second only to the USA. This situation has remained unchanged in 2021, with the USA and China, respectively, being the first and the second largest cosmetic markets in the world. The size of the Chinese cosmetic market keeps enlarging from 2011 to 2020, with the compound annual growth rate (CAGR) attaining 9.22% in this decade.

The coronavirus crisis from 2020 to 2022 does not impede the expansion of the Chinese cosmetic market. Figure 8.1 compares the growth rate of the cosmetic industry and the Chinese GDP growth from 2020 to 2022, the three-year period in which the coronavirus crisis halts the economic development in China. As Fig. 8.1 indicates, the cosmetic industry has expanded significantly faster than the Chinese economy as a whole. It can be inferred that the cosmetic industry is a net contributor to the growth of the Chinese economy in this period.

The contrast between the growth rate of the Chinese GDP and that of the cosmetic industry from 2019 to 2022 is explained above all by the use of innovative sales methods that cannot be easily disrupted by the difficulties caused by the coronavirus crisis. China is among the earliest countries in the world to have tried live commerce (*zhibo daihuo*). 2019,

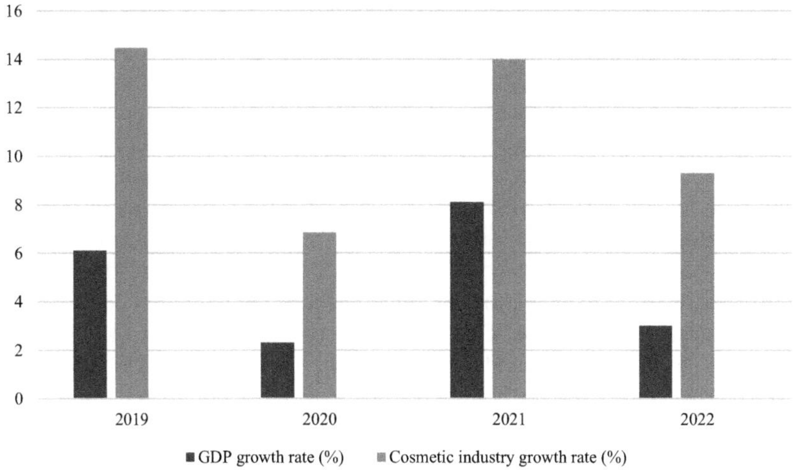

Fig. 8.1 Comparison of the GDP growth and the cosmetic industry growth 2019–2022 (*Source* National Bureau of Statistics [2019, 2020, 2021, 2022] and Forward-The Economist [2022])

the year before the outbreak of the coronavirus crisis, marks the start of live commerce. While the confinement paralyses people inside their apartments from 2020 to 2022, hosts on livestreaming platforms are particularly active in promoting online sales. Take 2021 as an example. The total size of the retail sales in that year exceeds 44 trillion yuan, and the sales of live commerce represent 5%. On 11 November 2021, Li Jiaqi and Weiya, then the two most influential hosts in live commerce in China, respectively, achieve 10.6 billion yuan and 8.2 billion yuan of transactions on the platform Tmall.[1]

The *2021 Annual Report on the Development of National Brands on Douyin* shows that the sales of products wearing national brands on Douyin (Chinese version of TikTok) increase by 667% in comparison with 2020. The increase of the sales for cosmetic products wearing national brands is 696%. In the first half of 2021, Linqingxuan, one of the domestic cosmetic brands, starts principally relying on live commerce for

[1] Meng, Qian, "Li Jiaqi and Weiya Earned More Than 10 Billion Yuan Overnight Through Live Commerce: Whose Banquet?," Newsweek, 3 November 2021, http://news.inewsweek.cn/survey/2021-11-03/14261.shtml.

sales. The volume of realised transactions increases by 12% in April 2021 and 15.9% in May 2021. Another cosmetic brand, SPES, sees its sales increase by 7461% compared with 2020. Given the shock of the coronavirus, domestic brands have opened their own live commerce studios and sell directly to consumers. It is estimated that in 2021, the number of domestic brands that created their own live commerce studios increases by 33% in comparison with 2020. The data by iimedia, an influential consulting firm in China, show that 72.6% of Chinese consumers tend to buy cosmetic products on E-commerce platforms. That the sales of cosmetics have massively moved online reduces the dependence of brands on physical infrastructure. The rise of live commerce goes to the interest of national cosmetic brands in particular, for they are less able than well-known international brands to rent stores in well-located areas.

Political Regulation Versus Economic Regulation

The regulatory framework of the cosmetic industry in China differs from that of the ISM industry. The regulation of private ISM firms in China is characterised by a bifurcated structure. While the CAC spearheads the political regulation, including above all the content regulation, regulatory agencies led by the MIIT and the SAMR oversee the economic regulation. The two agencies and their local outposts ensure the compliance of private ISM firms with the rules on consumer protection, investment, anti-trust, R&D of new products, and technical standards. They are also in charge of creating favourable policies to promote technological innovation and international competitiveness of Chinese firms. The bifurcation is absent in the regulation of the cosmetic industry, for there is no political regulation. Although the National Medical Products Administration (NMPA) is the regulator-in-chief of the cosmetic industry, the regulation of cosmetics is highly decentralised, with local authorities enjoying high discretion in making and implementing regulation rules. The absence of the political regulation explains that the first source of the party-state's double-bind logic, namely the contradiction between political and economic objectives, is non-existent in the regulation of cosmetics.

Domestic Level Versus Global Level

In the regulation of the ISM industry, the contradiction between the party-state's pursuit of global leadership and the continuity of domestic censorship constitutes one important source of the party-state's double-bind logic. When it comes to the regulation of the cosmetic industry, one legitimate question emerges: Is the party-state caught in the dilemma of having to achieve competing national and international objectives? Different from the ISM industry, the stakes of the cosmetic industry for the party-state are uni-dimensionally concentrated in the economic domain, not the political one. The contradiction between the domestic and international objectives in the regulation of the ISM sector is, at the end of the day, of a political nature. More precisely, the party-state's double-bind logic emerges, for the Chinese government simultaneously pursues domestic information control and global leadership. The two sides of objectives are fundamentally political. Since the cosmetic industry represents above all economic significance to the party-state, to argue that the latter faces the double-bind logic emanating from the pursuit of competing national and international goals does not make much sense.

Another factor also invalidates the contradiction between the domestic and international goals as the source of the party-state's double-bind logic in the regulation of the cosmetic industry: the Chinese ISM industry and the cosmetic industry do not have the same international clout. While China is widely recognised as the most connected country in the world, the Chinese cosmetic industry lacks international influence. At the Xiangshang Science Conference in October 2020, Huang Junhai, Professor at Shanghai Medicine Industry Research Institute, lamented that while China is the second largest country in the world in terms of cosmetic consumption, it is weak when it comes to the R&D of high-quality products. Together, five companies (l'Oréal, Procter & Gamble, Unilever, Estée Lauder, and Shiseido) represent around 52.4% of the global market shares in cosmetics. Among the 20 largest cosmetic firms operating in China, only 4 are national brands.

The contrasting influence of the Chinese ISM sector and the cosmetic sector on the international market is not unrelated to the policy design of the Chinese government. As several scholars highlighted (Yang, 2009; Jiang, 2012; Negro, 2017; Han, 2018), the Chinese government progressively pushed foreign ISM companies out of China after the ethnic riot in Xinjiang in July 2009. This policy left Chinese firms sufficient time to grow into international players and shielded them from international competition at the initial stage of their development. Google entered China in 2006. In early 2010, open conflicts broke out between Google and the Chinese government, for the latter accused Google of security breaches and cyber-attacks. Google had to move its search servers to Hong Kong. During its discussions with Chinese authorities in early 2010, the latter made it clear that moderation of "harmful" contents was a non-negotiable requirement if Google was to stay in mainland China.

In contrast, Chinese domestic cosmetic firms enjoyed less protection from the party-state. Due to poor living conditions in the first decade after the adoption of the Opening Up policy, the demand for cosmetic products was limited to a few rich families. Given widespread stigmatisation of make-up products, families that afforded to purchase cosmetics only consumed skin-protecting products, rather than make-up products. Weak demands did not incentivise domestic and international firms to invest in the cosmetic sector in general, and make-up products in particular. In the mid-1990s, after nearly two decades of economic liberalisation and social transformation, stigmatisation of citizens' expressions of individuality faded away. In parallel, the rising purchasing power of Chinese residents enabled them to consume cosmetics. Chinese cosmetic firms faced the competition of international firms from the outset. Due to frequent scandals related to the safety of domestic cosmetic products, international brands were much more appreciated, putting domestic brands at a disadvantaged position. This situation persists until today, because international brands are more competitive than national brands in the mid-end and high-end cosmetic market. Given the weakness of domestic brands, the party-state cannot rely on the cosmetic industry to achieve global leadership. Weak relevance of the cosmetic sector to the party-state's pursuit of global leadership further explains the absence of the contradiction between domestic and international regulation goals.

Fragmented Authoritarianism: Territorial/Vertical Fragmentation

Fragmented authoritarianism causes dilemmas for the party-state in the regulation of both the ISM sector and the cosmetic sector. The regulatory framework of cosmetics features both territorial fragmentation and bureaucratic fragmentation. Concerning territorial fragmentation, it is important to highlight the important discretionary power of local authorities. While the adoption of cosmetic security standards and licensing requirements of cosmetic manufacturers is centralised to the central level, local governments can fix their own supportive and restrictive regulatory policies, as long as the latter do not contradict the general regulatory framework fixed by central-level agencies. Therefore, it is no surprise to see provinces regulate their cosmetic firms differently.

The Chinese cosmetic industry is characterised by strong regional concentration. According to DBR, the regulation intensity in the localities relying on one specific sector to realise their economic growth is particularly strong. However, such regulations mostly take on a facilitative flavour rather than a restrictive one, the objective being to enable the concerned local government to outperform its neighbours in terms of economic performance. DBR also predicts that if national regulations impede the growth of local private firms with strategic economic values, it is not excluded that local authorities mobilise "feigned compliance" or even squarely disobey to national regulations.

Guangdong and Zhejiang have the strongest cosmetic industries in China. By January 2023, the number of registered cosmetic firms in Guangdong and Zhejiang is, respectively, 4698 and 1437.[2] At the press conference of Guangdong Medical Products Administration on 30 May 2023, one of the provincial outposts of the NMPA, the provincial regulator revealed that by April 2023, the gross industrial output value of the cosmetic industry in Guangdong was around 210 billion yuan, representing 60% of the national total. 56% of the cosmetic manufacturers in China, or 3100 manufacturers, operate in Guangdong. Nox Bellcow, one of the leading original design manufacturers (ODM) in China based in Guangdong, is well known for its capacity to produce cosmetic facial masks. It produces more than 6.5 million facial masks per day, bypassing all the other firms in the world. The cosmetic industry crucially underpins the economic growth of certain districts of Guangzhou, capital city of Guangdong. For instance, it is one of the pillar industries of Baiyun District. In 2020, the 130 cosmetic firms in this district generated 10.09 billion yuan of revenues, representing an increase of 25.7% in comparison with 2019.

The provincial and municipal authorities in Guangdong have been arduously promoting the cosmetic industry. On 9 January 2021, the

[2] Forward, "Regional Distribution of the Chinese Cosmetic Industry in 2023, the Development of the Cosmetic Industry in Guangdong and Zhejiang Province Is Faster than Elsewhere (illustrations)," sina.com, 24 March 2023, https://finance.sina.com.cn/roll/2023-03-24/doc-imymykha3002188.shtml.

General Office of the Guangdong provincial government issued the document "The Implementation Plan for Promoting High-Quality Development of the Cosmetic Industry in Guangdong Province" (*Guangdong sheng tuidong huazhuangpin chanye gaozhiliang fazhan shishi fang'an*). This document fixed ambitious objectives. By 2025, Guangdong will have formed 3–5 cosmetic companies the annual turnover of which will exceed 20 billion yuan. The province also aimed to form 10 well-known domestic brands the combined market shares of which in the Chinese national market would be over 50%. To achieve these goals, the document laid down generous policy support. The cosmetic industry figured among the industries that the provincial government supported in priority. Financial and budgetary policies favourably tilted towards the cosmetic sector to reduce firms' burden, including fiscal exemptions of the advertising fees and the communication fees.

Eager to reinforce the national and international influence of their cosmetic industry, local authorities are largely supportive of local firms. They go as far as to violate the orders of the central government, including lowering the safety standards of cosmetic products. This is the dark side of regulator-firm alliance at the local level: it nurtures corruption among local officials and abets immunity of firms. Eager to keep promising firms growing, local officials are inclined to covering up the scandals involving local firms. Huaxizi is an emergent domestic cosmetic brand registered in Hangzhou, Zhejiang province. It has undergone an astonishing growth trajectory in its rise to national stardom. Created in 2017, Huaxizi had already realised an annual turnover of 2 billion yuan in 2019 on the B2C platform Tmall. During the annual shopping festival of 18 June 2021, the size of the sales realised by Huaxizi on Tmall was the largest among all make-up brands, surpassing Perfect Diary which ranked third. The name Huaxizi is composed of two parts: "flower" and "Xizi". Xizi was a beautiful woman who lived in the turbulent period of Spring–Fall (BC770–BC476) under which China was divided into more than 100 small kingdoms. Citizen of Yue Kingdom and dotted with an exceptional beauty, she was sent as a spy to the enemy state Wu Kingdom to corrupt its king. The latter, indulged in debauchery, was defeated in a major war between Wu and Yue. The rich cultural connotations entailed into the brand Huaxizi have contributed to make it a cultural symbol of Hangzhou city.

It is not uncommon to hear livestreaming hosts and consumers complain about the poor quality of the make-up products of Huaxizi. For instance, while the eyeshadow of Huaxizi produces fairly plain colours, the eyeshadow palette is made with great care. Consumers also highlight that the sculptured lipstick and loose face powder of Huaxizi respectively cause peeling and allergy. Nevertheless, since Huaxizi is a star company of Hangzhou, the Hangzhou municipal government tends to protect the firm. As long as there are no serious scandals causing deaths or serious health problems for consumers, Hangzhou regulators turn a deaf ear to consumers' multiple complaints. In July 2022, the parent company of Huaxizi, Zhejiang Yige Enterprise Management Group Co., Ltd, inaugurated its Global Innovation and Industrial Centre in Qiantang District of Hangzhou. This move was intended to improve the R&D of Huaxizi's products. In January 2023, Zhejiang government even awarded the honour of "Remarkable Organisation in Security Control and Technological Empowerment of Medical Products in Zhejiang Province" to Huaxizi.

Fragmented Authoritarianism: Bureaucratic/Horizontal Fragmentation

Bureaucratic fragmentation is very evident in the regulation of the cosmetic industry. The NMPA is the chief regulator of the cosmetic sector. Its competences include the distribution of licences to producers of cosmetic products (*huazhuangpin shengchan qiye xuke*), the quality inspection, and the approval and prohibition of cosmetic ingredients. Besides NMPA, several other agencies are involved in the national-level regulatory process, too. The General Administration of Customs adjusts the tax rate of imported cosmetic products. The NIFDC (National Institutes for Food and Drug Control) introduces and modifies the rules on the registration and notification management of cosmetics. The SAMR investigates illegal behaviour of cosmetic firms and decides administrative sanctions to them. Since private firms are registered at the local level, most competences of the SAMR are assumed by its local outposts.

Bureaucratic fragmentation also exists at the local level. The division of labour indicated in the aforementioned *Implementation Plan for Promoting High-Quality Development of the Cosmetic Industry in Guangdong Province* well illustrates lack of a streamlined hierarchical bureaucratic structure. This policy document lays down 23 supportive policies for promoting the cosmetic industry. Table 8.3 resumes the information on the involved agencies and their respective competences. As it shows, overlapping competences are ubiquitous. Not only one policy area is taken in charge by more than one agency, but also one agency oversees more than one policy area.

As Table 8.3 shows, except one policy area, each of the other 22 policies are assumed by at least 2 agencies. To clarify the hierarchical relations of different agencies, Guangdong government often designates one agency as the leading agency, and the other agencies as collaborating agencies. For instance, the 3rd policy aims to facilitate the land use of cosmetic firms. Municipal governments are allowed to provide differentiated guarantees of land use to ensure high-quality development of cosmetic firms. They can rent land to cosmetic firms in the first place and sell it to firms subsequently. For this policy area, the Department of Natural Resources is designated as the leading agency, and the Department of Industry and Information Technology as the collaborating agency. The division of labour helps ensure clarity of responsibilities.

However, bureaucratic turf battles risk emerging under two circumstances. First, no agency is designated as the leading agency. Therefore, all the involved agencies have to collaborate. This is the case of policies 1, 4, 13, 19, and 20. Given lack of hierarchy among them, it is very likely that the policy implementation is hindered by bureaucratic turf battles. Second, although there is one leading agency, there exists at least 2 collaborating agencies. This is the case for policies 2, 5, 6, 9, 10, 12, 14, 15, 16, 18, 21, and 23. In this case, unless the leading agency manages to clearly define the turf of each collaborator, bureaucratic battles can hardly be avoided. Since the agencies involved in the same policy area are often located at the same administrative level, the leading agency has no direct oversight over its collaborators. This implies that bureaucratic turf battles will probably happen.

What worsens bureaucratic fragmentation in Guangdong is that one agency is often in charge of at least 2 policy areas. This increases the possibilities of frictions among agencies. Again, I take the example of the division of labour laid down in the *Implementation Plan for Promoting High-Quality Development of the Cosmetic Industry in Guangdong Province*. Based on my counting, this document mentions in total 16 agencies. Except the municipal government of Guangzhou, all the other 15 agencies are provincial-level ones. Figure 8.2 indicates the number of policy areas each agency is in charge of. Since the municipal government of Guangzhou is not a provincial-level regulatory agency, Fig. 8.2 does not include it.

Table 8.3 Distribution of regulation competences in Guangdong in cosmetics

Policy Number	*Competences*	*Number of involved agencies*
1	Promote the construction of cosmetic industrial parks in Guangdong	6 (collaboration)
2	Build industrial clusters of research institutions and firms; improve the capacity to design, develop, and research on high-end cosmetic products	3 (1 leads, 2 follow)
3	Provide guarantees of land use to cosmetic firms	2 (1 leads, 1 follows)
4	Provide budgetary and fiscal policy support	2 (collaboration)
5	Encourage private capital to invest in cosmetics	4 (1 leads, 3 follow)
6	Encourage cosmetic firms to get listed on the stock exchange market	3 (1 leads, 2 follow)
7	Optimise product verification and validation; accelerate the commercialisation of new products	1
8	Fight fake products; protect intellectual property rights	2 (1 leads, 1 follows)
9	Put the cosmetic industry on the list of priority industries of Guangdong; deploy provincial research projects on cosmetics	3 (1 leads, 2 follow)
10	Promote firm-university cooperation; accelerate industrial applications of academic research findings	4 (1 leads, 3 follow)
11	Encourage cosmetic firms to move their headquarters to Guangdong; incubate new brands	2 (1 leads, 1 follows)
12	Encourage cosmetic professional guilds to elaborate evaluation criteria of high-quality products	3 (1 leads, 2 follow)
13	Encourage chemical firms to deploy fundamental research into the raw materials of cosmetic products	3 (collaboration)
14	Build unified purchase platforms; apply collective price bargaining with international providers of raw materials	3 (1 leads, 2 follow)
15	Build influential marketing platforms to reduce costs	3 (1 leads, 2 follow)
16	Create research teams in cosmetics; organise academic and talent exchanges	4 (1 leads, 3 follow)
17	Encourage universities to create courses on cosmetics; increase the mobility of talents between universities, firms, and research institutions	2 (1 leads, 1 follow)

(continued)

Table 8.3 (continued)

Policy Number	*Competences*	*Number of involved agencies*
18	Support key firms to introduce high-level talents in design and marketing	3 (1 leads, 2 follow)
19	Promote the integration between cosmetics, culture, tourism, arts, and entertainment	2 (collaborate)
20	Organise diverse activities like Make-up Fashion Week or Guangdong Live Commerce Festival	2 (collaborate)
21	Open a special tourism route to increase the influence of the Guangdong cosmetic industry	4 (1 leads, 3 follow)
22	Reinforce positive communication of domestic brands; create a positive image for the cosmetic brands of Guangdong	2 (1 leads, 1 follow)
23	Encourage professional guilds of the cosmetic industry to create professional norms and standards	3 (1 leads, 2 follow)

Figure 8.2 shows that, except the Fiscal Administration of Guangdong, the Department of Civil Affairs, and the Department of Natural Resources, each of the other 12 provincial agencies is involved in at least 2 policy areas. The Department of Medical Products of Guangdong is the most deeply involved agency in the regulation of the cosmetic industry. Involved in 16 of the 23 policy areas, it is the most likely to have interactions with other agencies in the regulatory process. The Department of Market Regulation is the second most involved agency, participating in 9 of the 23 policy areas. Therefore, it also has ample chances of having to collaborate (or bargain) with other agencies. The regulation of the cosmetic industry does not have any streamlined regulatory framework. Instead, regulatory responsibilities are much subdivided to different agencies.

To some extent, bureaucratic fragmentation in the regulation of the cosmetic industry is a more serious issue than that in the regulation of the ISM industry. Although the regulation of ISM firms is taken in charge by two different groups of regulators—political and economic ones—the political regulation of the ISM industry is spearheaded by one meta-regulator, the CAC. On the one hand, the competences of the local outposts of the CAC are much limited due to the centralisation of the political regulation. Therefore, in the central-local power relations, the CAC has clear advantages over its local outposts. On the other hand,

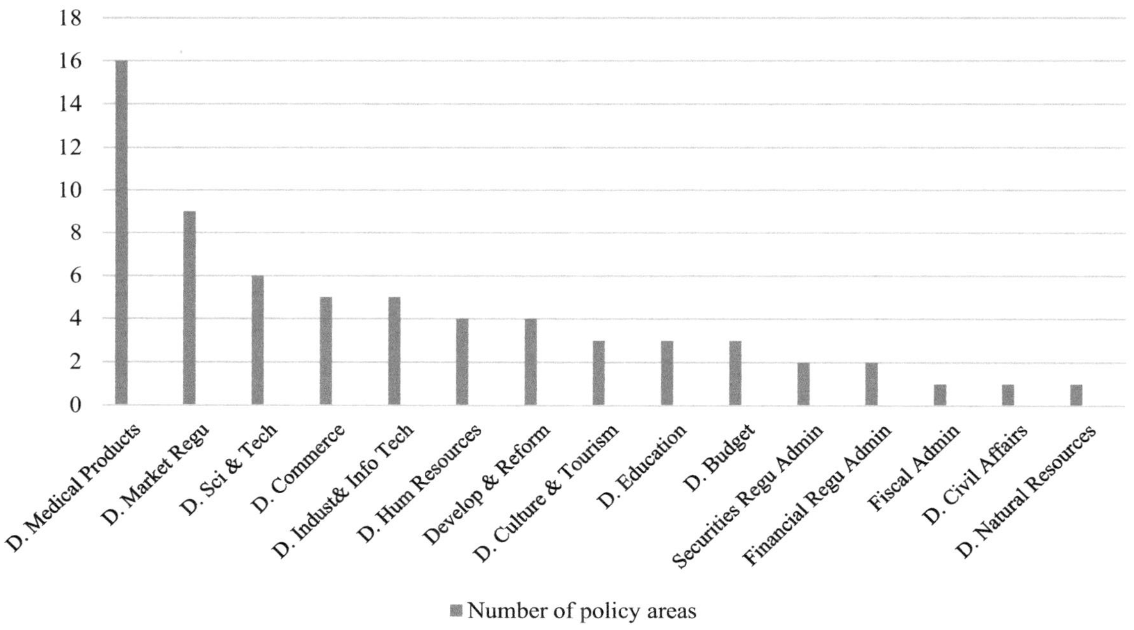

Fig. 8.2 Number of policy areas that each agency oversees in Guangdong (*Note* "D." means "Department of" (*ting*, in Chinese). It is the provincial equivalent of the ministries)

among the regulatory agencies at the same administrative level as the CAC, the latter is granted a sufficiently privileged position that it can easily ensure the coordination between different agencies (while ministries report to the State Council, the CAC, as a party institution, directly reports to President Xi Jinping). This is not the case of the Department of Medical Products of Guangdong. The relations between it and other provincial regulators (e.g., Department of Market Regulation, Department of Commerce, etc.) are not hierarchical. Therefore, it has much less capacities to ensure their collaboration in the regulatory process. Bureaucratic fragmentation and inter-agency battles constitute one defining characteristic of the regulation of cosmetics in China.

Implications on the Margins of Manoeuvring of Private Firms

It has been shown thus far that in the regulation of the cosmetic industry, the party-state faces a smaller number of contradictions than in the regulation of the ISM sector. Since the cosmetic industry represents above all economic importance for the party-state, the contradiction between the political and economic regulations is non-existent for the regulation of the cosmetic industry. In addition, the party-state does not pursue competing national and international objectives when regulating cosmetics. What is common to the ISM industry and the cosmetic industry is that the regulation of both sectors is marked by "fragmented authoritarianism" which unfolds alongside the territorial/vertical and the bureaucratic/horizontal dimensions. "Fragmented authoritarianism" also constitutes the source of the party-state's double-bind logic in the regulation of the cosmetic industry. While the previous section demonstrates the existence of the party-state's predicament, this part will analyse the extent to which such predicament can be transformed into firms' margins of manoeuvring.

To do so, I develop the case study on Yatsenglobal, one of the most influential domestic cosmetic firms in China. As one of the qualitative research methods, case studies help draw conclusive findings by presenting the facts, analysing the origins, and identifying the explanations with certain generalisability. Case studies are often mobilised when large-scale empirical studies are impossible to conduct. They are adapted for accomplishing theory-building and falsifiability identification tasks. It is not uncommon to see influential scholars in Chinese internet studies mobilise case studies in their research (see Zhao, 2008; Negro, 2017; Meng, 2011). The case study below illustrates the extent to which vertical and horizontal fragmentations in China have enabled private internet firms to seek margins of manoeuvring in the wrestling with regulators.

Exploitation of the Vertical Fragmentation by Cosmetic Firms

Headquartered in Haizhu district of Guangzhou, capital city of Guangdong province, Yatsenglobal is a landmark E-commerce platform. It is particularly known for having incubated the viral cosmetic brand Perfect Diary in 2017. The firm is created in 2016 by Huang Jinfeng, Chen Yuwen and Lü Jianhua, all of whom are alumni of Sun Yat-Sen University in Guangzhou. Yatsenglobal is listed in New York Stock Exchange on 19 November 2020. With the listing of its parent company, Perfect Diary becomes the first listed domestic make-up brand.

The growth trajectory of Yatsenglobal resembles that of private ISM firms, for it also heavily relies on the support of private capital. Jia and Winseck (2018) nuance the view according to which the rise of Chinese firms is voluntarily nurtured by the party-state. They argue that similar to their American counterparts, Chinese internet firms could not rise to international stardom without the backing of the financial market. In their words, "we must grasp not just the tight relationship between the state and business, but the emergent three-way ties between the state, internet companies, and financial capital" (2018: 31–32). Before Yatsenglobal illustrated its amazing growth potential in 2019,[3] it was mainly backed up by three private capital: Hillhouse, ZhenFund, and Banyan Partners.

Yatsenglobal quickly became the *chouchou* of the Guangzhou municipal government and the Guangdong provincial government. In December 2019, Guangzhou government put Yatsenglobal on the list of "Unicorn Innovative Companies" of the city. In December 2020, the firm was granted the honorary title of "Advanced Group of Guangdong" (*Guangdong sheng xianjin jiti*). Being recognised as a unicorn firm brought firms tangible benefits, including greater chances to get listed on stock

[3] The growth potential of Yatsenglobal in 2019 is spectacular. The sales of make-up products of Perfect Diary of Yatsenglobal on Tmall during the shopping festival of 11 November surpassed all its competitors. In addition, from January to November 2019, among all make-up brands on Tmall, domestic and international brands confounded, Perfect Diary realised the highest sales volume and reached 1.5 billion yuan. See the news at https://www.yatsenglobal.com/newsinfo/1968027.html.

exchange markets. In addition, Guangzhou government granted a one-off bonus of around 500 000–1 million yuan to its unicorn firms. The sum of the bonus varied according to the districts where firms were headquartered.

Interviewed by *Guangzhou Daily* on 27 October 2019 at the margins of Understanding China Conference (Guangzhou), Mr Huang Yigeng, Deputy CEO of Yatsenglobal, explained how Guangzhou government lent support to his company. National-level policies stipulate that each cosmetic product must have a label before being put on the market. Since firms have to go through lengthy administrative procedures to get the label ready, it happens that they cannot commercialise their products in time. The delay is likely to weaken the competitiveness of the concerned firm, because competing firms may very well commercialise similar products before the concerned firm obtains the label. Yatsenglobal obtained targeted help of its chief service official (CSO), sent by the district government of Haizhu. The CSO ensured the coordination between firms and different regulators, the objective being to ensure that firms obtained the label as soon as possible.

The CSO is a pioneering institution that the Haizhu district of Guangzhou started experimenting in 2018. The Investment Promotion Administration (IPA) of Haizhu sends a CSO, normally a functionary at the IPA, to each key private firm selected by the district government for attracting external investment. The CSO provides services to firms in an all-out manner: selection of headquarters, registration to market regulators, application for operating licences, registration to fiscal authorities. The creation of the CSO allows private firms like Yatsenglobal to spend less time on red tapes. It illustrates the agility of local regulators in balancing the needs of local private firms and national regulators. While they do ensure that private firms in their jurisdiction comply with national regulations, they take diverse initiatives to reduce the costs of corporate compliance.

Territorial fragmentation in the Chinese economic regulation has generated what Jean Oi termed as "local state corporatism" (1992). This term implies that the competences delegated to local authorities enable them to run as autonomous corporations competing with each other for resources and influence. Promising private firms like Perfect Diary and Huaxizi capable of attracting capital, technology, and talents constitute important trumps enabling local authorities to outperform their neighbours. For this reason, they are spoiled, not coerced. Local governments' dependence on them turns into their bargaining power for seeking preferential treatment. The regulation of the ISM industry thus resembles that of the cosmetic industry, for, in both cases, territorial fragmentation plays in favour of private firms. However, the autonomy of ISM firms and cosmetic firms means differently for citizens. While greater autonomy granted to ISM firms may allow the latter to satisfy citizens' needs for high-quality information, greater autonomy to cosmetic firms frequently contradicts citizens' interests, for it incentivises firms to lower the safety standards of cosmetic products. I will turn to this point immediately.

Exploitation of the Bureaucratic Fragmentation by Cosmetic Firms

Bureaucratic fragmentation does not always open extra margins of manoeuvring for private cosmetic firms. In Hangzhou, Guangzhou, and other Chinese cities with influential cosmetic companies, overlaps in regulators' competences have not enabled cosmetic industries to be shielded from administrative sanctions when they violate quality standards. Facing the ferocious competition of international cosmetic brands, national brands often try to increase their sales by heavily investing in marketing. It is estimated that Chinese domestic brands invest around 50–60% of their revenues in marketing. As a result, investment into the R&D of new products is often insufficient. This partly explains Chinese consumers' complaints about the quality of domestic products. Some Chinese consumers ironically declare that domestic brands "undertake marketing with heart, manufacture products with feet" (*yongxin zuo yingxiao, yongjiao zuo chanpin*).

In China, the NMPA and its local outposts oversee the safety of cosmetic products. At the outbreak of safety-related scandals, the NMPA-led agencies often intervene with an iron fist. In the decentralised regulatory framework, the provincial outposts of the NMPA fight at the frontline against fake cosmetic products. According to the *Regulation Guidelines on the Supervision of Cosmetic Products*, while the NMPA regulates the cosmetic industry at the national level, the medical products administration at the county level and above oversee cosmetics in their respective jurisdiction. The same regulation charges provincial and national regulators with conducting sampling inspections in cosmetic firms. The frequency of sampling inspections is not fixed. Provincial regulators must make the inspection results available on their website. If the complaints about specific types of cosmetics are intensive, provincial regulators must conduct targeted inspections.

On 6 September 2022, Guangdong cosmetic regulators released their inspection results of 31 firms in Guangzhou, Foshan, and other cities of the province. While 7 firms respected relevant regulation rules, the other 24 were urged to take corrective measures within a limited period of time. The firms criticised by regulators included some influential players: Sephora and Lebee. The equation of local cosmetic regulators differs from that of local ISM regulators, for the wrongdoing of cosmetic firms endangers citizens' health. Therefore, cosmetic firms must strike a right balance between defending local economic interests and protecting citizens' health. Regulators often have to defend the latter, for damaging citizens' health exposes them to high public pressure.

Grassroots regulators at the municipal and county levels sanction wrongdoing firms with an iron fist, too. It is especially so when the scandal obtains wide media attention. In such cases, local regulators' dilemmas weaken, for their range of choices narrows down. The paradox in the regulation of the cosmetic industry is that the more a brand is well-known, the more local regulators are under the public pressure to sanction it for the violation of regulation rules. Several well-known international brands have been sanctioned for having failed to comply. In October 2021, Shanghai Administration for Market Regulation ruled that the subsidiary of Sephora in Shanghai paid the fine of 400 000 yuan for fake advertisements. Sephora is a leading French cosmetic brand. Born in 1969, it was acquired by LVMH in 1997. Sephora opened its first Chinese shoppe in Shanghai in April 2005. The counterpart to the international fame of Sephora was the high public attention to the quality of its products. It is estimated that by 22 June 2021, Sephora had been sanctioned 16 times by Chinese market regulators.[4]

When it comes to sanction wrongdoing cosmetic firms, regulators at the same administrative level have been fairly quick to reach consensus. The reason for which horizontal fragmentation has generated limited margins of manoeuvring for private cosmetic firms is threefold: centrality of health concerns, lack of public support towards wrongdoing cosmetic firms, and selectiveness in bureaucratic fragmentation. These factors

[4] "Sephora Was Sanctioned Again for Misleading Advertising," China Quality Net, 21 October 2021, https://www.cqn.com.cn/ms/content/2021-10/21/content_8743957.htm.

crucially explain that to understand the implications of fragmented authoritarianism on private economic entities, it is insufficient to investigate whether the party-state, or regulators in general, is caught in dilemmas. It also matters to clarify the configurations of the alliances among three types of actors: regulators, private firms, and citizens. First, the cosmetic industry differs from the ISM industry in one crucial dimension: cosmetic products directly influence the physical health of citizens. The centrality of health concerns in citizens' consumption of cosmetics explains that when scandals break out, citizens are more than likely to position with regulators, not firms. Bearing the brunt of public pressure, whatever the overlaps in their respective responsibilities, cosmetic regulators quickly form alliances, the objective being to punish wrongdoing firms to calm down public anger. The particularity of the cosmetic industry facilitates consensus-making among regulators. Since it is certain that citizens welcome regulators' decision of severely sanctioning wrongdoing firms, involved regulators are inclined to adopting conservative positions.

Second, since the cosmetic industry represents above all economic interests, the wrongdoing of cosmetic firms is mostly motivated by economic reasons. If the wrongdoing of internet firms can be associated with their carrying of bold speeches, the wrongdoing of private cosmetic firms is seldom motivated by such lofty considerations. The collective representations of the motivations of dishonest cosmetic firms in China are that firms violate rules to pursue commercial profits. Given that consumers seldom support cosmetic firms that produce fake products or make false advertisements, regulators are reckless in punishing them. In addition, the more private cosmetic firms are severely sanctioned, the more regulators garner public support. Interest alignment between cosmetic regulators neutralises the hindrance that horizontal fragmentation represents to bureaucratic coordination.

Third, it is important to take account of the selectiveness in bureaucratic fragmentation in the institutional layout for the cosmetic regulation. While bureaucratic fragmentation is clear among the institutions responsible for promoting the growth of the cosmetic industry, it weakens when it comes to ensure corporate compliance with established regulations. As stated above, in Guangdong, 16 agencies are in charge of planning and promoting the cosmetic industry. The fact that more than one agency is involved in the same policy area is conducive to turf battles. However, the NMPA, the SAMR, and their local outposts are specifically designated to oversee corporate compliance with the cosmetic safety standards. Since the latter are fixed by the NMPA, bureaucratic fragmentation in terms of overseeing the implementation of safety standards is weak.

I call this bifurcated situation as *selective bureaucratic fragmentation*. It implies that depending on the issue areas, the intensity of bureaucratic fragmentation varies for the regulators at the same administrative level. In the cosmetic regulation, for instance, the regulators in charge of promoting the growth of the cosmetic industry are more likely to be troubled by bureaucratic fragmentation than the regulators ensuring corporate compliance with safety standards. With selective bureaucratic fragmentation, regulators at the same administrative level and who work on different issues of the same policy area are not equally caught within the double-bind logic. Accordingly, regulatees are not equally capable of obtaining margins of manoeuvring by allying with one group of regulators to resist other groups. It is certain that, when caught in mediatised scandals related to the quality of cosmetic products, cosmetic firms, however economically influential they are, can hardly obtain the support of any regulator.

Academic Contributions and Generalisability of DBR

To test how far DBR can travel, this chapter has conducted dual robustness tests: external robustness test and sectoral robustness test. It has found that DBR is well present in other authoritarian settings than China and in other economic sectors than the ISM industry, although some caveats are necessary to keep in mind. Overall, this chapter has shown that DBR is a useful approach for studying the dynamics behind state-firm interactions in autocracies. DBR reveals autocrats' existential dilemmas between self-serving and public-serving. This is the root reason behind the compatibility between DBR and authoritarian regulatory contexts.

It is high time to summarise the conditions under which DBR can be set in motion. The external robustness test in this chapter lends further support to the claim of this book: the application scope of DBR is limited to autocracies, not democracies. The detailed study on the Singaporean regulation of the ISM industry reveals that DBR is well present in Singapore. In a similar way as in China, the PAP-led Singaporean government is caught in the double-bind logic. What distinguishes China from Singapore is that the Singaporean government faces a smaller number of contradictions than the Chinese government. While both governments seek competing political and economic goals, fragmented authoritarianism is absent in Singapore. Likewise, that Singapore has limited resources and influence enabling it to occupy the centre of the global stage spares it the third source of the double-bind logic: the contradiction between the pursuit of global leadership and the continuity of domestic censorship.

Similar to China, the predicament of the Singaporean government has enlarged the margins of manoeuvring of private ISM firms. The semi-authoritarian Singaporean political system features greater openness than the Chinese one, for it allows private firms to seek autonomy in more diverse ways than in China: mobilising media resources, contesting controversial regulations in courtrooms, and voting other political parties than the PAP during general elections. Nevertheless, these alternative channels in Singapore have shown very limited efficiency. The entrenched dominance of the PAP in the legislative, executive, and judiciary arms of Singapore in the past 60-odd years has weakened counter-power groups of the country. The implication of the PAP's political monopoly for the margins of manoeuvring of private businesses is clear: institutionalised mechanisms for undertaking contestation activities are more often than not window-dressing institutions. Therefore, the political opportunity structure of private ISM firms in Singapore is that they also have to play with the double-bind logic of regulators to obtain autonomy.

The analysis of the Singaporean case shows the validity of DBR in authoritarian settings. The caveat to bear in mind is that the sources of authoritarian regulators' dilemmas vary, and need to be analysed on a case-by-case basis. For instance, the PAP-led Singaporean government faces a smaller number of dilemmas than China does. Concerning the cross-national validity of DBR, the contradiction between the political and economic regulatory goals is, among all the sources of contradictions, the fundamental one that sets DBR in motion. The political and economic goals of the ISM regulation, respectively, match the self-serving and public-serving goals of autocrats. The more autocrats seek to simultaneously realise the two sets of objectives, the more they are caught in the double-bind logic.

Gilson and Milhaupt (2011) advanced the term "economically benevolent dictators" to explain why authoritarian countries like China and South Korea could realise exceptional economic performance. This term captures well the quintessence of DBR, for it reveals the dilemma between self-serving and public-serving goals in some authoritarian polities. Autocracies that maintain themselves in power exclusively or principally by repression or manipulation are expected to suffer less from the double-bind logic. This is in particular the case of the entrenched autocracies like Iran or North Korea that voluntarily curtail the development of the ICTs to nip the bud of destabilising factors.

DBR also shows sectoral robustness and can explain state-firm interactions in other economic sectors than the ISM. Depending on the stakes that the concerned sector represents to the regulators of a country, the sources of dilemmas vary. This chapter has tested the sectoral robustness of DBR by applying it to the regulation of the cosmetic industry in China. Given that cosmetics represents above all economic importance to the party-state, the latter does not seek competing economic and political objectives in the regulation process. In addition, since the "beautiful industry" is not among the sectors with the potential to decisively influence the Chinese position in the global competition, the dilemma of the party-state between pursuing global leadership and continuing domestic information control in the ISM regulation is barely visible in the regulation of cosmetics. Nevertheless, fragmented authoritarianism is well present in the cosmetic regulation, and it has helped enlarge the autonomous sphere of private cosmetic firms in China, as shown by the robustness test in this chapter.

Territorial fragmentation and bureaucratic fragmentation do not equally enable private cosmetic firms to pursue margins of manoeuvring. The extent to which they generate autonomy for cosmetic firms depends on the regulated issue areas. As analysed in this chapter, when it comes to non-sensitive issues (e.g., investment policies, fiscal policies), private cosmetic firms can very well play the division between central and local regulators to seek favourable policies. Nevertheless, as far as sensitive issues (e.g., safety of cosmetic products) are concerned, cosmetic regulators are cornered by public pressure and can seldom make compromises to wrongdoing firms. High political and reputational risks related to the cover-up of cosmetic scandals disincentivise regulators from siding with wrongdoing firms, even though the latter realise huge annual turnovers and generate large fiscal incomes for local governments. These risks work as a facilitator of consensus-building among regulators, regardless of being national or local ones.

The last factor that hinders the creation of regulator-firm alliances in the cosmetic regulation is the overwhelmingly negative public opinion towards wrongdoing cosmetic firms. Here lies a significant difference between ISM firms and cosmetic firms. It is common to see Chinese citizens demonstrate sympathy towards the ISM firms that are punished for having carried boundary-spanning speeches. However, citizens seldom do so towards punished cosmetic firms, perceiving the latter as deserving the inflicted punishment after cheating consumers or harming their health. That citizens lack sympathy towards wrongdoing cosmetic companies encourages regulators to sanction firms recklessly. Regulators' eagerness in winning public support, reflected for instance in their mediatisation of exemplary sanctions to dishonest cosmetic manufacturers, does not allow private firms to ally themselves with one group of regulators to resist other groups.

Limits and New Paths for Future Research

Since DBR is conceptualised on the basis of the Chinese regulation of the ISM industry from 1987 to 2022, and that it is tested in one other authoritarian context of Singapore, the cross-country comparison is relatively limited in this book. What this book has accomplished is that it makes clear that DBR hardly appears in democratic settings. Unlike their authoritarian counterparts, politicians and political parties in democracies cannot maintain themselves in power indefinitely. They can only keep themselves in power by winning elections. Power legitimisation and power preservation are not contradictory in democracies, because the power cannot be preserved without legitimising it beforehand.

The equations of autocrats are different, for lack of rule of law allows autocrats to preserve their power without legitimising it by electoral means. Should there be fair and open elections, autocrats may well be ousted. Therefore, power preservation and power legitimisation are contradictory in autocracies. However, what is particularly hypothesis-generating is that to ensure that power does not change hands, autocracies must enjoy a certain form of legitimacy. They cannot become fully legitimate, but they cannot afford to be perceived as totally illegitimate either, for fear of provoking regime instability. Chinese experiences with the ISM regulation show that in exchange for legitimacy, the party-state does accept to grant certain autonomy to private ISM firms. The status of the civil society as the marker of the regime's legitimacy enables them to "hack" the state, obliging it to unleash autonomy for firms. Although this book has explained in great detail how the double-bind logic has emerged in China and Singapore and how it has enabled private ISM firms to seek autonomy, there is no in-depth cross-country comparison among a greater number of autocracies. Does DBR exist in all autocracies or only in certain autocracies? I recommend future researchers to conduct large-scale comparative studies on China and authoritarian regimes in other regions of the world. These projects will help pinpoint the generalisability of DBR with greater precision.

Does the Double-Bind Logic Produce a Tipping Point?

The second research path is to investigate the directions that political authoritarianism under Xi Jinping will head for. Since DBR portrays the party-state as moving between RFOs and SFOs, it considers the tightening social control under Xi Jinping as one ephemeral phenomenon. This book argues that the current tightening illustrates the party-state's moving to RFOs. It predicts that the party-state will move to SFOs again when the regime's legitimacy falls to a dangerously low level or when the tight control hinders economic objectives. Such theoretical predictions have been corroborated several times in China, with the alternation between more control and less control having happened at several crucial turning points: the 1999 handling of the Falun Gong Movement, the 2009 crackdown of blogs, and the 2013 crackdown of microblogs.

The abandoning of the zero-covid policy at the beginning of December 2022 further illustrates the explanatory power of DBR. From 2020 to 2022, the zero-covid policy was applied in China at the expense of huge sacrifices of Chinese citizens: transfer of covid-positive individuals to centralised confinement centres, difficulties in getting treatment for other diseases than covid, covid-tests every two days (even one test per day in covid-stricken areas), etc. The "zero covid" policy lasted for so long that speculations mounted that it might well change into a permanent mechanism of social control. The tricolour code and the itinerary record, two landmark inventions in China during the coronavirus crisis, were used to check each citizen's possibilities of being infected. Among others, they informed of the places that the concerned individual visited, the frequency of covid-tests, and whether this person contacted another infected individual.

After the 20th Party Congress in October 2022, which was the most sensitive political event in China in that year, the zero-covid policy was progressively abandoned due to citizens' multiple protests that took on increasingly violent forms. In the middle of December in Beijing, residents who took bus and metro were no longer required to show the covid-test results of 48 hours or 72 hours. It sufficed that they scanned the health code attached inside the public transportation means. At the same time, Tsinghua University and Peking University announced that they no longer published the number of the daily covid cases on campus. Other universities in China quickly followed suit. After the abandoning of the zero covid in the end of December 2022, both the health code and the itinerary record stopped being used in China.

Nevertheless, the DBR approach has not provided predictions on whether a tipping point could turn up in China. At the tipping point, the double-bind logic is expected to stop working, and power preservation and power legitimisation are no longer contradictory for the party-state. The suppression of the double-bind logic goes through the pursuit of one side of interests and the neglecting of the other. Therefore, the tipping point matches the point of regime change: if the party-state solely pursues SFOs or puts them at a considerably more important position than RFOs, China will become a democracy. In contrast, if the party-state cares about keeping itself in power and neglects SFOs, China will become a totalitarian state.

This book has not been as ambitious as to speculate on the conditions where the double-bind logic stops casting the spell over the governance of the party-state. However, this does not mean that the reflections on the tipping point are without importance. Xi Jinping, often referred to as the most powerful man in China after Mao Zedong, differs from his predecessors in terms of power concentration. Breaking down Deng's implicit rule on presidential limits, Xi peacefully starts his third presidential term in March 2023. This has been unseen in the post-Deng epoch. Should the signs of power concentration under Xi be interpreted as the signs of the breaking-down of the double-bind logic? This question deserves in-depth investigations by future researchers.

To sum up, this book answers one fundamental question: To what extent are Chinese private ISM firms autonomous from the authoritarian party-state? Drawing on the insights from regulation studies, communication studies, Chinese political autonomy, and the political sociology of social movements, I have argued that the seemingly strong capacities of the party-state are not unlimited and that its capacity to use them is not unlimited either. Investigating the regulation of the Chinese ISM industry from 1987 to 2022, I have argued that the party-state is constrained by a "double-bind logic". Pursuing competing RFOs and SFOs, the party-state does not afford to privilege one set of objectives only and has to switch between the two. By exploiting the party-state's predicament, private ISM firms in China have been able to reclaim autonomy against the Chinese authoritarian backdrop. They have succeeded in rising to international stardom, to such a point that they are even perceived as a threat to the party-state's authority nowadays. DBR takes on particular importance in China. The manoeuvring opportunities that DBR opens for ISM firms compensate for their lack of formal channels to contest unfavourable regulations. In this sense, DBR repairs, to some extent, the disadvantaged position of private ISM firms in the Chinese political economic landscape.

The findings of this book contribute to sharpen readers' comprehension towards two interrelated theories: "dictators' dilemma" and "authoritarian resilience". By unearthing the dynamics enabling Chinese private ISM firms to push the boundaries of their autonomous sphere, DBR shows that although China is the largest autocracy in the world, the double-bind logic works as a self-correcting mechanism of the regime. It ensures that authoritarianism does not reach a dangerously high level in China. These theoretical claims are particularly relevant for understanding the situation that China finds itself in today: the increasingly authoritarian stance of the party-state has made more than one observer predict the country's totalitarian turn. DBR sheds light on this issue: by dismissing the party-state's unconstrained power, DBR gives reasons for the Chinese civil society to remain optimistic about their potential to continue to seek autonomy.

References

Carpenter, Daniel P. 2001. "The Political Foundations of Bureaucratic Autonomy: A Response to Kernell." *Studies in American Political Development* 15: 113–122

Cunningham, Edward, Tony Saich, and Jesse Turiel. 2020. "Understanding CCP Resilience: Surveying Chinese Public Opinion Through Time." Ash Center of Harvard Kennedy School. https://ash.harvard.edu/publications/understanding-ccp-resilience-surveying-chinese-public-opinion-through-time.

George, Cherian. 2007. "Consolidating Authoritarian Rule: Calibrated Coercion in Singapore." *The Pacific Review* 20(2): 127–145.

Gerring, John. 2001. *Social Science Methodology: A Criterial Framework*. Cambridge: Cambridge University Press.

Gilson, Ronald, and Curtis Milhaupt. 2011. "Economically Benevolent Dictators: Lessons for Developing Democracies." *The American Journal of Comparative Law* 59(1): 227–288.

Han, Rongbin. 2018. *Contesting Cyberspace in China: Online Expression and Authoritarian Resilience*. New York: Columbia University Press.

Hay, Colin. 1999. "Marxism and the State." In *Marxism and Social Science*, edited by A. Gamble, D. Marsh, and T. Tant, 152–174. Champaign, IL: University of Illinois Press.

Jia, Lianrui, and Dwayne Winseck. 2018. "The Political Economy of Chinese Internet Companies: Financialization, Concentration, and Capitalization." *International Communication Gazette* 80(1): 30–59.

Jiang, Min. 2012. "Internet Companies in China: Dancing Between the Party Line and the Bottom Line." *Asie Vision* 47: 1–50.

Kalathil, Shanthi, and Taylor C. Boas. 2003. *Open Networks, Closed Regimes: The Impact of the Internet on Authoritarian Rule*. Washington, DC: Carnegie Endowment for International Peace.

Lagerkvist, Johan. 2011. "New Media Entrepreneurs in China: Allies of the Party-State or the Civil Society?" *Journal of International Affairs*. Fall/Winter 2011: 169–182.

Li, Yanwei and Genea Canelles. 2021. "Governing Airbnb in Amsterdam and Singapore: A Comparative Study on Governance Strategies and Styles." *Sage Open*, October–December: 1–11.

Lin, Li-wen, and Curtis Milhaupt. 2013. "We Are the (National) Champions: Understanding the Mechanisms of State Capitalism in China". *Stanford Law Review* 65(4): 697–759.

Ma, Aifang. 2022. *Double Bind Regulation of the Internet and Social Media in Contemporary China*. PhD dissertation defended on 28 June 2022 at Sciences Po, Paris.

Meng, Bingchun. 2011. "From Steamed Bun to Grass Mud Horse: E Gao as Alternative Political Discourse on the Chinese Internet." *Global Media and Communication* 7(1): 33–51.

Mill, Stuart. 1843/2011. *A System of Logic: Ratiocinative and Inductive*. New York: Columbia University Press.

Mitnick, Barry M. 1980. *The Political Economy of Regulation: Creating, Designing and Removing Regulatory Forms*. New York: Columbia University Press.

Negro, Gianluigi. 2017. *The Internet in China: From Infrastructure to a Nascent Civil Society*. Cham: Palgrave Macmillan.

North, Douglass. 1990. *Institutions, Institutional Change and Economic Performance*. New York: Cambridge University Press.

O'Brien, Kevin, and Lianjiang Li. 2006. *Rightful Resistance in the Chinese Countryside*. New York: Cambridge University Press.

Oi, Jean C. 1992. "Fiscal Reform and the Economic Foundations of Local State Corporatism in China." *World Politics* 45: 99–126.

Ortmann, Stephan. 2011. Singapore: Authoritarian but Newly Competitive. *Journal of Democracy* 22(4): 153–164.

Ringen, Stein. 2016. *The Perfect Dictatorship: China in the 21st Century*. Hong Kong: HKU Press.

Yang, Guobin. 2009. *The Power of the Internet in China: Citizen Activism Online*. New York: Columbia University Press.

Yang, Long. 2010. "Fa Lü Yuan Sheng Tai De Sha Shou: Dao De Fan Fa Lü Hua He Fa lü Fan Dao De Hua" [The Killer of the Original Ecology of Law: The Pan-Legalization of Morality and the Pan-Moralization of Law]. *Xue Shu Lun Tan* [Academic Forum] 7: 158–163.

Yeo, Yukyung. 2020. *Varieties of State Regulation: How China Regulates Its Socialist Market Economy*. Cambridge, MA: Harvard University Press.

Zhao, Yuezhi. 2008. *Communication in China: Political Economy, Power, and Conflict*. Lanham, MD: Rowman & Littlefield Publishers.

Annex: Principal Events in the Coronavirus Crisis from January to June 2020

Dates	*Events*
2019/12/30	Li Wenliang posted on a WeChat group that the hospital where he worked identified 7 cases of SARS. He asked his former classmates not to spread the information
2020/01/01	Li Wenliang was summoned by Wuhan local police for having spread "unverified information" on social media. Seven other people were also summoned
2020/01/17	Baibuting District of Wuhan held the Banquet for Ten Thousand Families (wan jia yan)
2020/01/19	The Chinese leading pulmonologist Zhong Nanshan confirmed that the new virus was transmissible from human to human
2020/01/23	First day of lockdown in Wuhan
2020/01/24	Chinese Lunar Calendar New Year Festival
2020/02/07	Li Wenliang died of coronavirus at Wuhan Central Hospital; wide public outcry in Chinese cyberspace
2020/02/13	Jiang Chaoliang was replaced by Ying Yong as the new Party Secretary of Hubei province
2020/02/26	Daily increase of new corona cases in foreign countries exceeded that in China
2020/02/28	World Health Organisation raised the global risk assessment of the covid-19 from "high" to "very high"

(continued)

A. Ma, *China as a Double-Bind Regulatory State*,
https://doi.org/10.1007/978-981-99-8857-0

(continued)

Dates	*Events*
2020/03/10	President Xi Jinping visited Wuhan; on the same day, the article about the interview of Doctor Ai Fen, one of the earliest doctors who revealed the coronavirus, was censored in cyberspace
2020/03/18	For the first time, Wuhan authorities announced zero increase of new cases
2020/03/26	The funeral parlors in Wuhan were successively opened. Local residents lined up to collect the ashes of their family members who died of coronavirus. Censorship of the posts on this issue
2020/03/28	Several metro lines re-opened in Wuhan
2020/04/04	National commemoration ceremony of those who died of coronavirus
2020/04/08	The 76-day lockdown in Wuhan was lifted
2020/04/16	French President Emmanuel Macron explained to *Financial Times* his concerns about the Chinese government's lack of transparency on the origins of the coronavirus
2020/04/17	The Chinese government revised the number of the death toll and increased the number by 1300. Therefore, the total number of the Chinese dying of covid-19 rose to 4632
2020/04/29	Australian Prime Minister Scott Morrison called for an independent investigation into the origins of the coronavirus
2020/05/07	Shulan, a county-level city of Jilin province, reported new corona cases
2020/05/13	Gai Dongping, Deputy Mayor of Jilin, declared a lockdown of the city. The lockdown was lifted on 7 June
2020/05/19	194 member states of the World Health Organisation agreed with Australia to organise independent investigations of the origins of coronavirus in China. China opposed the initiative
2020/05/27	The USA became the first country in the world where the death toll of coronavirus surpassed 100 000
2020/06/12	Regional outbreak of coronavirus in the Xinfadi wholesale market of Beijing
2020/06/14	The State Council instructed to launch the "strictest investigations in Xinfadi and the surrounding areas" to identify and cut the transmission of the virus
2020/06/24	Beijing declared that the outbreak of covid-19 in the Xinfadi market was brought under control

Source Compiled by the author based on Fang (2020), Yang (2022), and BBC[1]

[1] "First Anniversary of the Outbreak of Coronavirus in Wuhan: A Review of the Key Moments," BBC, 28 December 2020, https://www.bbc.com/zhongwen/simp/chinese-news-55433966.

Bibliography

Allison, Graham. 1971. *Essence of Decision: Explaining the Cuba Missile Crisis*. Boston: Little Brown.

Allison, Graham and Morton H. Halperin. 1972. "Bureaucratic Politics: A Paradigm and Some Policy Implications." *World Politics* 24: 40–79.

Arsène, Séverine. 2011. "De l'autocensure aux mobilisations: prendre la parole en ligne en contexte autoritaire." *Revue Française de Science Politique* 61(5): 893–915.

Arsène, Séverine. 2012a. "Protester sur le web chinois (1994-2011)." *Le temps des médias* 18(1): 99–110.

Arsène, Séverine. 2012b. "Chine: Internet, levier de puissance nationale." *Politique Etrangère* 2: 291–303.

Austin, Greg. 2014. *Cyber Policy in China*. Cambridge, UK: Polity.

Baldwin, Robert, Martin Cave, and Martin Lodge. 2010. "Introduction: Regulation—The Field and the Developing Agenda." In *The Oxford Handbook of Regulation*, edited by R. Baldwin, M. Cave, and M. Lodge, 3–16. Oxford: Oxford University Press.

Ball-Rokeach, Sandra J., and Melvin L. Defleur. 1976. "A Dependency Model of Mass Media Effects." *Communication Research* 3(1): 3–31.

Ball-Rokeach, Sandra J., and Melvin L. Defleur. 1979. "A Dependency Model of Mass Media Effects." In *Inter/Media*, edited by G. Gumpert, 81–96. New York: Free Press.

Bandurski, David. 2018. "Can the Internet and Social Media Change the Party?" In *Routledge Handbook of the Chinese Communist Party*, edited by W.W.-L. Lam, 372–390. Oxon, UK: Routledge.

A. Ma, *China as a Double-Bind Regulatory State*,
https://doi.org/10.1007/978-981-99-8857-0

Barber, Benjamin. 1984. *Strong Democracy: Participatory Politics for a New Age*. Berkeley, CA: University of California Press.

Barber, Benjamin. 1998. "Three scenarios for the Future of Technology and Strong Democracy." *Political Science Quarterly* 113(4): 573–589.

Barrow, Clyde W. 2002. "The Miliband-Poulantzas Debate: An Intellectual History." In *Paradigm Lost: State Theory Reconsidered*, edited by S. Aranowitz and P. Bratsis, 3–52. Minneapolis, MN: University of Minnesota Press.

Béjà, Jean-Philippe. 2019. "Xi Jinping's China: On the Road to Neo-Totalitarianism." *Social Research* 86(1): 203–230.

Bell, Daniel. 1976. *The Coming of Post-Industrial Society*. New York: Basic Books

Benoît, Cyril. 2018. "Régulation." In *Dictionnaire d'économie politique: capitalisme, institutions, pouvoir*, edited by C. Hay and A. Smith, 415–420. Paris: Presses de Sciences Po.

Benoît, Cyril. 2020. *Réguler l'accès aux médicaments*. Grenoble: Presses Universitaires de Grenoble.

Birney, Mayling. 2013. "Decentralisation and Veiled Corruption Under China's 'Rule of Mandates'." *World Development* 53: 55–67.

Boullenois, Camille. 2020. "Poverty Alleviation in China: The Rise of State-Sponsored Corporate Paternalism." *China Perspectives* 3: 47–56.

Bourdieu, Pierre. 1979. "Public Opinion Does Not Exist." In *Communication and Class Struggle 1*, edited by A. Mattelart and S. Siegerlaub, 124–130. New York: International General.

Brady, Anne-Marie. 2018. "'The New (Old) Normal': The Party's Propaganda System Under Jiang, Hu, and Xi." In *Routledge Handbook of the Chinese Communist Party*, edited by W.W.-L. Lam, 165–181. Oxon, UK: Routledge.

Cabestan, Jean-Pierre. 2018. *Demain la Chine: démocratie ou dictature?* Paris: Gallimard.

Cabestan, Jean-Pierre. 2017. "The Party Runs the Show: How the CCP Controls the State and Towers over the Government, Legislature and Judiciary." In *Routledge Handbook of the Chinese Communist Party*, edited by W.W.-L. Lam, 75–91. Oxon, UK: Routledge.

CAICT. 2020. "Quan Qiu Shu Zi Jing Ji Xin Tu Jing 2020 Nian" [New Perspectives of the Global Digital Economy in 2020]. http://www.caict.ac.cn/kxyj/qwfb/bps/202010/P020201014373499777701.pdf

Cairns, Christopher. 2016. "Fragmented Authoritarianism? Reforms to China's Internet Censorship System Under Xi Jinping." Presented at the American Political Science Association Annual Meeting in Philadelphia, PA on 3 September 2016.

Calvert, Randall L., Mathew D. McCubbins, and Barry R. Weingast. 1989. "A Theory of Political Control and Agency Discretion." *American Journal of Political Science* 33(3): 588–611.

Carpenter, Daniel P. 2001. "The Political Foundations of Bureaucratic Autonomy: A Response to Kernell." *Studies in American Political Development* 15: 113–122

Carpenter, Daniel. 2010. *Reputation and Power: Organisational Image and Pharmaceutical Regulation at the FDA*. Princeton, NJ: Princeton University Press

Carpenter, Daniel, and David Moss. 2014. *Preventing Regulatory Capture: Special Interest Influence and How to Limit It*. New York: Cambridge University Press.

Carter, Nancy, Denise Bryant-Lukosius, Alba Dicenso, Jennifer Blythe, and Alan J. Neville. 2014. "The Use of Triangulation in Qualitative Research." *Oncology Nursing Forum* 41(5): 545–547.

Chai, Yanfei. 2018. *Zhejiang Gaige Kaifang 40 Nian Koushu Lishi (1978–2018)* (Oral History of 40 Years of Reform and Opening-Up in Zhejiang). Hangzhou: Zhejiang Science and Technology Press.

Chen, Deren. 2017. "Cong Shi Jie Wang Shang Da Hui de Ba Da Bian Hua Tan Jiu Xin Shi Ti Jing Ji" (Explore New Real Economy Based on the Eight Changes of Ten Editions of Global Netrepreneur Conference). Ali Research Institute, https://read01.com/zh-my/oLLo6NJ.html#.Yga_B9_MLIU

Chen, Feng, and Ting Gong. 1997. "Party Versus Market in Post-Mao China: The Erosion of the Leninist Organisation from Below." *Journal of Communist Studies and Transition Politics* 13: 148-166.

Chen, Hangping. 2021. "Qi Di 2020 Nian Teng Xun 82 Fen Pan Jue Shu: Nan Shan Bi Sheng Ke Shi Zhen De Ma?" [82 Court Rulings Since 2020: Is the Phenomenon of 'Nanshan Must-Be-Victorious Quest' True?]. 2 March. WeChat public account "Zhong Guo Fa Lü Ping Lun" (China Legal Review).

Chen, Shuo, and Zhitao Zhu. 2019. "Xiang Dui Ji Xiao Kao He Yu Zi You Cai Liang Quan: Zhong Guo Xian Ji Guan Yuan Zhi Li Kao Cha" [Relative Performance Evaluation and Discretionary Power: Investigations into Chinese County-Level Cadres' Governance]. Fudan University Working Paper, 1–24.

Cheng, Manli. 2017. "Zhong Guo Dui Wai Chuan Bo De Li Shi Hui Gu Yu Zhan Wang 2009–2017 Nian" [History and Perspectives of Chinese External Propaganda]. *Xin Wen Yu Xie Zuo* [Journalism and Writing] 8: 5–9

Clark, Duncan. 2016. *Alibaba: The House That Jack Ma Built*. New York: HarperCollins Publishers.

CNNIC. February 2021. *The 47th China Statistical Report on Internet Development*.

Coglianese, Cary and David Lazer. 2003. "Management-Based Regulation: Prescribing Private Management to Achieve Public Goals." *Law & Society Review* 37: 691–730.

Crete-Nishihata, Masashi, Andrew Hilts, Jeffrey Knockel, Jason Q. Ng, Lotus Ruan, and Greg Wiseman. 2016. "Harmonized Histories? A Year of Fragmented Censorship Across Chinese Live Streaming Applications." Citizen Lab, 1–13, https://citizenlab.ca/2016/11/harmonized-histories-year-fragmented-censorship-across-chinese-live-streaming-applications/.

Crandell, Jedidiah R., Masashi Crete-Nishihata, Jeffrey Knockel, Sarah McKune, Adam Senft, Diana Tseng, and Greg Wiseman. 2013. "Chat Program Censorship and Surveillance in China: Tracking Tom-Skype and Sina UC." *First Monday* 18(7).

Creemers, Rogier. 2015. "The Pivot in Chinese Cyber Governance: Integrating Internet Control in Xi Jinping's China." *China Perspectives* 4: 5–13.

Cui, Ying, and Aijun Zhang. 2019. *Wei Bo Yu Lun Dao Xiang Yan Jiu* [Analysis of the Public Opinion on Weibo]. Tianjin: Tian Jin Ren Min Chu Ban She [Tianjin People's Publishing House].

Cullen, Richard and D. W. Choy. 2005. "China's Media: The Impact of the Internet." *San Diego International Law Journal* 6(323): 323–339.

Culpepper, Pepper D., and Kathleen Thelen. 2020. "Are We All Amazon Primed? Consumers and the Politics of Platform Power." *Comparative Political Studies* 53(2): 288–318.

Cunningham, Edward, Tony Saich, and Jesse Turiel. 2020. "Understanding CCP Resilience: Surveying Chinese Public Opinion Through Time." Ash Center of Harvard Kennedy School, https://ash.harvard.edu/publications/understanding-ccp-resilience-surveying-chinese-public-opinion-through-time.

Defleur, Melvin L., and Sandra J. Ball-Rokeach. 1989. *Theories of Mass Communication*. Fifth Edition. New York: Longman.

Deng, Jinting, and Pinxin Liu. 2017. "Consultative Authoritarianism: The Drafting of China's Internet Security Law and E-Commerce Law." *Journal of Contemporary China* 26(107): 679–695.

Dickson, Bruce. 2003. *Red Capitalists in China: The Party, Private Entrepreneurs, and Prospects for Political Change*. Cambridge, UK: Cambridge University Press.

Dickson, Bruce. 2007. "Integrating Wealth and Power in China: The Communist Party's Embrace of the Private Sector." *The China Quarterly* 192: 827–854.

Dickson, Bruce. 2016. *The Dictator's Dilemma: The Chinese Communist Party's Strategy for Survival*. Oxford: Oxford University Press.

Dixit, Avinash. 2010. "Democracy, Autocracy and Development." *Journal of Globalisation and Development* 1(1): 1–45.

Dollar, David. 2018. "The Highs and Lows of Chinese Decentralisation." East Asia Forum, Brookings Institution, 1 November.

Donald, Stephanie Hemelryk, Michael Keane, and Yin Hong (eds.). 2002. *Media in China: Consumption, Content and Crisis*. New York: Routledge.

Dworkin, Gerald. 1976. "Autonomy and Behavioural Control." *Hastings Center Report* 6: 23–28.

Economy, Elisabeth. 2018. *The Third Revolution: Xi Jinping and the New Chinese State*. Oxford: Oxford University Press.

Eggertsson, Thrainn. 1990. *Economic Behaviour and Institutions*. Cambridge: Cambridge University Press.

Ekman, Alice. 2020. *Rouge Vif: l'Idéal Communiste Chinois*. Paris: Editions de l'Observatoire.

Esarey, Ashley, and Qiang Xiao. 2011. "Digital Communication and Political Change in China." *International Journal of Communication* 5: 298–319.

European Central Bank. 2020. "The Digital Economy and the Euro Area." *Economic Bulletin* 8.

Fang, Fang. 2020. *Wuhan: Ville Close*. Paris: Stock.

Finez, Jean. 2014. "La construction des prix à la SNCF une socio-histoire de la tarification: de la péréquation au yield management (1938–2012)." *Revue Française de Sociologie* 55(1): 5–39.

Foster, William, and Seymour Goodman. 2000. "The Diffusion of the Internet in China." Report of the Centre for International Security and Cooperation at Stanford University, https://cisac.fsi.stanford.edu/publications/diffusion_of_the_internet_in_china_the.

Francisco, Ronald A. 2005. "The Dictator's Dilemma." In *Repression and Mobilisation*, edited by C. Davenport, H. Johnson, and C. Mueller, 58–81. Minneapolis, MN: University of Minnesota Press.

Freedom House. 2017. "Zhong Guo Ling Hun Zheng Duo Zhan: Xi Jin Ping Zhi Xia De Zong Jiao Fu Xing Ya Zhi He Di Kang" [Fight for Souls in China: Religious Revival, Repression and Resistance under Xi].

Fu, Diana. 2018. *Mobilizing without the Masses: Control and Contention in China*. Cambridge: Cambridge University Press.

Fuchs, Christian. 2016. "Baidu, Weibo and Renren: The Global Political Economy of Social Media in China." *Asian Journal of Communication* 26(1): 14–41.

Gailmard, Sean. 2009. "Multiple Principals and Oversight of Bureaucratic Policy-Making." *Journal of Theoretical Politics* 21(2): 161–186.

George, Cherian. 2007. "Consolidating Authoritarian Rule: Calibrated Coercion in Singapore." *The Pacific Review* 20(2): 127–145.

Gerring, John. 2001. *Social Science Methodology: A Criterial Framework*. Cambridge: Cambridge University Press.

Gerschewski, Johannes. 2018. "Legitimacy in Autocracies: Oxymoron or Essential Feature." *Perspectives on Politics* 16(3): 652–665.

Gewirtz, Julian. 2019. "The Futurists of Beijing: Alvin Toffler, Zhao Ziyang, and China's 'New Technological Revolution'." *The Journal of Asian Studies* 78(1): 115–140.

Giddens, Anthony. 1981. *A Contemporary Critique of Historical Materialism*. London: Macmillan.

Gilson, Ronald, and Curtis Milhaupt. 2011. "Economically Benevolent Dictators: Lessons for Developing Democracies." *The American Journal of Comparative Law* 59(1): 227–288.

Glassman, Ronald M. 1991. *China in Transition: Communism, Capitalism, and Democracy*. New York: Praeger.

Goldstone, Jack. 1995. "The Coming Chinese Collapse." *Foreign Policy* 99: 35–52.

Gramsci, Antonio. 1979. *Letters from Prison*. London: Quartet Books

Gray, Garry C., and Susan S. Silbey. 2014. "Governing Inside the Organisation: Interpreting Regulation and Compliance." *American Journal of Sociology* 120(1): 96–145.

Greenlee, Michael. 2006. "A King Who Devours His People: Jiang Zemin and the Falun Gong Crackdown: A Bibliography." *International Journal of Legal Information* 34(3): 556–584.

Grossman, Emiliano, and Sabine Saurugger. 2012. *Les Groupes d'intérêt: action collective et stratégies de représentation*. Paris: Armand Colin.

Guriev, Sergei, and Daniel Treisman. 2019. "Informational Autocrats." *Journal of Economic Perspectives* 33(4): 100–127.

Hammond, Thomas, and Jack Knott. 1996. "Who Controls the Bureaucracy? Presidential Power, Congressional Dominance, Legal Constraints, and Bureaucratic Autonomy in a Multi-Institutional Model of Policy Making." *Journal of Law, Economics, and Organization* 12: 119–166.

Han, Dong. 2016. "From Vagueness to Clarity? Articulating Legal Criteria of Digital Content Regulation in China." *Global Media and Communication* 12(3): 211–227.

Han, Rongbin. 2015a. "Manufacturing Consent in Cyberspace: China's Fifty-Cent Army." *Journal of Current Chinese Affairs* 44(2): 105–134.

Han, Rongbin. 2015b. "Defending the Authoritarian Regime Online: China's 'Voluntary Fifty-Cent Army'." *The China Quarterly* 224: 1006–1025.

Han, Rongbin. 2018. *Contesting Cyberspace in China: Online Expression and Authoritarian Resilience*. New York: Columbia University Press.

Han, Rongbin. 2019. "Patriotism Without State Blessing: Chinese Cyber Nationalists in a Predicament." In *Handbook of Protest and Resistance in China*, edited by T. Wright, 346–360. Cheltenham: Edward Elgar Publishing.

Harding, Harry. 1987. *China's Second Revolution*. Washington DC: Brookings Institution.

Harwit, Eric, and Duncan Clark. 2001. "Shaping the Internet in China: Evolution of Political Control over Network Infrastructure and Content." *Asian Survey* 41(3): 377–408.

Haworth, Lawrence. 1991. "Dworkin on Autonomy". *Ethics* 102(1): 129–139.

Hay, Colin, and Daniel Wincott. 1998. "Structure, Agency and Historical Institutionalism." *Political Studies* 46: 951–957.

Hay, Colin. 1999. "Marxism and the State." In *Marxism and Social Science*, edited by A. Gamble, D. Marsh, and T. Tant, 152–174. Champaign, IL: University of Illinois Press.

Hay, Colin. 2002. *Political Analysis: A Critical Introduction*. Basingstoke: Palgrave.

Hay, Colin. 2009. "King Canute and the 'Problem' of Structure and Agency: On Times, Tides and Heresthetics." *Political Studies* 57: 260–279.

Hay, Colin. 2011. "Ideas and the construction of interests." In *Ideas and Politics in Social Science Research*, edited by D. Béland and R. H. Cox, 65–82, Oxford: Oxford University Press.

Hay, Colin, and Andy Smith. 2018. "Introduction : Le Rapport Capitalisme-Politique." In *Dictionnaire d'économie politique*, edited by C. Hay and A. Smith, 19–49. Paris: Presses de Sciences Po.

He, Baogang, and Mark Warren. 2011. "Authoritarian Deliberation: The Deliberative Turn in Chinese Political Development." *Perspectives on Politics* 9(2): 269–289.

Hibou, Béatrice. 2011. *Anatomie Politique de la Domination*. Paris: La Découverte.

Hintze, Otto. 1975. *The Historical Essays of Otto Hintze*. New York: Oxford University Press.

Hirschman, Albert. 1970. *Exit, Voice and Loyalty: Responses to Declines in Firms, Organisations, and States*. Cambridge, MA: Harvard University Press.

Howell, Jude, and Tim Pringle. 2018. "Shades of Authoritarianism and State-Labour Relations in China." *British Journal of Industrial Relations* 57(2): 223–246.

Hsueh, Roselyn. 2011. *China's Regulatory State: A New Strategy for Globalisation*. Ithaca, NY: Cornell University.

Hu, Shuli. 2011. "The Rise of the Business Media in China." In *Changing Media, Changing China*, edited by S. Shirk, 77–90. New York: Oxford University Press.

Hu, Yong. 2008. *Zhong Sheng Xuan Hua: Wang Luo Shi Dai De Ge Ren Biao Da Yu Gong Gong Tao Lun* [The Rising Cacophony: Personal Expression and Public Discussion in the Internet Age]. Guilin: Guang Xi Shi Fan Da Xue Chu Ban She [Guangxi Normal University Press].

Huang, Haifeng. 2015. "Propaganda as Signalling." *Comparative Politics* 47(4): 419–437.

Huang, Haifeng. 2018. "The Pathology of Hard Propaganda." *The Journal of Politics* 80(3): 1034–1038.

Huang, Yasheng. 2008. *Capitalism with Chinese Characteristics: Entrepreneurship and the State*. Cambridge: Cambridge University Press.

Huntington, Samuel P. 1991. *The Third Wave: Democratisation in the Late Twentieth Century*. Norman, OK: University of Oklahoma Press.

Jessop, Bob. 1977. "Recent Theories of the Capitalist State." *Cambridge Journal of Economics* 1: 353–373.

Jevons, W. Stanley. 1879. *The Theory of Political Economy*. Second Edition. London: Macmillan and Co.

Jia, Kai and Martin Kenney. 2022. "The Chinese Platform Business Group: An Alternative to the Silicon Valley Model." *Journal of Chinese Governance* 7(1): 58–80.

Jia, Lianrui, and Dwayne Winseck. 2018. "The Political Economy of Chinese Internet Companies: Financialization, Concentration, and Capitalization." *International Communication Gazette* 80(1): 30–59.

Jiang, Fei, and Nan Zhang. 2019. "Zhong Guo Dui Wai Chuan Bo De San Ci Lang Chao (1978–2019)" [Three Waves of International Communication in China (1978–2019)]. *Quan Qiu Chuan Mei Xue Kan* [Global Media Journal] 2: 39–58.

Jiang, Min. 2012. "Internet Companies in China: Dancing Between the Party Line and the Bottom Line." *Asie Vision* 47: 1–50.

Jiang, Min. 2014. "The Business and Politics of Search Engines: A Comparative Study of Baidu and Google's Search Results of Internet Events in China." *New Media & Society* 16(2): 212–233.

Jiang, Min. 2015. "Managing the Micro-Self: The Governmentality of Real Name Registration Policy in Chinese Microblogosphere." *Information, Communication, and Society* 19(2): 203–220.

Jin, Hehui, Yingyi Qian, and Barry R. Weingast. 2005. "Regional Decentralization and Fiscal Incentives: Federalism, Chinese Style." *Journal of Public Economics* 89(9): 1719–1742.

Johansson, S. Ryan. 1988. "The Computer Paradigm and the Role of Cultural Information in Social Systems." *Historical Methods* 21: 172–188.

Kalathil, Shanthi, and Taylor C. Boas. 2003. *Open Networks, Closed Regimes: The Impact of the Internet on Authoritarian Rule*. Washington, DC: Carnegie Endowment for International Peace.

Kennedy, Scott. 2009. *The Business of Lobbying in China*. Cambridge, MA: Harvard University Press.

Kennedy, Scott. 2011. "Fragmented Influence: Business Lobbying in China in Comparative Perspectives." In *Beyond the Middle Kingdom: Comparative Perspectives on China's Capitalist Transformation*, edited by S. Kennedy, 113–135. Stanford, CA: Stanford University Press.

King, Gary, Jennifer Pan, and Margaret E. Roberts. 2013. "How Censorship in China Allows Government Criticism But Silences Collective Expression." *American Political Science Review* 107(2): 1–18.

King, Gary, Jennifer Pan, and Margaret E. Roberts. 2017. "How the Chinese Government Fabricates Social Media Posts for Strategic Distraction, Not Engaged Argument." *American Political Science Review* 111(3): 484–501.

Knockel, Jeffrey, Lotus Ruan, and Mashashi Crete-Nishihata. 2017. "Measuring Decentralization of Chinese Keyword Censorship via Mobile Games." Presented at the 7th USENIX Workshop on Free and Open Communications on the Internet (FOCI 17). Vancouver, BC: USENIX Association.

Koop, Christel, and Martin Lodge. 2017. "What Is Regulation? An Interdisciplinary Concept Analysis." *Regulation & Governance* 11: 95–108.

Krastev, Ivan. 2011. "Paradoxes of the New Authoritarianism." *Journal of Democracy* 22(2): 5–16.

Kuran, Timur. 1997. *Private Truths, Public Lies: The Social Consequences of Preference Falsification*. Cambridge, MA: Harvard University Press.

Laffont, Jean-Jacques, and Jean Tirole. 2012. *Théories des incitations et réglementations*. Paris: Economica.

Lagerkvist, Johan. 2008. "Internet Ideotainment in the PRC: National Responses to Cultural Globalization." *Journal of Contemporary China* 17(54): 121–140.

Lagerkvist, Johan. 2010. *After the Internet, Before Democracy: Competing Norms in Chinese Media and Society*. Bern, Switzerland: Peter Lang Verlag.

Lagerkvist, Johan. 2011. "New Media Entrepreneurs in China: Allies of the Party-State or the Civil Society?" *Journal of International Affairs*. Fall/ Winter 2011: 169–182.

Lagerkvist, Johan. 2012. "Principal-Agent Dilemma in China's Social Media Sector? The Party-State and Industry Real-Name Registration Waltz." *International Journal of Communication* 6: 2628–2646.

Landry, Pierre F. 2008. *Decentralised Authoritarianism in China: The Communist Party's Control of Local Elites in the Post-Mao Era*. Cambridge: Cambridge University Press.

Lee, Ching Kwan, and Zhang, Yonghong. 2013. "The power of instability: unravelling the microfoundations of bargained authoritarianism in China." *American Journal of Sociology* 118(6): 1475–1508.

Légaré, Kathia. 2007. "L'influence du système international sur l'évolution des régimes semi- autoritaires: quatre approches des révolutions de couleur (2003–2005)." *Etudes Internationales* 38(4): 501–522.

Lei, Ya-wen. 2018. *The Contentious Public Sphere: Law, Media and Authoritarian Rule in China*. Princeton, NJ: Princeton University Press.

Leung, Beatrice. 2002. "China and Falun Gong: Party and Society Relations in the Modern Era." *Journal of Contemporary China* 11(33): 761–784.

Li, Hongbin and Li-An Zhou. 2005. "Political Turnover and Economic Performance: The Incentive Role of Personnel Control in China." *Journal of Public Economics* 89(9): 1743–1762.

Li, Yanwei and Genea Canelles. 2021. "Governing Airbnb in Amsterdam and Singapore: A Comparative Study on Governance Strategies and Styles." *Sage Open*, October–December: 1–11.

Liao, Zhixiong. 2016. "An Economic Analysis on Internet Regulation in China and Proposals to Policy and Law Makers." *International Journal of Technology Policy and Law* 2 (2–4): 242–256.

Libman, Alexander, and Michael Rochlitz. 2019. *Federalism in China and Russia*. Cheltenham: Edward Elgar Publishing.

Lieberthal, Kenneth, and David M. Lampton. 1992. *Bureaucracy, Politics, and Decision Making in Post-Mao China*. Berkeley, CA: University of California Press.

Lieberthal, Kenneth, and Michel Oksenberg. 1988. *Policy Making in China: Leaders, Structures and Processes*. Princeton, NJ: Princeton University Press.

Lin, Justin Yifu, and Zhiqiang Liu. 1998. "Zhong Guo De Cai Zheng Fen Quan Yu Jing Ji Zeng Zhang" [Shared Budgetary System and Economic Growth in China]. Presented at the Third Pan-Pacific Joint Conference of Economic Organizations, 13–18 January, Bangkok, Thailand, http://vod6.pkudl.cn/xzgl/czss/contents/thesis/the_010_01.pdf

Lin, Li-wen, and Curtis Milhaupt. 2013. "We Are the (National) Champions: Understanding the Mechanisms of State Capitalism in China". *Stanford Law Review* 65(4): 697–759.

Liu, Mianmian. 2014. *Zhong Guo Niu Ren* [Chinese Superman]. New York: Mirror Books.

Lodge, Martin. 2002. *On Different Tracks: Designing Railways Regulation in Britain and Germany*. Westport: Praeger.

Lu, Wei, Jia Du, Jin Zhang, Feicheng Ma, and Taowen Le. 2002. "Internet Development in China". *Journal of Information Science* 28(3): 207–233.

Lukes, Steven. 2005. *Power: A Radical View*. Second Edition. Basingstoke: Palgrave Macmillan.

Lynch, Daniel. 1999. "Dilemmas of 'Thought Work' in Fin-De-Siècle China." *The China Quarterly* 157: 173–201.

Ma, Aifang. 2018. *L'intelligence artificielle en Chine: Un état des lieux*. Paris: Fondation pour l'Innovation Politique.

Ma, Aifang. 2022. *Double Bind Regulation of the Internet and Social Media in Contemporary China*. PhD dissertation defended on 28 June 2022 at Sciences Po, Paris.

Ma, Aifang. 2023a. "China Internet Regulations 1987–2022", https://doi.org/10.21410/7E4/R6NLOH.

Ma, Aifang. 2023b. "Geographical Distribution of the Top 100 Internet Firms in China from 2013 to 2022", https://doi.org/10.21410/7E4/PWQ6WF.

Ma, Aifang. 2023c. *La Régulation du numérique: Chine, Etats-Unis, France*. Paris: Fondation pour l'Innovation Politique.

Ma, Aifang, and Yong Hu. 2024. "Zhong Guo Dui Min Ying Hu Lian Wang Qi Ye De Fei Zheng Shi Jian Guan: Yi Min Ying Hu Lian Wang Qi Ye Dang Jian Wei Li" [Informal Regulation of Private Internet Firms in China: A Case Study of Party-Building]. *Twenty-First Century*. Forthcoming.

MacKinnon, Rebecca. 2009. "China's Censorship 2.0: How Companies Censor Bloggers." *First Monday* 14(2), https://firstmonday.org/ojs/index.php/fm/article/download/2378/2089.

MacKinnon, Rebecca. 2011. "Liberation Technology: China's 'Networked Authoritarianism'." *Journal of Democracy* 22(2): 32–46.

Mann, Michael. 1984. "The Autonomous Power of the State: Its Origins, Mechanisms and Results." *European Journal of Sociology* 25(2): 185–213.

Mann, Michael. 2012. *The Sources of Social Power 2: The Rise of Classes and Nation-States, 1760–1914*. New York: Cambridge University Press.

Marx, Karl. 1926. *The Eighteenth Brumaire of Louis Bonaparte*. P. Eden, trans. London: Allen & Unwin.

Marx, Karl, and Frederick Engels. 1848. *Manifesto of the Communist Party*.

Maskin, Eric, Yingyi Qian, and Chenggang Xu. 2000. "Incentives, Scale Economies and Organization Forms." *Review of Economic Studies* 67: 359–378.

Mayer, Nonna. 1995. "L'entretien selon Pierre Bourdieu, analyse critique de La misère du monde." *Revue Française de Sociologie* 36(2): 355–370.

Mayer, Nonna. 2009. "Opinion Publique." In *Dictionnaire des mouvements sociaux*, edited by Olivier Fillieule, Lilan Mathieu and Cécile Péchu, 429–435. Paris: Presses de Sciences Po.

Mayer, Nonna. 2018. "Qualitatif ou quantitatif ? Plaidoyer pour l'éclectique méthodologique." *Bulletin de Méthodologie Sociologique* 139: 7–33.

Mayntz, Renate. 2009. "The Changing Governance of Large Technical Infrastructure Systems." In *Über Governance: Institutionen und Prozesse politischer Regelung*, edited by Renate Mayntz, 121–150. Frankfurt: Campus Verlag.

McCubins, Mathew D., Roger G. Noll, and Barry R. Weingast. 1987. "Administrative Procedures as Instruments of Political Control." *Journal of Law, Economics, and Organization* 3: 243–277.

McGeary, Aaron D. 2001. "China's Great Balancing Act: Maximizing the Internet's Benefits While Limiting Its Detriments." *The International Lawyer* 35(1): 219–230.

McGregor, Richard. 2010. *The Party: The Secret World of China's Communist Rulers*. New York: HarperCollins Publishers.

McNight, Scott, Martin Kenney, and Dan Breznitz. 2021. "Platformizing the Economy? Building and Regulating Chinese Digital Platforms." Available at SSRN, https://papers.ssrn.com/sol3/papers.cfm?abstract_id=3885190

Mendras, Henri, and Jean Etienne. 1996. *Les Grands Auteurs de la Sociologie: Tocqueville, Marx, Durkheim, Weber*. Paris: Hatier.

Meng, Bingchun. 2011. "From Steamed Bun to Grass Mud Horse: E Gao as Alternative Political Discourse on the Chinese Internet." *Global Media and Communication* 7(1): 33–51.

Mertha, Andrew. 2005a. "China's Soft Centralization: Shifting Tiao/Kuai Authority Relations." *The China Quarterly* 184: 792–810.

Mertha, Andrew. 2005b. *The Politics of Piracy: Intellectual Property in Contemporary China*. Ithaca, NY: Cornell University Press.

Mertha, Andrew. 2009. "'Fragmented Authoritarianism 2.0': Political Pluralization in the Chinese Policy Process." *The China Quarterly* 200: 995–1012.

Miao, Di. 2011. "Between Propaganda and Commercials: Chinese Television Today." In *Changing Media, Changing China*, edited by S. Shirk, 91–114. New York: Oxford University Press.

Miao, Weishan, Min Jiang, and Yunxia Pang. 2021. "Historicizing Internet Regulation in China: A Meta-Analysis of Chinese Internet Policies (1994–2017)". *International Journal of Communication* 15: 2003–2026.

Milhaupt, Curtis, and Wentong Zheng. 2015. "Beyond Ownership: State Capitalism and the Chinese Firm." *The Georgetown Law Journal* 103: 665–722.

Mill, Stuart. 1843/2011. *A System of Logic: Ratiocinative and Inductive*. New York: Columbia University Press.

Miller, Blake. 2018. "The Limits of Commercialized Censorship in China." Working Paper, http://eprints.lse.ac.uk/101417/1/miller_limits_of_commercialized_censorship_in_china.pdf.

Mitnick, Barry M. 1980. *The Political Economy of Regulation: Creating, Designing and Removing Regulatory Forms*. New York: Columbia University Press.

Montinola, Gabriella, Yingyi Qian, and Barry R. Weingast. 1995. "Federalism, Chinese Style: The Political Basis for Economic Success in China." *World Politics* 48(1): 50–81.

Morgan, David L. 1996. "Focus Groups." *Annual Review of Sociology* 22: 129–152.

Morozov, Evgeny. 2011. *The Net Delusion: The Dark Side of Internet Freedom*. New York: Public Affairs.

Mueller, Milton, and Tan Zixiang. 1997. *China in the Information Age: Telecommunications and the Dilemmas of Reform*.Westport: Praeger.

Nathan, Andrew. 2009. "China Since Tiananmen: Authoritarian Impermanence." *Journal of Democracy* 20(3): 37–40.

Nathan, Andrew. 2003. "China's Changing of the Guard: Authoritarian Resilience." *Journal of Democracy* 14(1): 6–17.

Naughton, Barry. 1995. *Growing Out of the Plan: Chinese Economic Reform 1978–1993*. Cambridge: Cambridge University Press.

Negro, Gianluigi. 2017. *The Internet in China: From Infrastructure to a Nascent Civil Society*. Cham: Palgrave Macmillan.

Nicholson, Jessica R. 2020. "New Digital Economy Estimates." Bureau of Economic Analysis, U.S. Department of Commerce, 1–32, https://www.bea.gov/system/files/2020-08/New-Digital-Economy-Estimates-August-2020.pdf.

Noesselt, Nele. 2014. "Microblogs and the Adaptation of the Chinese Party-Stat's Governance Strategy." *Governance* 27(3): 449–468.

Noll, Roger G. 1980. "What Is Regulation?" Social Science Working Paper 324, California Institute of Technology.

Norris, Pippa. 2001. *Digital Divide : Civic Engagement , Information Poverty, and the Internet Worldwide*. New York: Cambridge University Press.

North, Douglass. 1990. *Institutions, Institutional Change and Economic Performance*. New York: Cambridge University Press.

Nye, Joseph. 2004. "The Decline of America's Soft Power: Why Washington Should Worry." *Foreign Affairs* 83(3): 16–20.

Obar, Jonathan A. and Steve Wildman. 2015. "Social Media Definition and the Governance Challenge: An Introduction to the Special Issue". *Telecommunications Policy* 39(9): 745–750.

Oberdorff, Henri. 2010. *La démocratie à l'ère numérique*. Grenoble: Presses Universitaires de Grenoble.

Oberschall, Anthony. 1996. "Opportunities and Framing in the Eastern European Revolts of 1989." In *Comparative Perspectives on Social Movements. Political Opportunities, Mobilizing Structures, and Cultural Framings*, edited by D. McAdam, J.D. McCarthy, and M.N. Zald, 93–121. Cambridge: Cambridge University Press.

O'Brien, Kevin. 1994. "Implementing Political Reform in China's Villages." *The Australian Journal of Chinese Affairs* 32: 33–59.

O'Brien, Kevin, and Lianjiang Li. 2006. *Rightful Resistance in the Chinese Countryside*. New York: Cambridge University Press.

Oi, Jean C. 1992. "Fiscal Reform and the Economic Foundations of Local State Corporatism in China." *World Politics* 45: 99–126.

Olson, Mancur. 1993. "Dictatorship, Democracy and Development." *American Political Science Review* 87(3): 567–576.

Ortmann, Stephan. 2011. Singapore: Authoritarian but Newly Competitive. *Journal of Democracy* 22(4): 153–164.

Pairault, Thierry. 2013. "Les entreprises chinoises sous la tutelle directe du gouvernement illustré par leur investissement en Afrique." *Revue de la Régulation* 13, https://journals.openedition.org/regulation/10195.

Pearson, Margaret M. 2005. "The Business of Governing Business in China: Institutions and Norms for the Emerging Regulatory State." *World Politics* 57: 296–322.

Perry, Elisabeth J. 2007. "Study Chinese Politics: Farewell to Revolution?" *China Journal* 57(1): 1–22.

Pierson, Paul. 2005. "The Study of Policy Development." *The Journal of Policy History* 17(1): 34–51.

Polanyi, Karl. 2001. *The Great Transformation of Our Time: The Political and Economic Origins of Our Time*. Boston: Beacon Press.

Poulantzas, Nicolas and Ralph Miliband. 1972. "The Problem of the Capitalist State". In *Ideology in Social Science: Readings in Critical Social Theory*, edited by Robin Blackburn, 238–262. New York: Pantheon Books.

Przeworski, Adam, and Fernando Limongi. 1993. "Political Regimes and Economic Growth." *The Journal of Economic Perspectives* 7(3): 51–69.

Qi, Yanli. 2000. "The Current Situation and Prospect of Chinese Information Resources on the Web." *Social Science Computer Review* 18(4): 484–489.

Qian, Gang, and David Bandurski. 2011. "China's Emerging Public Sphere: The Impact of Media Commercialization, Professionalism, and the Internet in an Era of Transition." In *Changing Media, Changing China*, edited by S. Shirk, 38–76. Oxford: Oxford University Press.

Qian, Yingyi, and Chenggang Xu. 1993. "Why China's Economic Reforms Differ: The M-Form Hierarchy and Entry/Expansion of the Non-State Sector." *Economics of Transition* 1(2): 135–170.

Qian, Yingyi, and Barry Weingast. 1996. "China's Transition to Markets: Market-Preserving Federalism, Chinese Style." *Journal of Economic Policy Reform* 1(2): 149–185.

Qiao, Kunyuan. 2013. "Wo Guo Guan Yuan Jin Sheng Jin Biao Sai Ji Zhi: Li Lun Yu Zheng Ju" [Tournament Mechanism in the Promotion of Chinese Government Officials: Theories and Proves]. *Jing Ji Ke Xue* [Economic Science] 1: 88–98.

Rahman, K. Sabeel, Thelen, Kathleen. 2019. "The Rise of the Platform Business Model and the Transformation Twenty-First-Century Capitalism." *Politics & Society* 47(2): 177–204.

Rauchfleischer, Adrian and Schäfer, Mike. 2015. "Multiple Public Spheres of Weibo: A Typology of Forms and Potentials of Online Public Spheres in China." *Information, Communication & Society* 18(2): 139–155.

Ringen, Stein. 2016. *The Perfect Dictatorship: China in the 21st Century*. Hong Kong: HKU Press.

Riskin, Carl. 1987. *China's Political Economy: The Quest for Development Since 1949*. New York: Oxford University Press.

Roberts, Margaret. 2018. *Censored: Distraction and Diversion Inside China's Great Firewall*. Princeton: Princeton University Press.

Rocca, Jean-Louis. 2006. La Condition chinoise: la mise au travail capitaliste à l'âge des réformes (1978- 2004). Paris: Karthala.

Sautman, Barry. 1992. "Sirens of the Strongman: Neo-Authoritarianism in Chinese Political Theory." *The China Quarterly* 129: 72–102.

Scott, C. James. 1985. *Weapons of the Weak: Everyday Forms of Peasant Resistance*. New Haven: Yale University Press.

Segal, Gerald. 1994. "China Changes Shape: Regionalism and Foreign Policy: Introduction." The Adephi Papers 34(287): 3–5.

Shambaugh, David. 2008. *China's Communist Party: Atrophy and Adaptation*. Washington, DC: Woodrow Wilson Center Press.

Shirk, Susan. 1993. *The Political Logic of Economic Reform in China*. Berkeley, CA: University of California Press.

Shirk, Susan. 1985. "The Politics of Industrial Reform." In *The Political Economy of Reform in Post-Mao China*, edited by E.J. Perry, and C. Wong, 195–221. Cambridge, MA: Harvard Council on East Asian Studies.

Shirk, Susan. 2011a. "Chapter 1: Changing Media, Changing China." In *Changing Media, Changing China*, edited by S. Shirk, 1–37. Oxford: Oxford University Press.

Shirk, Susan. 2011b. "Changing Media, Changing Foreign Policy." In *Changing Media, Changing China*, edited by S. Shirk, 225–252. Oxford: Oxford University Press.

Shirk, Susan. 2018. "China in Xi's 'New Era': The Return to Personalistic Rule". *Journal of Democracy* 29(2): 22–36.

Shirky, Clay. 2008. *Here Comes Everybody: The Power of Organizing Without Organizations*. New York: The Penguin Press.

Skocpol, Theda. 1979. *States and Social Revolutions*. Cambridge: Cambridge University Press.

Solinger, Dorothy J. 1987. "The 1980 Inflation and the Politics of Price Control in the PRC." In David M. Lampton (ed.), Policy Implementation in Post-Mao China. Berkeley: University of California Press, pp. 106–110.

Stern, Rachel E., and Kevin O'Brien. 2012. "Politics at the Boundary: Mixed Signals and the Chinese State." *Modern China* 38(2): 174–198.

Stigler, George J. 1971. "The Theory of Economic Regulation." *Bell Journal of Economics and Management Science* 2: 3–21.

Stockmann, Daniela. 2011. "What Kind of Information Does the Public Demand? Getting the News During the 2005 Anti-Japanese Protests". In *Changing Media, Changing China*, edited by S. Shirk, 175–201. Oxford: Oxford University Press.

Strittmatter, Kai. 2020. *Dictature 2.0 : Quand la Chine surveille son people (et demain le monde)*. Translation by Olivier Mannoni. Paris: Tallandier.

Sullivan, Jonathan. 2014. "China's Weibo: Is Faster Different?" *New Media & Society* 16(1): 24–37.

Surowiec, Pawel, and Vaclav Stetka. 2018. "Introduction: Social Media, Politics and Democracy in Post-Transition Central and Eastern Europe." In *Social Media and Politics in Central and Eastern Europe*, edited by V. Stetka and P. Surowiec, 1–20. London: Routledge.

Svensson, Marina. 2014. "Voice, Power and Connectivity in China's Microblogosphere: Digital Divides on Sina Weibo." *China Information* 28(2): 168–188.

Tai, Zixue. 2006. *The Internet in China: Cyberspace and Civil Society*. New York: Routledge.

Tai, Zixue, and Tao Sun. 2007. "Media Dependencies in a Changing Media Environment: The Case of the 2003 SARS Epidemics in China." *New Media & Society* 9(6): 987–1009.

Tang, Wenfang. 2016. *Populist Authoritarianism: Chinese Political Culture and Regime Sustainability*. Oxford: Oxford University Press.

Teets, Jessica C. 2013. "Let many civil societies bloom: the rise of consultative authoritarianism in China." *The China Quarterly* 213: 19–38.

Teng, Biao. 2012. "Rights Defence (weiquan), Microblogs (weibo) and the Surrounding Gaze (weiguan): The Rights Defence Movement Online and Offline." *China Perspectives* 3: 29–39.

Tilly, Charles. 1981. *As Sociology Meets History*. New York: Academic Press.

Ting, Michael M. 2003. "A Strategic Theory of Bureaucratic Redundancy." *American Journal of Political Science* 47(2): 274–292.

Toffler, Alvin. 1980. *The Third Wave*. New York: William Morrow Publishing House.

Tong, Yanqi. 2011. "Morality, Benevolence, and Responsibility: Regime Legitimacy in China from Past to the Present." *Journal of Chinese Political Science* 16: 141–159.

Tsai, Kelley S. 2005. "Capitalists Without a Class: Political Diversity Among Private Entrepreneurs in China." *Comparative Political Studies* 38(9): 1130–1158.

Tsebelis, George. 1995. "Decision Making in Political Systems: Veto Players in Presidentialism, Parliamentarism, Multicameralism and Multipartyism." *Bristish Journal of Political Science* 25(3): 289–325.

Tsui, Lokman. 2003. "The Panopticon as the Antithesis of a Space of Freedom: Control and Regulation of the Internet in China." *China Information* 17(2): 65–82.

Umesao, Tadao. 1963. "Information Industry Theory: Dawn of the Coming Era of the Ectodermal Industry." Hoso Asahi, January: 4–17.

Unger, Jonathan. 1996. "'Bridges': Private Business, the Chinese Government and the Rise of New Associations." *The China Quarterly* 147: 795–819.

Villeneuve, Nart. 2008. "Sear Monitor: Toward a Measure of Transparency." Citizen Lab Occasional Paper#1: 1–29.

Wang, Hongyu, and Fayong Shi. 2018. "Weibo Use and Political Participation: The Mechanism Explaining the Positive Effect of Weibo Use on Online Political Participation Among College Students in Contemporary China." *Information, Communication & Society* 21(4): 516–530.

Weber, Max. 1946. *Politics as a Vocation*. New York: Oxford University Press.
Weber, Max. 1978. *Economy and Society*. New York: Bedminster Press.
Weber, Max. 1998. "Class, Status and Party." In *From Max Weber: Essays in Sociology*, edited and translated by H.H. Gerth, and C.W. Mills, 180–195. New York: Oxford University Press.
White, Gordon. 1994. "Democratization and Economic Reform in China." *Australian Journal of Chinese Affairs* 31: 73–92.
Wintrobe, Ronald. 1998. *The Political Economy of Dictatorship*. Cambridge: Cambridge University Press.
Wolchok, Scott, Randy Yao, and J. Alex Halderman. 2009. "Analysis of the Green Dam Censorware System." https://jhalderm.com/pub/gd/.
Woll, Cornelia. 2008. *Firm Interests: How Governments Shape Business Lobbying on Global Trade*. Ithaca, NY: Cornell University Press.
Wong, Christine. 1985. "Material Allocation and Decentralization: Impact of the Local Sector on Industrial Reform." In *The Political Economy of Reform in Post-Mao China*, edited by E. Perry, and C. Wong, 253–278. Cambridge, MA: Harvard University Press.
Wu, Angela Xiao. 2020. "The Evolution of Regime Imaginaries on the Chinese Internet." *Journal of Political Ideologies* 25(2): 139–161.
Xi, Jinping. 2021. "Nu Li Cheng Wei Shi Jie Zhu Yao Ke Xue Zhong Xin He Chuang Xin Gao Di" [Try Our Best to Become a Principle Global Science and Technology Centre and a Global Innovation Highland]. *Fen Dou* [Hard-Working] 6: 4–11.
Xiao, Lihui. 2008. "Xian Wei Shu Ji Yan Zhong De Zhong Yang Yu Di Fang Guan Xi" [Central-Local Relationship in the Eyes of the Party-Secretaries of the Counties]. *Jing Ji She Hui Ti Zhi Bi Jiao* [Comparison of Economic and Social Systems] 4: 145–150.
Xiao, Qiang. 2011. "The Rise of Online Public Opinion and Its Political Impact." In *Changing Media, Changing China*, edited by S. Shirk, 202–224. Oxford: Oxford University Press.
Xue, Xiaorong. 2018. *Hu Lian Wang Qi Ye Dang Jian: Ji Shu Zi Ben Yu Zheng Zhi Luo Ji Zhang Li Xia De Zheng Dang Zhi Li* [Party-Building in Internet Firms: Party Governance under the Logics of Technology, Capital and Politics]. Beijing: Current Affairs Press.
Yang, Dali. 1997. *Beyond Beijing: Liberalization and the Regions in China*. New York: Routledge.
Yang, Guobin. 2003. "The Co-Evolution of the Internet and Civil Society in China." *Asian Survey* 43(3): 405–422.
Yang, Guobin. 2009a. *The Power of the Internet in China: Citizen Activism Online*. New York: Columbia University Press.
Yang, Guobin. 2009b. "China Since Tiananmen: Online Activism." *Journal of Democracy* 20(3): 33–36.

Yang, Long. 2010. "Fa Lü Yuan Sheng Tai De Sha Shou: Dao De Fan Fa Lü Hua He Fa lü Fan Dao De Hua" [The Killer of the Original Ecology of Law: The Pan-Legalization of Morality and the Pan-Moralization of Law]. *Xue Shu Lun Tan* [Academic Forum] 7: 158–163.

Ye, Shan. 2017. "Di Fang Zheng Fu Shui Shou Jian Mian Cheng Nuo De Xiao Li Xia Ci" [Flaws in Local Governments' Promises to Reduce and to Exempt Taxations]. *Dang Dai Fa Xue* [Contemporary Law Studies] 6: 116–125.

Yeo, Yukyung. 2020. *Varieties of State Regulation: How China Regulates Its Socialist Market Economy*. Cambridge, MA: Harvard University Press.

Yin, Laisheng, and Bangyan Feng. 2014. "Cong Cheng Shi Jing Zheng Dao Qu Yu He Zuo" [From Competition between Cities to Cooperation between Regions]. Jing Ji Ti Zhi Gai Ge [Reform of Economic Systems] 5.

Zhan, Xuying. 2015. "Di Fang Zheng Fu Ji Xiao Ping Gu De Bei Lun Jie Xi: Re Dian Ju Jiao" [Analysis of the Paradoxes in Evaluating Local Governments' Performance: Focus on Hot Points]. *Zhong Guo Xing Zheng Guan Li* [Public Administration in China] 11: 45–49.

Zhang, Jun, Haichao Fan, Zhiwei Xu, and Longfei Zhou. 2020. "GDP Zeng Su De Jie Gou Xing Xia Tiao: Ji Yu Guan Yuan Kao He Ji Zhi De Shi Jiao" [Structural Downward Adjustment of the GDP Growth: From the Perspective of the Evaluation System of Officials]. *Jing Ji Yan Jiu* [Economic Studies] 55(5): 31–48.

Zhao, Yuezhi. 1998. *Media, Market, and Democracy in China: Between the Party Line and the Bottom Line*. Urbana, IL: University of Illinois Press.

Zhao, Yuezhi. 2000. "From Commercialization to Conglomeration: The Transformation of the Chinese Press Within the Orbit of the Party State." *Journal of Communication* 50(2): 3–26.

Zhao, Yuezhi. 2008. *Communication in China: Political Economy, Power, and Conflict*. Lanham, MD: Rowman & Littlefield Publishers.

Zhao, Ziyang. 2016. *Zhao Zi Yang Wen Ji 1980–1989* [Collected Works of Zhao Ziyang 1980–1989]. Hong Kong: Chinese University Press.

Zheng, Wentong. 2010. "Transplanting Antitrust in China: Economic Transition, Market Structure, and State Control." *University of Pennsylvania Journal of International Law* 32(2): 643–721.

Zheng, Yongnian. 2006. *De facto Federalism in China: Reforms and Dynamics of Central-Local Relations*. Singapore: World Scientific Publishing.

Zheng, Yongnian. 2007. *Technological Empowerment: The Internet, State and Society in China*. Stanford, CA: Stanford University Press.

Index

A. Ma, *China as a Double-Bind Regulatory State*,
https://doi.org/10.1007/978-981-99-8857-0

GPSR Compliance

The European Union's (EU) General Product Safety Regulation (GPSR) is a set of rules that requires consumer products to be safe and our obligations to ensure this.

If you have any concerns about our products, you can contact us on ProductSafety@springernature.com

In case Publisher is established outside the EU, the EU authorized representative is:

Springer Nature Customer Service Center GmbH
Europaplatz 3
69115 Heidelberg, Germany

Batch number: 10725849

Printed by Printforce, the Netherlands